Ibogaine

Proceedings of the First International Conference

This book is a volume of The Alkaloids, edited by Geoffrey Cordell, College of Pharmacy, University of Illinois at Chicago, Chicago, Illinois

Ibogaine

Proceedings of the First International Conference

Edited by

Kenneth R. Alper
New York University School of Medicine
New York, New York

Stanley D. Glick
Albany Medical College
Albany, New York

ACADEMIC PRESS
A Harcourt Science and Technology Company
San Diego San Francisco New York Boston London Sydney Tokyo

This book is printed on acid-free paper. ∞

Academic Press
A Harcourt Science and Technology Company
525 B Street, Suite 1900, San Diego, California 92101-4495, USA
http://www.academicpress.com

Academic Press
Harcourt Place, 32 Jamestown Road, London NW1 7BY, UK
http://www.academicpress.com

Library of Congress Catalog Card Number: 2001092788

International Standard Book Number: 0-12-053206-9

PRINTED IN THE UNITED STATES OF AMERICA
01 02 03 04 05 06 MM 9 8 7 6 5 4 3 2 1

CONTENTS

Drug Discrimination Studies with Ibogaine

SCOTT HELSLEY, RICHARD A. RABIN, AND J. C. WINTER

Comparative Neuropharmacology of Ibogaine and Its *O*-Desmethyl Metabolite, Noribogaine

MICHAEL H. BAUMANN, JOHN PABLO, SYED F. ALI, RICHARD B. ROTHMAN, AND DEBORAH C. MASH

Characterization of Multiple Sites of Action of Ibogaine

HENRY SERSHEN, AUDREY HASHIM, AND ABEL LAJTHA

Changes in Gene Expression and Signal Transduction Following Ibogaine Treatment

EMMANUEL S. ONAIVI, B. EMMANUEL AKINSHOLA, SYED F. ALI

Ibogaine in the Treatment of Heroin Withdrawal

DEBORAH C. MASH, CRAIG A. KOVERA, JOHN PABLO, RACHEL TYNDALE, FRANK R. ERVIN, JEFFREY D. KAMLET, AND W. LEE HEARN

Sigma Receptors and *Iboga* Alkaloids

WAYNE D. BOWEN

Ibogaine Neurotoxicity Assessment: Electrophysiological, Neurochemical, and Neurohistological Methods

ZBIGNIEW K. BINIENDA, ANDREW C. SCALLET, LARRY C. SCHMUED, AND SYED F. ALI

Modulation of the Effects of Rewarding Drugs by Ibogaine

LINDA A. PARKER AND SHEPARD SIEGEL

CONTRIBUTORS

Numbers in parentheses indicate the pages on which the authors' contributions begin.

B. Emmanuel Akinshola (135), Department of Pharmacology, College of Medicine, Howard University, Washington, DC 20059, eakinshola@howard.edu

Norma E. Alexander (293), NDA International, 46 Oxford Place, Staten Island NY 10301, nmitsogo@aol.com

Syed F. Ali (79, 135), Division of Neurotoxicology, National Center for Toxicological Research, 3900 NCTR Road, Jefferson, AR 72079, sali@nctr.fda.gov

Kenneth R. Alper (1, 249), Departments of Psychiatry and Neurology, New York University School of Medicine, 550 First Avenue, New York, NY 10016, kra1@nyu.edu

Michael H. Baumann (79), Clinical Psychopharmacology Section, Intramural Research Program, NIDA, National Institutes of Health, Baltimore, MD 21224, mbaumann@irp.nida.nih.gov

Dana Beal (249), Cures-not-Wars, 9 Bleecker Street, New York, NY 10012, dana@cures-not-wars.org

Zbigniew K. Binienda (193), Division of Neurotoxicology, National Center for Toxicological Research, 3900 NCTR Road, Jefferson, AR 72079, zbinienda@nctr.fda.gov

Wayne D. Bowen (173), Laboratory of Medicinal Chemistry, NIDDK, NIH, Building 8 B1-23, 8 Center Drive, MSC 0820, Bethesda, MD 20892, bowenw@bdg8.niddk.nih.gov

Frank R. Ervin (155), Department of Psychiatry and Human Genetics, McGill University, Montreal, Quebec H3A 2T5, Canada, md18@musica.mcgill.ca

James W. Fernandez (235), Department of Anthropology, University of Chicago, 1126 E. 59th Street, Chicago, IL 60637, jwfi@midway.uchicago.edu

RENATE L. FERNANDEZ (235), Department of Anthropology, University of Chicago, 1126 E. 59th Street, Chicago, IL 60637, rlf2@midway.uchicago.edu

GEERTE FRENKEN (283), INTASH, P.O. Box 426113, San Francisco, CA 94142, straightup27@yahoo.com

STANLEY D. GLICK (39), Center for Neuropharmacology and Neuroscience, Albany Medical College, Mail Code 136, Albany, NY 12208, glicks@mail.amc.edu

AUDREY HASHIM (115), Nathan S. Kline Institute for Psychiatric Research, 140 Old Orangeburg Road, Building 35, Orangeburg, NY 10962, hashim@nki.rfmh.org

W. LEE HEARN (155), Metro-Dade County Medical Examiner Department, Number One Bob Hope Road, Miami, FL 33136-1133, WLH@co.miami-dade.fl.us

SCOTT HELSLEY (63), Department of Anesthesiology, Duke University Medical Center, Box 3094 Med. Ctr., Durham, NC 27710, scotthelsley@hotmail.com

JEFFREY D. KAMLET (155), Addiction Treatment Program, Mt. Sinai Medical Center, 300 Arthur Godfrey Road, Suite 200, Miami, FL 33140, kamletmd@aol.com

CHARLES D. KAPLAN (249), Department of Psychiatry and Neuropsychology, Maastricht University, P. O. Box 616, Maastricht 6200MD, The Netherlands, Ch.Kaplan@SP.UNIMAAS.NL

HERBERT KLEBER (xv), Columbia University College of Physicians and Surgeons, 1051 Riverside Drive, New York, NY 10032, hdk3@columbia.edu

CRAIG A. KOVERA (155), Departments of Neurology and Pharmacology, University of Miami School of Medicine, Miami, FL 33124, ckovera@brandinst.com

ABEL LAJTHA (115), Nathan S. Kline Institute for Psychiatric Research, 140 Old Orangeburg Road, Building 35, Orangeburg, NY 10962 lajtha@nki.rfmh.org

HOWARD S. LOTSOF (293), NDA International, 46 Oxford Place, Staten Island, NY 10301, HSL123@aol.com

ISABELLE M. MAISONNEUVE (39), Center for Neuropharmacology and Neuroscience, Albany Medical College, Mail Code 136, Albany, NY 12208, maisoni@mail.amc.edu

Deborah C. Mash (79, 155), Departments of Neurology and Pharmacology, University of Miami School of Medicine, Miami, FL 33124, dmash@ miami.edu

Emanuel S. Onaivi (135), Departments of Psychiatry and Pharmacology, Vanderbilt University School of Medicine, 5428 Country Drive, Nashville, TN 37211, onaivies@ctrvax.vanderbilt.edu

John Pablo (79, 155), Departments of Neurology and Pharmacology, University of Miami School of Medicine, Miami, FL 33124, jpp71@ hotmail.com

Linda A. Parker (211), Department of Psychology, Wilfrid Laurier University, Waterloo, Ontario N2L 3C5, Canada, lparker@machl.wlu.ca

Piotr Popik (227), Institute of Pharmacology, Polish Academy of Sciences, 12 Smetna Street, Krakow 31-343, Poland, nfpopik@cyf-kr.edu.pl

Richard A. Rabin (63), Department of Pharmacology and Toxicology, SUNY Buffalo, 132 Farber Hall, South Campus, Buffalo, NY 14214, rarabin@buffalo.edu

Richard B. Rothman (79), Clinical Psychopharmacology Section, Intramural Research Program, NIDA, National Institutes of Health, Baltimore, MD 21224, rrothman@irp.nida.nih.gov

Andrew C. Scallet (193), Division of Neurotoxicology, National Center for Toxicological Research, 3900 NCTR Road, Jefferson, AR 72079, ascallet@nctr.fda.gov

Larry C. Schmued (193), Division of Neurotoxicology, National Center for Toxicological Research, 3900 NCTR Road, Jefferson, AR 72079, lschmued@nctr.fda.gov

Henry Sershen (115), Nathan S. Kline Institute for Psychiatric Research, 140 Old Orangeburg Road, Building 35, Orangeburg, NY 10962, sershen@nki.rfmh.org

Shephard Siegel (211), Department of Psychology, McMaster University, 1280 Main Street West, Hamilton, Ontario L8S 4K1, Canada, siegel@mcmaster.ca

Phil Skolnick (55), Lilly Research Laboratories, Lilly Corporate Center, Drop Code 0510, Indianapolis, IN 46285, skolnick_phil@lilly.com

Karen K. Szumlinski (39), Center for Neuropharmacology and Neuroscience, Albany Medical College, Mail Code 136, Albany, NY 12208, szumlink@musc.edu

RACHEL TYNDALE (155), Center for Addiction and Mental Health, University of Toronto, Toronto, Ontario M5S 1A1, Canada, r.tyndale@utoronto.ca

J. C. WINTER (63), Department of Pharmacology and Toxicology, SUNY Buffalo, 118a Farber Hall, South Campus, Buffalo, NY 14214, jcwinter@acsu.fjbuffalo.edu

MALGORZATA WRÓBEL (227), Institute of Pharmacology, Polish Academy of Sciences, 12 Smetna Street, Krakow 31-343, Poland, widla@if-pan.krakow.pl

FOREWORD

The use of, abuse of, and dependence on a variety of licit and illicit substances constitute the major public health problem facing the United States and many other countries. Drug abuse is the leading cause of new HIV infections, a major cause of cancer deaths as well as automobile and boat accidents, the largest contributor to our burgeoning prison population, and the largest cause of crime, violence, domestic and child abuse, and community destruction. In the United States there are more than 50 million nicotine addicts, at least 15 million alcoholics and problem drinkers, more than 3 million marijuana addicts, 2 to 3 million cocaine addicts, and more than a million heroin addicts. The number of "hardcore" addicts to illicit drugs is well over 6 million. It is not surprising, given these numbers and social costs, that theories of ways to improve the situation abound. From total prohibition to total legalization and numerous steps in between, arguments rage over the best approach. For experienced observers of the situation, not blinded by partisan beliefs and rhetoric, it appears clear that there is no one answer.

Mencken's observation about simple solutions, "there is always a well-known solution to every human problem—neat, plausible, and wrong," is as true now as in 1920 when he made it. Pure "supply reduction" models founder on the rocks of "need" and "greed"—the desire for the euphoric effects of these agents, and the willingness of individuals to take risks to provide them because of the large profits. Likewise, the pure "demand reduction" model shows its inadequacy by the lack of interest of many addicts in stopping and the failures of our current prevention and treatment programs to either prevent or treat sufficiently. We need both a balanced model and better prevention and treatment methods.

The current view of addiction is a marriage of brain and behavior. Sophisticated imaging procedures and basic science research into the neurobiology of reward have identified key elements in the reinforcing effects of various psychoactive substances. Motivational circuits underlie the desirability of abused drugs. Brain changes after prolonged use help keep the habit going, as well as increase the likelihood of relapse after hard-won abstinence. In one sense, the reward circuitry has been "hijacked" by the rapid intense effects of chemicals at the expense of the more usual rewarding behaviors. Successful treatment thus often requires both medications—to help addicts cope with the brain's changes and urges—and relapse prevention techniques and learning—to help addicts regain the ability to get rewards from nonchemical means. The failure of many treatment attempts is a testimony to the difficulty of the task.

Because of this difficulty, there is a constant search for new methods—better, faster, easier. The search for a "quick fix" is not limited to addicts—researchers, treatment providers, family members, friends, and policy makers share it at times as well. The fact that it hasn't yet been found doesn't mean it can't be found, so efforts continue. The story of ibogaine for addiction is part of that search.

One hundred years ago, as well as recently, treatment of withdrawal was often seen as synonymous with treatment of addiction. Numerous drugs and techniques—some innocuous, some lethal, most in between—were tried to improve withdrawal. None were successful for the larger task of healing addiction, although some have worked reasonably well in treating withdrawal. We still cannot successfully treat a substantial number of addicted individuals. The difficulty may lie both in the persistence of brain changes and the difficulty of making lifestyle changes. The search has been hampered by the intensive warfare between those who believe any medication is a "crutch," and those who view addiction as a medical disorder that may ultimately yield to a combination of medications and behavioral techniques, as employed in other chronic medical conditions. It has also been hampered by the lack of interest of major pharmaceutical firms in devoting resources to the search. Stigma connected to addiction and a perceived lack of possible profitability in a medication have contributed to this unwillingness. Medications could have a variety of roles, some more likely to be found than others: providing a rapid, safe effective withdrawal; decreasing craving; providing a "window of opportunity" for the individual to develop relapse prevention skills and alternative reinforcers; reversing brain changes; blocking or ameliorating the effects of the abused substances; and providing a cost-effective way of reaching larger numbers of individuals.

The diffusion of psychedelic drugs into the larger culture in the 1960s led to a variety of uses. While some people used them for "recreational" purposes, escapism, and altered sensory experiences, others used them in religious activities, serious exploration of altered states, and, at times, formal therapy. LSD, for example, was used in the treatment of alcoholism. Although initially it appeared to yield promising results, manifested by a high percentage of abstinence, follow-up studies demonstrated no sustained efficacy, and efforts were mainly dropped. The rise in the street use of the drug among the young may have contributed in part to the loss of interest among researchers, but lack of efficacy appears to have been a major factor. In contrast, the use of peyote to treat alcoholism in some Native American groups has persisted for decades, perhaps because of its restriction to clear religious ceremonial occasions.

Ibogaine appears to have followed a therapeutic path similar to LSD, but it did not become a street drug, probably because of some unpleasant side effects and possibly weak reinforcing effects. Initially it was touted as both a rapid effective withdrawal method and a cure for heroin and cocaine addiction. Later, as relapses became apparent, it was labeled as an "addiction interrupter," and still later as useful mainly for a small group of "motivated" individuals. In contrast to the situation with LSD, a variety of groups with very different agendas pushed ibogaine's use for therapy—as described succinctly in the chapter by Alper *et al.* in this volume. Because, as noted earlier, commercial interest in addiction treatment medications was minimal, pressure by these disparate groups was aimed at government agencies—especially the National Institute on Drug Abuse (NIDA) and the Food and Drug Administration (FDA)—and individuals, including this author, who were mistakenly, or more likely deliberately for reasons such as their position on other issues of interest to certain groups, targeted for coercive actions. Whether the actions against NIDA were ultimately helpful, harmful, or insignificant in getting the desired results is not totally clear. My own view is there may have been a short-term gain, but a long-term loss, because of the perceived marginalization of the drug.

More important for the long-term goal of developing new medications for addiction was the persistence of scholarly research on ibogaine in both animals and humans. Such research laid out possible mechanisms of action and found metabolites or congeners that may be of more interest than the parent compound. Ultimately the usefulness, or lack thereof, of ibogaine and related compounds in the treatment of addiction will rise or fall on such research. If the drug does have useful effects, it may be possible to develop synthetic agents that produce desired actions on addiction without undesirable effects. In any event, Alper is to be congratulated for both the enormous effort to put together the scientific conference on which this book is based and the book itself, which can bring the findings to a larger audience than was present at the meeting. The need for new medications to treat addiction is as great as ever. Whether or not ibogaine is useful is a scientific question that can be answered neither by street demonstrations nor by avoiding careful, controlled research. As scientists, our obligation is to keep looking for safe and effective methods to prevent and treat this great international scourge.

Herbert Kleber
Columbia University College of Physicians and Surgeons

PREFACE

The responsibilities of an editor for an established book series such as *The Alkaloids: Chemistry and Biology* are twofold: to attract first-rate reviews in well-known areas of alkaloid chemistry and biology and to offer new insights into the breadth and depth of the field as it evolves. Over the years, with chapters on alkaloid biosynthesis and enzymology, the use of alkaloids as chiral auxiliaries, and the therapeutic aspects of alkaloids, the boundaries of the series have been continuously challenged. This volume moves the boundaries even wider, as we examine the social and psychological as well as the chemical, biological, and clinical issues regarding a single alkaloid, ibogaine. No previous volume has been devoted to a single alkaloid or to the proceedings of a conference, but then no previous alkaloid has engendered so much controversy over how to handle its potent biological effects.

This volume brings together sixteen chapters from presentations made at the First International Conference on Ibogaine, held in New York in November 1999. They cover the gamut from indigenous ethnomedical experiences in tropical Africa to diverse clinical trials in Europe and the United States, with much of the extensive chemistry, biology, and pharmacology also described. The volume is compelling reading as one contemplates the vast social impact of various addictive agents (most of which themselves are alkaloids!) and what can be achieved in science to alleviate the personal and societal pressures of profound addiction.

Is ibogaine an alkaloid that can save the world from drug addiction? Probably not. Should it hold a prominent position in the list of antiaddictive strategies under investigation? The evidence reviewed in this book would appear to suggest so. As a paradigm for medication development, with its multifold receptor effects, ibogaine may change the way physicians consider the biological complexities of treating addiction.

With over 25% of the U.S. population addicted to some form of drug (nicotine, alcohol, cocaine, marihuana, and heroin) one must conclude that there is significant financial, not to mention moral and ethical, incentive to examine alternative strategic efforts in alleviating drug addiction. In the future, one hopes that many *iboga* alkaloid derivatives, including 10-hydroxyibogamine and 18-methoxycoronaridine, will be made available for a wider and deeper exploration of the power of this alkaloid skeleton to modulate those receptors in the brain relating to addiction, a bane that causes such interminable suffering to individuals, families, and societies around the world.

Geoffrey A. Cordell
University of Illinois at Chicago

PREFACE

The First International Conference on Ibogaine, held at the New York University School of Medicine on November 5 and 6, 1999, was remarkable for its blend of the instrumental and the expressive. On the instrumental side, presentations of those from the academic research community, the United States Food and Drug Administration (FDA), and the National Institute on Drug Abuse (NIDA) included preclinical data on ibogaine's pharmacological actions, evidence of efficacy and safety in animal models, and case reports in humans. On the expressive side was representation of the sacred culture of Bwiti in Africa and the medical subculture of ibogaine in the United States and Europe. Charts and graphs of data were a significant aspect of the fabric of the experience of the event, as were the emotionally salient presentations of Bwiti initiates and the attestations of those who reported having benefited from ibogaine in the treatment of opioid dependence and who advocated earnestly for its development.

The conference was organized in the belief that the *iboga* alkaloids are an interesting pharmacologic paradigm for the development of the treatment of addiction. If this indeed turns out to be the case, then the optimal clinical approach to their use will demand integration of the instrumental and expressive, an imperative of importance in medicine, and particularly in the treatment of addiction. The medical dictum "our patients are our greatest teachers," is a statement about the instrumental importance of observation and experience and the expressive importance of empathy. This dictum may be of particular relevance to ibogaine, in which a significant portion of the collective clinical experience has originated from an addict self-help context involving individuals without formal medical credentials, and the addicts themselves.

Deep gratitude is extended to the participants of this First International Conference on Ibogaine for having listened to one another and for their contributions to this volume.

Kenneth Alper
New York University School of Medicine

Stanley Glick
Albany Medical College

A NOTE CONCERNING THE NUMBERING OF *IBOGA* ALKALOIDS

Two systems exist for numbering the carbon and nitrogen atoms within the diversity of the monoterpenoid indole alkaloids, which currently comprise some 5000 structures. Among these structures are the derivatives of ibogamine, the parent structure of ibogaine. One numbering system applied is that used by *Chemical Abstracts,* in which numbers are assigned systematically to the various alkaloid skeleta on an individual skeleton basis. As a result, corresponding carbons in slightly different skeleta can have quite different numbers. The other approach, a biogenetic one, was developed by Le Men and Taylor (*1*). In this numbering method, the monoterpenoid indole alkaloids are numbered uniformly and systematically based on a parent carbon skeleton. Consequently, corresponding carbon atoms in very different alkaloid structures, such as quinine, ibogaine, and camptothecin, can be directly related. Historically, the Le Men and Taylor system is used by natural product chemists and many biologists, while the *Chemical Abstracts* approach is found in aspects of the medical and biological literature.

The literature on the *iboga* alkaloids reflects the use of both the Le Men and Taylor and *Chemical Abstracts* systems and is therefore a potential source of confusion. For example, it has been common in the current medical literature to refer to ibogaine as 12-methoxyibogamine, and to the metabolite noribogaine as 12-hydroxyibogamine. In the Le Men and Taylor system these alkaloids are 10-methoxyibogamine and 10-hydroxyibogamine, respectively. On the other hand, 18-methoxycoronaridine (18-MC), a synthetic *iboga* alkaloid derivative of current interest, is named according to the Le Men and Taylor system. In this volume, the use of the Le Men and Taylor system was recommended, but the choice was left to the discretion of the individual contributors.

The premise that underlies this volume and the First International Conference on Ibogaine is that the *iboga* alkaloids are pharmacologically interesting and potentially clinically valuable. If this is accepted, then synthetic natural product chemists will likely produce a profusion of *iboga* alkaloid derivatives for biological evaluation. These compounds will be numbered utilizing the Le Men and Taylor system, as was 18-MC. It is therefore recommended that the Le Men and Taylor system be adopted as the normative approach to the chemical nomenclature of the *iboga* alkaloids in both the chemical and the biological literature.

1. J. Le Men and W. I. Taylor, *Experientia* **21,** 508 (1965).

—CHAPTER 1—

IBOGAINE: A REVIEW

KENNETH R. ALPER

Departments of Psychiatry and Neurology
New York University School of Medicine
New York, NY 10016

THE ALKALOIDS, Vol.56
0099-9598/01 $35.00

I. Introduction, Chemical Properties, and Historical Time Line

A. Introduction

Ibogaine, a naturally occurring plant alkaloid with a history of use as a medicinal and ceremonial agent in West Central Africa, has been alleged to be effective in the treatment of drug abuse. The National Institute on Drug Abuse (NIDA) has given significant support to animal research, and the U.S. Food and Drug Administration (FDA) has approved Phase I studies in humans. Evidence for ibogaine's effectiveness includes a substantial preclinical literature on reduced drug self-administration and withdrawal in animals, and case reports in humans. There is relatively little financial incentive for its development by the pharmaceutical industry because ibogaine is isolated from a botanical source in which it naturally occurs, and its chemical structure cannot be patented. This has left the academic community and the public sector with a crucial role in research on ibogaine, which was a major reason for organizing the First International Conference on Ibogaine.

A major focus of the Conference was the possible mechanism(s) of action of ibogaine. Ibogaine is of interest because it appears to have a novel mechanism of action distinct from other existing pharmacotherapeutic approaches to addiction, and it potentially could provide a paradigm for understanding the neurobiology of addiction and the development of new treatments. Another important focus of the Conference was to review human experience with ibogaine and preclinical and clinical evidence of efficacy and safety. The Conference also featured presentations related to the sociological and anthropological aspects of the sacramental context of the use of iboga in Africa and the distinctive ibogaine subculture of the United States and Europe.

B. Chemical Structure and Properties

Ibogaine (10-methoxyibogamine) (Figure 1) is an indole alkaloid with molecular formula $C_{20}H_{26}N_2O$ and molecular weight 310.44. Ibogaine is the most abundant alkaloid in the root bark of the Apocynaceous shrub *Tabernanthe iboga*, which grows in West Central Africa. In the dried root bark, the part of the plant

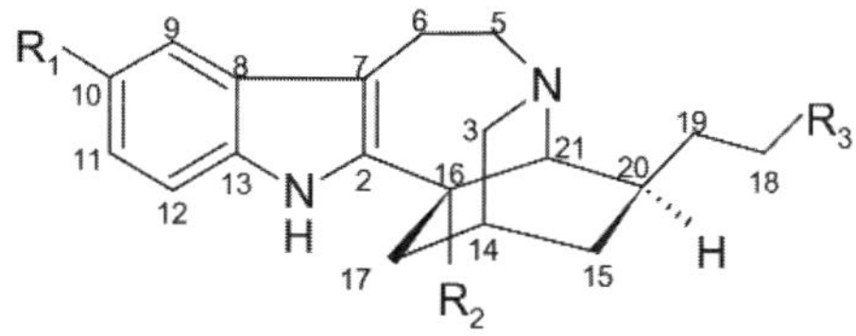

Alkaloid	R_1	R_2	R_3
Ibogaine	OCH_3	H	H
Noribogaine	OH	H	H
(±)-18-Methoxycoronaridine	H	CO_2CH_3	OCH_3

FIGURE 1. CHEMICAL STRUCTURES OF IBOGAINE, NORIBOGAINE, AND 18-METHOXYCORONARIDINE. The ibogamine skeleton above is numbered using the LeMen and Taylor system in which ibogaine is designated as 10-methoxyibogamine and noribogaine as 10-hydroxyibogamine. Alternatively, according to the Chemical Abstracts numbering system for the ibogamine skeleton which is frequently encountered in the biological and medical literature, ibogaine and noribogaine have respectively been referred to as 12-methoxyibogamine and 12-hydroxyibogamine.

in which alkaloid content is highest, total alkaloid content is reportedly 5 to 6% (*1*).

Ibogaine has a melting point of 153°, a pK_a of 8.1 in 80% methylcellosolve, and it crystallizes as prismatic needles from ethanol. Ibogaine is levorotatory $[\alpha]_D$ –53° (in 95% ethanol), soluble in ethanol, ether, chloroform, acetone and benzene, but it is practically insoluble in water. Ibogaine is decomposed by the action of heat and light. Ibogaine hydrochloride decomposes at 299°, is also levorotatory $[\alpha]_D$ –63° (ethanol), $[\alpha]_D$ –49° (H_2O), and is soluble in water, methanol, and ethanol, slightly soluble in acetone and chloroform, and practically insoluble in ether (*2*). The X-ray crystal analysis that confirmed the structure of ibogaine has been described (*3*). The literature provides references to the mass spectrum of ibogaine (*4*), and the proton (*5,6*) and the ^{13}C (*7-9*) NMR spectra of ibogaine and other *iboga* alkaloids. Analytic chemical methods for extraction, derivatization, and detection of ibogaine utilizing combined gas chromatography-mass spectometry have been described (*10-13*).

Ibogaine undergoes demethylation to form its principal metabolite, noribogaine, also known as *O*-desmethylibogaine or 10-hydroxyibogamine. 18-methoxycoronaridine (18-MC, see Glick *et al.* in this volume) is an ibogaine congener that appears to have efficacy similar to ibogaine in animal models of drug dependence with evidence of less potential toxicity.

C. Historical Time Line

The following time line outlines the historical events relating to the development of ibogaine as a treatment for drug dependence. Elsewhere in this volume, Alper *et al.* provide a more detailed contemporary history of ibogaine in the United States and Europe.

1864: The first description of *T. iboga* is published. A specimen is brought to France from Gabon. A published description of the ceremonial use of *T. iboga* in Gabon appears in 1885 (*14*).

1901: Ibogaine is isolated and crystallized from *T. iboga* root bark (*15-17*).

1901-1905: The first pharmacodynamic studies of ibogaine are performed. During this period ibogaine is recommended as a treatment for "asthenia" at a dosage range of 10 to 30 mg per day (*14*).

1939-1970: Ibogaine is sold in France as Lambarène, a "neuromuscular stimulant," in 8 mg tablets, recommended for indications that include fatigue, depression, and recovery from infectious disease (*14*).

1955: Harris Isbell administers doses of ibogaine of up to 300 mg to eight already detoxified morphine addicts at the U.S. Addiction Research Center in Lexington, Kentucky (*18*).

1957: The description of the definitive chemical structure of ibogaine is published. The total synthesis of ibogaine is reported in 1965 (*19-21*).

1962-1963: In the United States, Howard Lotsof administers ibogaine to 19 individuals at dosages of 6 to 19 mg/kg, including 7 with opioid dependence who note an apparent effect on acute withdrawal symptomatology (*22,23*).

1967-1970: The World Health Assembly classifies ibogaine with hallucinogens and stimulants as a "substance likely to cause dependency or endanger human health." The U.S. Food and Drug Administration (FDA) assigns ibogaine Schedule I classification. The International Olympic Committee bans ibogaine as a potential doping agent. Sales of Lambarène cease in France (*14*).

1969: Dr. Claudio Naranjo, a psychiatrist, receives a French patent for the psychotherapeutic use of ibogaine at a dosage of 4 to 5 mg/kg (*24*).

1985: Howard Lotsof receives a U.S. patent for the use of ibogaine in opioid withdrawal (*22*). Additional patents follow for indications of dependence on cocaine and other stimulants (*23*), alcohol (*25*), nicotine (*26*), and polysubstance abuse (*27*).

1988-1994: U.S. and Dutch researchers publish initial findings suggestive of the efficacy of ibogaine in animal models of addiction, including diminished opioid self-administration and withdrawal (*28-30*), as well as diminished cocaine self-administration (*31*).

1989-1993: Treatments are conducted outside of conventional medical settings in the Netherlands involving the International Coalition of Addict Self-Help (ICASH), Dutch Addict Self Help (DASH), and NDA International (*22,32-35*).

1991: Based on case reports and preclinical evidence suggesting possible efficacy, NIDA Medication Development Division (MDD) begins its ibogaine project. The major objectives of the ibogaine project are preclinical toxicological evaluation and development of a human protocol.

August 1993: FDA Advisory Panel meeting, chaired by Medical Review Officer Curtis Wright, is held to formally consider Investigational New Drug Application filed by Dr. Deborah Mash, Professor of Neurology at the University of Miami School of Medicine. Approval is given for human trials. The approved ibogaine dosage levels are 1, 2, and 5 mg/kg. The Phase I dose escalation study begins December 1993, but activity is eventually suspended (*36*).

October 1993-December 1994: The National Institute on Drug Abuse (NIDA) holds a total of four Phase I/II protocol development meetings, which include outside consultants. The resulting draft protocol calls for the single administration of fixed dosages of ibogaine of 150 and 300 mg versus placebo for the indication of cocaine dependence (*37*).

March 1995: The NIDA Ibogaine Review Meeting is held in Rockville, Maryland, chaired by the MDD Deputy Director, Dr. Frank Vocci. The possibility of NIDA funding a human trial of the efficacy of ibogaine is considered. Opinions of representatives of the pharmaceutical industry are mostly critical, and are a significant influence in the decision not to fund the trial. NIDA ends its ibogaine project, but it does continue to support some preclinical research on *iboga* alkaloids.

Mid 1990s-2001: Ibogaine becomes increasingly available in alternative settings, in view of the lack of approval in the Europe and the United States. Treatments in settings based on a conventional medical model are conducted in

Panama in 1994 and 1995 and in St. Kitts from 1996 to the present. Informal scenes begin in the United States, Slovenia, Britain, the Netherlands, and the Czech Republic. The Ibogaine Mailing List (*38*) begins in 1997 and heralds an increasing utilization of the Internet within the ibogaine medical subculture.

II. Mechanisms of Action

A. Neurotransmitter Activities

1. General Comments

Elsewhere in this volume, Glick *et al.*, Sershen *et al.*, and Skolnick review the mechanism of action of ibogaine. Popik and Skolnick (*39*) provide a recent, detailed review of ibogaine's receptor activities. Ibogaine appears to have a novel mechanism of action that differs from other existing pharmacotherapies of addiction, and its mechanism of action does not appear to be readily explained on the basis of existing pharmacologic approaches to addiction. Ibogaine's effects may result from complex interactions between multiple neurotransmitter systems rather than predominant activity within a single neurotransmitter system (*39-42*).

Several laboratories have reported on the results of pharmacological screens of the receptor binding profile of ibogaine (*40,43-45*). Ibogaine has low micromolar affinities for multiple binding sites within the central nervous system, including *N*-methyl-D-aspartate (NMDA), kappa- and mu-opioid and sigma$_2$ receptors, sodium channels, and the serotonin transporter. Although not apparent in binding studies, functional studies indicate significant activity of ibogaine as a noncompetitive antagonist at the nicotinic acetylcholine receptor (*46-50*).

Although *in vitro* activities in the micromolar range are often described as ancillary in attempting to characterize a drug's *in vivo* mechanism of action, micromolar activity may be pharmacologically important with regard to ibogaine or noribogaine due to the relatively high concentrations reached in the brain (*40,44,51*). Hough *et al.* (*51*) noted a brain level of ibogaine of 10 μM in female rats at 1 hour after the administration of 40 mg/kg ibogaine intraperitoneally (i.p.), which is the usual dosage, animal, gender and route of administration used in that laboratory to investigate ibogaine's effects on drug self-administration and withdrawal. Brain levels of ibogaine, and its major metabolite noribogaine, ranged from 1 to 17 μM between 15 minutes and 2 hours in male rats following the oral administration ibogaine at a dose of 50 mg/kg (*44*).

2. Glutamate

Elsewhere in this volume, Skolnick reviews the possible relevance of ibogaine's activity as a glutamate antagonist to its putative effects in drug dependence. There is evidence that suggests that antagonists of the *N*-methyl-D-aspartate (NMDA) subtype of glutamate receptor are a potentially promising class of agents for the development of medications for addiction (*52-54*). Ibogaine's apparent activity as a noncompetitive NMDA antagonist has been suggested to be a possible mechanism mediating its putative effects on drug dependence (*39,41,55-58*).

Ibogaine competitively inhibits the binding of the NMDA antagonist MK801 to the NMDA receptor complex, with reported affinities in the range of 0.02 to 9.8 μM (*40,45,55-57,59,60*). Functional evidence supporting an antagonist action of ibogaine at the NMDA receptor includes observations of reduced glutamate-induced cell death in neuronal cultures, reduction of NMDA-activated currents in hippocampal cultures (*55,58*), prevention of NMDA-mediated depolarization in frog motoneurons (*59*), and protection against NMDA-induced convulsions (*61*). Glycine, which acts as an NMDA co-agonist by binding at the NMDA receptor, attenuates ibogaine's effect of blocking naloxone-precipitated jumping (*58*). MK801 and ibogaine do not produce identical effects, as evidenced by the observation that in the rat brain ibogaine lowered the concentration of dopamine while increasing the level of its metabolites, whereas MK801 did not have these effects (*62,63*).

3. *Opioid*

It has been suggested that ibogaine's or noribogaine's activity as a putative agonist at mu-opioid receptors might explain ibogaine's apparent efficacy in opioid withdrawal (*36,64,65*). Ibogaine binds to mu-opioid receptors with reported binding affinities in the range of 0.13 to 26 μM (*40,45,64,66*), with one study reporting a result in excess of 100 μM (*43*). Ibogaine behaves as an agonist in a functional assay for mu-opioid receptors, the binding of [^{35}S]-GTPγS (*65*). However, some observations are difficult to reconcile with a mu-agonist action of ibogaine. Ibogaine did not behave as a mu-opioid agonist in assays with isolated smooth muscle preparations (*67*). Unlike mu-opioid agonists, ibogaine (*68-70*) and noribogaine (*71*) do not appear by themselves to have antinociceptive effects.

Some findings suggest the intriguing possibility that ibogaine may act at the level of second messenger signal transduction to enhance the functional activity of mu-opioid receptors independently of any direct agonist interaction at opioid receptors. Both ibogaine and noribogaine reportedly potentiated morphine-induced inhibition of adenylyl cyclase *in vitro* with opioid receptors already occupied by the maximally effective concentration of morphine, but did not affect adenylyl cyclase in the absence of morphine (*72*). A similar interpretation might also explain the finding that ibogaine inhibited the development of tolerance to the antinociceptive effect of morphine in mice, without by itself affecting nociception (*73*).

Ibogaine binds to kappa-opioid receptors with reported binding affinities in the range of 2.2 to 30 μM (*43,45,56,66*). Evidence consistent with a kappa-opioid action of ibogaine includes the observation that the kappa-opioid antagonist, norbinaltorphimine antagonized some of the effects of ibogaine in morphine-treated rats (*74,75*). Kappa-opioid agonists reportedly can imitate certain effects of ibogaine, such as reduced cocaine and morphine self-administration (*76*), and reduction in locomotor activation to morphine accentuated by prior morphine exposure (*77*). Sershen, on the other hand, attributes a kappa-opioid antagonist action to ibogaine, based on the observation that stimulated dopamine efflux from mouse brain slices was decreased by a kappa-opioid agonist, and the decrease was offset by the addition of ibogaine (*78*). However, ibogaine's interactions with multiple neurotransmitter systems raises the possibility that the finding could be accounted for by mechanisms that do not involve the kappa-opioid receptor, as dopamine efflux is modulated by multiple neurotransmitters.

4. Serotonin

Ibogaine and serotonin both contain an indole ring in their structure, and ibogaine has been shown to bind to the serotonin transporter and to increase serotonin levels in the nucleus accumbens (NAc) (*41,79,80*). The demonstration that ibogaine blocks serotonin uptake (*81*) suggests that the effect of ibogaine on extracellular serotonin levels may be mediated by uptake inhibition, in addition to release (*80*). The reported affinity of ibogaine for the serotonin transporter ranges from 0.55 to 10 μM (*39,44,45,79,81*), and the affinity of noribogaine for the serotonin transporter is approximately 10-fold stronger (*45,79*). The magnitude of the effect of ibogaine on serotonin release is reportedly large and is comparable to that of the serotonin releasing agent fenfluramine, with noribogaine having a lesser effect, and 18-MC no effect (*80*). Some authors suggest a role for modulatory influence of serotonin in ibogaine's effects on dampening dopamine efflux in the NAc (*41,80*).

Ibogaine's hallucinogenic effect has been suggested to involve altered serotonergic neurotransmission (*42,80*). Ibogaine is reported in some studies to bind the 5-HT_{2A} receptor, which is thought to mediate the effects of "classical" indolealkylamine and phenethylamine hallucinogens (*82*), with three studies reporting affinities in the range of 4.1 to 12 μM (*40,45,83*), one reporting a value of 92.5 μM (*84*), and with two other studies reporting no significant affinity (*43,44*). Drug discrimination studies provide some functional evidence for the action of ibogaine as an agonist at the 5-HT_{2A} receptor, which is apparently a significant, although nonessential, determinant of the ibogaine stimulus (*84*) (see Section II.B, "Discrimination Studies"). Ibogaine binds to the 5-HT_3 receptor with reported affinities of 2.6 and 3.9 μM (*40,45*), and it was without significant affinity in two other studies (*43,83*). The 5-HT_3 receptor is apparently not involved in the ibogaine discriminative stimulus (*85*).

5. Dopamine

Ibogaine does not appear to significantly affect radioligand binding to D_1, D_2, D_3, or D_4 receptors (*40,43,44*) and is a competitive blocker of dopamine uptake at the dopamine transporter with affinities in the range of 1.5 to 20 μM (*81*). Where affinities for the serotonin and dopamine transporter have been estimated within the same study, the reported affinity of ibogaine for the serotonin transporter has generally been 10 to 50 times stronger than its affinity for the dopamine transporter (*44,79,81*). Ibogaine does not apparently affect norepinephrine reuptake (*44,45*).

French *et al.* (*86*) studied the electrophysiological activity of dopamine neurons in the ventral tegmental area (VTA) of rats given up to 7.5 mg/kg ibogaine intravenously and found a significant increase in firing rate. Ibogaine given i.p. at a dose of 40 mg/kg did not affect the spontaneous firing of VTA dopamine neurons or the response of VTA dopamine neurons to cocaine or morphine. Ibogaine reportedly lowers the concentration of dopamine, while increasing the level of its metabolites, indicating diminished release of dopamine in the brain of the rat (*62,63*) and the mouse (*87*). Decreased release of dopamine could possibly explain the observation of increased prolactin release following ibogaine administration (*62,63,88*). Staley *et al.* (*44*) have suggested that ibogaine might act at the dopamine transporter to inhibit the translocation of dopamine into synaptic vesicles, thereby redistributing dopamine from vesicular to cytoplasmic pools. As a result, the metabolism of dopamine by monoamine oxidase could explain the observation of decreased tissue dopamine content with increased levels of its metabolites.

The effects of ibogaine on dopamine efflux in response to the administration of drugs of abuse are described in Section III.E, "Dopamine Efflux".

6. Acetylcholine

Ibogaine is a nonselective and weak inhibitor of binding to muscarinic receptor subtypes. Reported affinities are 7.6 and 16 μM and 5.9 and 31 μM, respectively, for the M_1 and M_2 muscarinic receptor subtypes (*40,45*), with another study reporting no significant affinity of ibogaine for muscarinic receptors (*43*). Functional evidence consistent with a muscarinic cholinergic agonist effect of ibogaine includes the observations of the elimination of ibogaine-induced EEG dyssynchrony by atropine in cats (*89*), decreased heart rate following ibogaine administration in rats (*90*), and the attribution of the effect of cholinesterase inhibition to ibogaine in the older literature (*1,91*). The affinity of noribogaine for muscarinic receptors is apparently similar to that of ibogaine (*44,45*).

Several laboratories have reported that ibogaine produces noncompetitive functional inhibition of the nicotinic acetylcholine receptor, apparently involving open channel blockade (*46,48-50*). As with a number of other channel blockers, binding studies involving channels associated with nicotinic receptors have been limited by the lack of appropriate ligands, and investigations of the affinity of ibogaine for the nicotinic acetylcholine receptor have mainly involved functional assays. Utilizing $^{86}Rb^{+}$ efflux assays, Fryer and Lukas (*50*) found that ibogaine inhibited human ganglionic and muscle-type nicotinic acetylcholine receptors with IC_{50} values of 1.06 and 22.3 μM, respectively. Badio *et al.* (*48*) found that ibogaine inhibited $^{22}Na^{+}$ influx through rat ganglionic and human muscle-type nicotinic acetylcholine receptors with IC_{50} values of 0.020 μM and 2.0 μM, respectively. Noribogaine was 75-fold less active than ibogaine in the rat ganglionic cell assay. In mice, ibogaine at a dose of 10 mg/kg completely blocked the central antinociceptive nicotinic receptor-mediated response to epibatidine. Ibogaine has been associated with decreased acetylcholine-stimulated nicotinic receptor mediated catecholamine release in cultured cells (*49*) and decreased dopamine release evoked by nicotine in the NAc of the rat (*46,92*).

7. *Sigma Receptors*

Elsewhere in this volume, Bowen discusses ibogaine's action at the sigma receptor. The affinity of ibogaine for the $sigma_2$ receptor is strong relative to other known CNS receptors, and the reported range is 0.09 to 1.8 μM (*45,60,93,94*). The affinity of ibogaine for the $sigma_1$ receptor is reportedly on the order of 2 to 100 times weaker than its affinity for the $sigma_2$ receptor (*45,60,93,94*). The neurotoxic effects of ibogaine may involve activity at the $sigma_2$ receptor, which reportedly potentiates the neuronal response to NMDA (*95*).

8. *Sodium Channels*

The reported affinity of ibogaine for sodium channels ranges from 3.6 to 9 μM (*40,43*). There is apparently no experimental evidence regarding the functional significance of ibogaine's action at sodium channels.

B. Discrimination Studies

Elsewhere in this volume, Helsley *et al.* discuss the topic of ibogaine and drug discrimination. Drug discrimination studies offer a possible approach to the issue of ibogaine's mechanism of action and may help resolve the distinction between ibogaine's therapeutic and hallucinogenic effects. The $5\text{-}HT_{2A}$ receptor appears to be a significant, but nonessential, determinant of the ibogaine stimulus (*84,96*). The ibogaine stimulus is reportedly generalized to the indolealkylamine hallucinogen D-lysergic acid diethylamide (LSD) and the phenethylamine hallucinogen 2,5-dimethoxy-4-methylamphetamine (DOM), and this generalization is abolished by the addition of a $5\text{-}HT_{2A}$ receptor antagonist (*96*). The addition of a $5\text{-}HT_{2A}$ receptor antagonist did not attenuate stimulus control of ibogaine itself in the ibogaine-trained animals, indicating that $5\text{-}HT_{2A}$ is not an essential component of the ibogaine discriminative stimulus. The $5\text{-}HT_{2C}$

receptor, which plays a modulatory role in hallucinogenesis, is also involved, but is not essential to the ibogaine stimulus, and the 5-HT_{1A} and 5-HT_3 receptors are apparently not involved in the ibogaine stimulus (*85*). The ibogaine discriminative stimulus reportedly is potentiated by the serotonin reuptake inhibitor fluoxetine (*85*), and has an insignificant degree of generalization to the serotonin releaser D-fenfluramine (*97*).

Ibogaine showed a lack of substitution for phencyclidine (*98,99*), and substituted for MK 801 only at high (100 mg/kg) doses in mice (*58,61*), but not at lower (10 mg/kg) doses in rats (*99,100*), suggesting that the NMDA receptor is not a significant determinant of the ibogaine stimulus. Sigma_2, and mu- and kappa-opioid activity may be involved in the ibogaine discriminative stimulus (*99*). A high degree of stimulus generalization is reported between ibogaine and some of the Harmala alkaloids, a group of hallucinogenic beta-carbolines that are structurally related to ibogaine (*101,102*). While the discriminative stimulus for both the Harmala alkaloids and ibogaine apparently involves the 5-HT_2 receptor (*84,85,103*), it does not appear essential to generalization between ibogaine and harmaline, as generalization to the harmaline stimulus was unaffected by the addition of a 5-HT_2 antagonist in ibogaine-trained animals (*84*). Ibogaine-trained rats generalize to noribogaine (*100,104*), which in one study was more potent than ibogaine itself in eliciting ibogaine-appropriate responses (*100*).

C. Effects on Neuropeptides

Both ibogaine and cocaine given in multiple administrations over 4 days to rats reportedly increase neurotensin-like immunoreactivity (NTLI) in the striatum, substantia nigra, and NAc (*105*). However, unlike cocaine, which increased NTLI in the frontal cortex, ibogaine had no effect on frontal cortical NTLI. Ibogaine pretreatment prevented the increase of NTLI in striatum and substantia nigra induced by a single dose of cocaine. Substance P, like NTLI, was increased in the striatum and substantia nigra after either cocaine or ibogaine, with an increase in frontal cortex with cocaine and no effect with ibogaine (*106*). Ibogaine–induced increases in NTLI or substance P were blocked by administration of a D_1 antagonist.

Unlike the NTLI or substance P responses, ibogaine alone had no effect on dynorphin. However, ibogaine pretreatment dramatically enhanced cocaine-induced increases in dynorphin, a kappa-opioid agonist (*107*). The authors suggested that the increase in dynorphin related to cocaine's interaction with ibogaine could result in enhanced kappa-opioid activity. Kappa-opioid agonists reportedly decrease cocaine intake in animal models (*108,109*).

D. Possible Effects on Neuroadaptations Related to Drug Sensitization or Tolerance

There is some evidence to suggest that ibogaine treatment might result in the "resetting" or "normalization" of neuroadaptations related to drug sensitization or tolerance (*110*). Ibogaine pretreatment blocked the expression of sensitization-induced increases in the release of dopamine in the NAc shell in response to cocaine in cocaine-sensitized rats (*111*). The effect of ibogaine on diminished locomotor activity and dopamine efflux in the NAc in response to morphine is more evident in animals with prior exposure to morphine (*112,113*), which is consistent with a relatively selective effect of ibogaine on neuroadaptations acquired from drug exposure. Similarly, the observation that ibogaine inhibited the development of tolerance in morphine-tolerant mice, but had no effect on morphine nociception in morphine-naïve mice (*114*), suggests a selective effect on acquired neuroadaptations related to repeated morphine exposure.

Ibogaine appears to have persistent effects not accounted for by a metabolite with a long biological half-life (*29,115*). Ibogaine's action could possibly involve the opposition or reversal of persistent neuroadaptive changes thought to be associated with drug tolerance or sensitization. Such an action could be related to persistent effects on second messengers (*72,116*). For example, sensitization to both opiates and cocaine is thought to involve enhanced stimulation of cyclic AMP (*117*). Ibogaine has been reported to potentiate the inhibition of adenylyl cyclase by serotonin (*72*), an effect that would be expected to oppose the enhanced transduction of cyclic AMP that is reportedly associated with stimulant sensitization (*117*).

III. Evidence of Efficacy in Animal Models

A. Drug Self-Administration

Evidence for ibogaine's effectiveness in animal models of addiction includes observations of reductions in self-administration of morphine or heroin (*29,31,118-120*), cocaine (*29,31,119,121*), and alcohol (*122*), and reduced nicotine preference (*75*). According to some reports, effects of ibogaine on drug self-administration are apparently persistent. Sershen *et al.* (*121*) administered ibogaine i.p. to mice as two 40 mg/kg dosages 6 hours apart, and found a diminution of cocaine preference that was still evident after 5 days. Glick *et al.* (*29,119*) noted reductions in cocaine and morphine self-administration that persisted for at least 2 days and were dose dependent in the range of 2.5 to 80 mg/kg. ibogaine given i.p. The persistence of an effect beyond the first day suggests a specific action of ibogaine on drug intake, as water intake was also suppressed initially by ibogaine on the first, but not the second day. Cappendijk and Dzoljic (*31*) found reductions in cocaine self-administration that persisted for more than 48 hours in rats treated with ibogaine at a dose of 40 mg/kg i.p., given as a single administration, or repeatedly on 3 consecutive days or three consecutive weeks.

In the studies by Glick *et al.* there was variation between results in individual rats with some showing persistent decreases in morphine or cocaine intake for several days or weeks after a single injection and others only after two or three weekly injections. The authors noted evidence of a continuous range of individual sensitivity to ibogaine among the experimental animals and that it appeared as if adjustments of the dosage regimen could produce long-term reductions in drug intake in most animals (*29*). Similarly, Cappendijk and Dzoljic (*31*) found the largest effects on cocaine self-administration occurred when ibogaine was given weekly for three consecutive weeks. This result suggests the possibility that the optimal schedule of ibogaine administration to limit cocaine intake may involve modification of the single dose regimen which has been used for opioid detoxification (*32,123*).

Dworkin *et al.* (*118*) found that pretreatment with ibogaine at a dose of 80 mg/kg i.p. diminished the response for heroin and cocaine, and also for food, suggesting a nonspecific confound. A 40 mg/kg i.p. dose of ibogaine sharply reduced heroin self-administration in the absence of a significant effect on food response, although the effect did not persist beyond 24 hours (*118*). Dworkin *et al.* cited methodologic factors relating to differences in gender, strain, and reinforcement schedule to explain the apparent discrepancy between their results and other studies that reported persistent effects (*29,31,119,121*).

Noribogaine has also been reported to reduce cocaine and morphine self-administration (*124*). The effect of noribogaine on drug self-administration persisted for 2 days, after the response for water, which was initially suppressed on the first day, had returned to baseline. Other *iboga* alkaloids have also been reported to reduce morphine and cocaine self-administration in rats for a period of a day or longer following a single i.p dose (*119*). Some of the *iboga* alkaloids tested in this study produced tremors, which typically occurred for a period of 2 to 3 hours, and were independent of persistent effects of drug self-administration. An ibogaine congener, 18-methoxycoronaridine (18-MC) (*45*), reportedly reduces in rats the self-administration of cocaine (*120*), morphine and alcohol (*125*), and nicotine preference (*75*) without any apparent reduction in the response for water.

B. Acute Opioid Withdrawal

Dzoljic *et al.* (*28*) administered ibogaine in a dose range of 4 to 16 μg intracerebroventricularly to rats and observed a dose-dependent attenuation of naloxone-precipitated withdrawal signs. This same group also found an attenuation of morphine withdrawal signs in rats with 40 mg/kg ibogaine administered i.p., and also norharman, an endogenously occurring hallucinogenic beta-carboline and a structural relative of ibogaine (*126*). Glick *et al.* have reported dose-dependent reduction of the signs of naltrexone-precipitated morphine withdrawal in rats administered ibogaine at doses of 20, 40, or 80 mg/kg i.p (*127*) or 18-MC (*128*) at doses of 20 and 40 mg/kg i.p. Attenuation of withdrawal signs was reported in morphine-dependent monkeys given 2 or 8 mg/kg ibogaine subcutaneously (*129*). In their chapter in this volume, Parker and Siegel report that 40 mg/kg ibogaine administered i.p attenuated naloxone-precipitated morphine withdrawal in rats, as well as withdrawal-induced place aversion.

Sharpe and Jaffe (*130*) reported that ibogaine in dosages ranging between 5 and 40 mg/kg administered subcutaneously failed to attenuate naloxone-precipitated withdrawal in rats, although they did find that one sign (grooming) was reduced, and noted the possible effect of methodological issues such as morphine exposure and withdrawal procedures, or the route of administration of ibogaine. Popik *et al.* (*58*) and Layer *et al.* (*56*) found that ibogaine at doses ranging from 40 to 80 mg/kg i.p. reduced naloxone-precipitated jumping in morphine dependent mice, although Francés *et al.* (*69*) found the opposite effect with 30 mg/kg ibogaine administered i.p. in mice. As pointed out by Popik and Skolnik (*39*), the divergent results in morphine dependent mice might relate to ibogaine having been given prior to the administration of naloxone in the studies by Popik *et al.* (*58*) and Layer *et al.* (*56*), whereas ibogaine was administered after naloxone in the study by Francés *et al.*

C. Conditioned Place Preference

Parker and Siegel review ibogaine and place preference in this volume. Ibogaine is reported to prevent the acquisition of place preference when given 24 hours before amphetamine (*131*) or morphine (*132*). The effect of ibogaine on blocking the acquisition of place preference was diminished across multiple conditioning trials. Ibogaine given after morphine did not apparently attenuate the expression of previously established morphine place preference (*133*).

D. Locomotor Activity

Pretreatment with ibogaine and its principal metabolite, noribogaine reportedly diminishes locomotor activation in response to morphine (*74,112,113,124,134-136*). The effect of ibogaine in reducing locomotor activity in response to morphine is reportedly greater in female than in male rats, probably reflecting the relatively greater bioavailability of ibogaine in females (*135*). The literature on cocaine appears to be less consistent, with some reports of decreased locomotor activation (*87,137-139*), and others reporting increases (*127,137,140,141*). This apparent disparity may be related in part to the species of experimental animal that was used, as Sershen *et al.* (*137*) report increased locomotor activity in response to cocaine in the rat, with the opposite result in the mouse.

Stereotypy is a methodologic issue that might explain some of the disparate results regarding ibogaine's interaction with the locomotor response to cocaine. Higher doses of stimulants can produce strereotypy, which could decrease the amount of measured locomotion relative to an animal that is experiencing less locomotor stimulation at a lower stimulant dose. Thus, the potentiation by ibogaine of locomotor activity related to cocaine administration can result in less measured movement in animals experiencing locomotor stimulation to the point of stereotypy (*110*). Ibogaine pretreatment reportedly potentiates stereotypy in rats receiving cocaine or methamphetamine (*111,142*).

E. Dopamine Efflux

Reductions in dopamine efflux in the NAc in response to morphine have been reported in animals pretreated with ibogaine (*113,115,134*), noribogaine (*124*), or 18-MC (*120,143*). Similarly, reductions in dopamine efflux in the NAc in response to nicotine have been reported in animals pretreated with ibogaine (*46,92*) and 18-MC (*42*).

As with locomotor stimulation, methodological issues may have played a part in apparently divergent results regarding ibogaine's effect on dopamine efflux in the NAc in response to cocaine or amphetamine, which is reportedly increased as measured by microdialysis (*134*), although the opposite result was observed in a study on cocaine using microvoltammetry (*139*). Dosage is an additional consideration that might influence ibogaine's effect on dopamine efflux in the NAc in response to cocaine, with a larger ibogaine dose reportedly producing an increase and a smaller dose producing a decrease (*144*).

Dopamine efflux in response to cocaine may also depend on whether dopamine measurements are made in the NAc core versus shell. Szumlinski *et al.* (*111*) found that ibogaine pretreatment (given 19 hours earlier) abolished the sensitized dopamine efflux in response to cocaine in the NAc shell in rats that had been sensitized by repeated prior exposure to cocaine. The same ibogaine pretreatment had no apparent effect on dopamine efflux in the NAc shell in response to "acute" (administered without prior cocaine exposure) cocaine. The authors noted a prior study in their laboratory that found a potentiation by ibogaine pretreatment of dopamine efflux in response to acute cocaine in which the position of the recording probe spanned both the core and shell regions of the NAc (*134*). These results indicate the possibility of a differential effect of ibogaine on dopamine efflux in response to cocaine between the NAc shell, which is thought to play a relatively greater role in the motivational aspects of drugs of abuse, and the NAc core, which, in turn, is thought to play a relatively greater role in motor behavior (*145*). The authors suggested that the effect of ibogaine on reduced cocaine self-administration may be mediated by the observed reduction in dopamine efflux in response to cocaine in the NAc shell in cocaine-sensitized animals (*111*). On the other hand, the enhancement by ibogaine preatreatment of locomotor activity seen in response to acute or chronic cocaine administration may be mediated by increased dopamine efflux in the NAc core. The observed increase in dopamine efflux with ibogaine pretreatment in the NAc core in response to acute cocaine (*134*) is consistent with such a formulation, although this group has yet to report on the result of the same experiment in cocaine-sensitized animals.

Ibogaine and 18-MC reportedly decrease dopamine release evoked by nicotine in the NAc of the rat (*46,92*). In the study by Benwell *et al.* (*46*), the decreased NAc dopamine release following ibogaine was independent of any change in locomotor activity, which was viewed as notable given the usual association between NAc dopamine efflux and locomotor activity in response to nicotine. The authors cited previous work in which a similar dissociation between NAc dopamine efflux and locomotor activity in response to nicotine was produced by treatment with NMDA antagonists, and they suggested that their findings might be related to ibogaine's NMDA antagonist activity.

IV. Evidence of Efficacy and Subjective Effects in Humans

A. Evidence Of Efficacy

1. Acute Opioid Withdrawal

One line of clinical evidence suggesting ibogaine's possible efficacy are the accounts of the addicts themselves, whose demand has led to the existence of an "informal" treatment network in Europe and the United States. Opioid dependence is the most common indication for which addicts have sought ibogaine treatment, which has been typically administered as a single dose. Common reported features of case reports describing ibogaine treatment (*35,36,146-149*) are reductions in drug craving and opiate withdrawal signs and symptoms within 1 to 2 hours, and sustained, complete resolution of the opioid withdrawal syndrome after the ingestion of ibogaine. These case studies appear consistent with general descriptions of ibogaine treatment (*33,34,150*).

Alper *et al.* (*32*) summarized 33 cases treated for the indication of opioid detoxification in nonmedical settings under open label conditions. These cases are a subset of those presented at the NIDA Ibogaine Review Meeting held in March, 1995 (*151*). A focus on acute opioid withdrawal may offset some of the methodological limitations of the informal treatment context because the acute opioid withdrawal syndrome is a clinically robust phenomenon that occurs within a relatively limited time frame and yields reasonably clear outcome measures. Despite the unconventional setting and the lack of structured clinical rating instruments, the lay "treatment guides" who reported on the case series might reasonably be expected to be able to assess the presence or absence of the relatively clinically obvious and unambiguous features of opioid withdrawal.

The subjects in this series of cases reported an average daily use of heroin of 0.64 ± 0.50 g, primarily by the intravenous route, and received an average dose of ibogaine of 19.3 ± 6.9 mg/kg p.o. (range of 6 to 29 mg/kg). Resolution of the signs of opioid withdrawal without further drug seeking behavior was observed in 25 patients. Other outcomes included drug seeking behavior without withdrawal signs (four patients), drug abstinence with attenuated withdrawal signs (two patients), drug seeking behavior with continued withdrawal signs (one patient), and one fatality, possibly involving surreptitious heroin use (see Section VI, "Safety"). The reported effectiveness of ibogaine in this series suggests the need for a systematic investigation in a conventional clinical research setting.

In their chapter in this volume, Mash *et al.* report having treated more than 150 subjects for substance dependence in a clinic located in St. Kitts, West Indies. A subset of 32 of these subjects was treated with a fixed dose of ibogaine of 800 mg for the indication of opioid withdrawal. Physician ratings utilizing structured instruments for signs and symptoms of opioid withdrawal indicated resolution of withdrawal signs and symptoms at time points corresponding to 12 hours following ibogaine administration and 24 hours after the last use of opiates, and at 24 hours following ibogaine administration and 36 hours after the last use of opiates. The resolution of withdrawal signs and symptoms was sustained during subsequent observations over an interval of approximately one week following ibogaine administration. Reductions of measures of depression and craving remained significantly reduced one month after treatment (*123*). The authors noted that ibogaine appeared to be equally efficacious in achieving detoxification from either methadone or heroin. The reported efficacy of ibogaine for the opioid withdrawal syndrome observed in the St. Kitts facility appears to confirm the earlier impressions of the case study literature (*32-36,146-150*).

2. *Long-Term Outcomes*

There is very little data regarding the long-term outcomes in patients treated with ibogaine. Lotsof (*151*) presented a summary of 41 individuals treated between 1962 and 1993 at the NIDA Ibogaine Review Meeting held in March 1995. The data consisted of self-reports obtained retrospectively, which are essentially anecdotal, but apparently represent the only formal presentation of a systematic attempt to determine long-term outcomes in patients treated with ibogaine. Thirty-eight of the 41 individuals presented in the summary reported some opioid use, with approximately 10 of these apparently additionally dependent on other drugs, mainly cocaine, alcohol, or sedative-hypnotics. The use of tobacco or cannabis was not apparently assessed. Across the sample of 41 individuals, nine individuals were treated twice and one was treated three times for a total of 52 treatments. The interval of time following treatment was recorded for which patients reported cessation of use of the drug or drugs on which they were dependent. Fifteen (29%) of the treatments were reportedly followed by cessation drug use for less than 2 months, 15 (29%) for at least 2 months and less than 6 months, 7 (13%) for at least 6 months and less than one year, 10 (19%) for a period of greater than one year, and in 5 (10%) outcomes could not be determined.

B. Subjective Effects

There appear to be common elements to experiences generally described by patients treated with ibogaine. The "stages" of the subjective ibogaine experience presented below are a composite derived by the author from interviews with patients and treatment guides, and general descriptions and case studies provided by the literature (*33-35,146,150*). Ibogaine has generally been administered in non-hospital settings, as a single p.o. dose, usually given in the morning. Vomiting is reportedly common and usually occurs relatively suddenly as a single episode in the first several hours of treatment. Patients generally lie still in a quiet darkened room throughout their treatment, a practice that is possibly related to the cerebellar effects of ibogaine, and because vomiting tends to be more frequent with movement. Patients later in treatment often experience muscle soreness, possibly due to reduced motor activity earlier in treatment, that resolves with motion, stretching, or massage.

1. Acute

The onset of this phase is within 1 to 3 hours of ingestion, with a duration on the order of 4 to 8 hours. The predominant reported experiences appear to involve a panoramic readout of long-term memory (*152*), particularly in the visual modality, and "visions" or "waking dream" states featuring archetypal experiences such as contact with transcendent beings, passage along a lengthy path, or floating. Descriptions of this state appear more consistent with the experience of dreams than of hallucinations. Informants appear to emphasize the experience of being placed in, entering, and exiting entire visual landscapes, rather than the intrusion of visual or auditory hallucinations on an otherwise continuous waking experience of reality. Ibogaine-related visual experiences are reported to be strongly associated with eye closure and suppressed by eye opening. The term "oneiric" (Greek, *oneiros*, dream) has been preferred to the term "hallucinogenic" in describing the subjective experience of the acute state. Not all subjects experience visual phenomena from ibogaine, which may be related to dose, bioavailability, and interindividual variation.

2. Evaluative

The onset of this phase is approximately 4 to 8 hours after ingestion, with a duration on the order of 8 to 20 hours. The volume of material recalled slows. The emotional tone of this phase is generally described as neutral and reflective. Attention is still focused on inner subjective experience rather than the external environment, and it is directed at evaluating the experiences of the acute phase. Patients in this and the acute phase above are apparently easily distracted and annoyed by ambient environmental stimuli and prefer as little environmental sensory stimulation as possible in order to maintain an attentional focus on inner experience.

3. Residual Stimulation

The onset of this phase is approximately 12 to 24 hours after ingestion, with a duration in the range of 24 to 72 hours or longer. There is a reported return of normal allocation of attention to the external environment. The intensity of the subjective psychoactive experience lessens, with mild residual subjective arousal or vigilance. Some patients report reduced need for sleep for several days to weeks following treatment. It is not clear to what extent such reports might reflect a persistent effect of ibogaine on sleep or a dyssomnia due to another cause.

V. Pharmacokinetics

A. Absorption

Jeffcoat *et al.* (*153*) administered single oral doses of ibogaine of 5 mg/kg and 50 mg/kg to rats, and estimated oral bioavailabilities of 16 and 71% at the two dosages, respectively, in females, and 7 and 43% in males. The dose-dependent bioavailability was interpreted as suggesting that ibogaine absorption, and/or first pass elimination, is nonlinear, and the greater bioavailability in females was viewed as consistent with gender-related differences in absorption kinetics. Pearl *et al.* (*135*) administered ibogaine at a dose of 40 mg/kg i.p. and found whole brain levels at 1, 5, and 19 hours post-administration of 10, 1, and 0.7 μM in female rats, and 6, 0.9, and 0.2 μM in male rats, respectively. In the same study, brain levels of noribogaine at 1, 5, and 19 hours post-administration were 20, 10, and 0.8 μM in female rats, and 13, 7, and 0.1 μM and male rats respectively. In addition to gender differences in bioavailability, the data also provide evidence for the pharmacologic relevance of micromolar activities of ibogaine and noribogaine measured *in vitro* (*40,44*).

Upton (*154*) reported on observations in rats given ibogaine in the form of oral suspension, oral solution, or via IV or intraperitoneal routes, and also reviewed data obtained in beagle dogs, cynomologous monkeys, and human subjects. Absorption of the oral suspension in rats was noted to be variable and incomplete. As in the study cited above by Jeffcoat (*153*), peak levels and bioavailability were greater in female than in male rats.

B. Distribution

Hough *et al.* (*51*) administered 40 mg/kg ibogaine by the intraperitoneal and subcutaneous routes and evaluated its distribution in plasma, brain, kidney, liver, and fat at 1 and 12 hours post-administration. Ibogaine levels were higher following subcutaneous versus intraperitoneal administration, suggesting a substantial "first pass" effect involving hepatic extraction. The results were consistent with the highly lipophilic nature of ibogaine; ibogaine concentrations at 1 hour postadministration were 100 times greater in fat, and 30 times greater in brain, than in plasma. These authors suggested that the prolonged actions of ibogaine could relate to adipose tissue serving as a reservoir with release and metabolism to noribogaine over an extended period of time (*51*). The apparently greater levels of ibogaine in whole blood versus plasma suggests the possibility that platelets might constitute a depot in which ibogaine is sequestered (*42*). If there is conversion of ibogaine to noribogaine in the brain, then the significantly greater polarity of noribogaine relative to ibogaine could prolong the presence of the more polar metabolite in the CNS after conversion from ibogaine (*42*).

C. Metabolism

The major metabolite of ibogaine, noribogaine, is formed through demethylation, apparently via the cytochrome P-450 2D6 (CYP2D6) isoform (*155*). Consistent with first pass metabolism of the parent drug, noribogaine is reportedly detectable in brain tissue within 15 minutes after oral administration of 50 mg/kg ibogaine (*44*). Noribogaine is itself pharmacologically active and is discussed in this volume by Baumann *et al.*

In pooled human liver microsomes, Pablo *et al.* identified two kinetically distinguishable ibogaine *O*-demethylase activities which corresponded, respectively, to high and low values of the apparent Michaelis constant (K_{mapp}) (*155*). The low K_{mapp} ibogaine *O*-demethylase activity was attributable to CYP2D6 and accounted for greater than 95% of the total intrinsic clearance in pooled human liver microsomes. The authors noted that the apparent involvement of the CYP2D6 suggests possible human pharmacogenetic differences in the metabolism of ibogaine. "Poor metabolizers" who lack a copy of the CYP2D6 gene (*156*) would be expected to have relatively less CYP2D6-catalyzed activity to metabolize ibogaine to noribogaine. Consistent with such an expectation, a subject identified as a phenotypic CYP2D6 poor metabolizer possessed only the high K_{mapp} ibogaine *O*-demethylase activity, which had accounted for only a small fraction of the intrinsic clearance. In another study, analysis of ibogaine and noribogaine levels in human subjects yielded a distribution interpreted as indicating three groups of rapid, intermediate, and poor metabolizers (*157*), a pattern consistent with the observed pharmacogenetic polymorphism of CYP2D6 in human populations (*156*).

D. Excretion

Ibogaine has an estimated half-life on the order of 1 hour in rodents (*158*), and 7.5 hours in man (Mash *et al.*, this volume). Ibogaine and its principal metabolite, noribogaine, are excreted via the renal and gastrointestinal tracts. In rats, Jeffcoat *et al.* (*153*) noted 60 to 70% elimination in urine and feces within 24 hours, and Hough *et al.* (*51*) found plasma and tissue levels to be 10 to 20-fold lower at 12 hours versus 1 hour post dose.

Upton and colleagues (*154*) cited several pharmacokinetic issues of potential concern based on their analysis of data obtained from rats. These include evidence for presystemic clearance potentially resulting in low bioavailability and interpatient variability, and saturable first pass clearance, which could also generate intrapatient variability. The possibility of saturable systemic clearance was also noted. Mash *et al.* (*36*) suggested the possibility of species or strain differences in ibogaine metabolism and clearance rates and cited the rapid elimination of ibogaine from the blood of primates, as opposed to rats or humans, as an example.

In human subjects, 90% of a 20 mg/kg p.o. dose of ibogaine was reportedly eliminated within 24 hours (*36*). Noribogaine is apparently eliminated significantly more slowly than ibogaine, and observations in human subjects indicate persistently high levels of noribogaine at 24 hours (*36,79,123,* Mash *et al.* in this volume). The sequestration and slow release from tissues of ibogaine or noribogaine and the slow elimination of noribogaine have been suggested to account for the apparently persistent effects of ibogaine.

VI. Safety

A. Neurotoxicity

1. Neuropathology

Multiple laboratories have reported on the degeneration of cerebellar Purkinje cells in rats given ibogaine at a dose of 100 mg/kg i.p. (*159,160*). However, the available evidence suggests that the neurotoxic effects of ibogaine may occur at levels higher than those observed to have effects on opioid withdrawal and self-administration. Molinari *et al.* (*161*) found no evidence of cerebellar Purkinje cell degeneration with 40 mg/kg i.p. administered as a single dose, which is reported to reduce morphine or cocaine self-administration or morphine withdrawal in rats (*29,119,126,161*). Xu *et al.* (*162*) evaluated biomarkers of cerebellar neurotoxicity in rats treated with single doses of ibogaine of 25, 50, 75, and 100 mg/kg i.p. The biomarkers used in this study included the specific labeling of degenerating neurons with silver, and Purkinje neurons with antisera to calbindin. Astrocytes were identified with antisera to glial fibrillary acidic protein (GFAP), a marker of reactive gliosis, a general response of astrocytes to CNS injury. The 25 mg/kg dosage was found to correspond to a no-observable-adverse-effect-level (NOAEL). Helsley *et al.* (*102*) treated rats with 10 mg/kg ibogaine every other day for 60 days and observed no evidence of neurotoxicity.

Regarding the question of neurotoxicity in brain areas outside the cerebellum, O'Hearn and Molliver (*163*) have stated, "Evidence of neuronal injury following ibogaine administration in rats appears to be almost entirely limited to the cerebellum." While the cerebellum appears to be the brain region most vulnerable to neurotoxic effects of ibogaine, some research has addressed the issue of neurotoxicity in other brain regions. O'Callaghan *et al.* (*164*) examined GFAP in male and female rats exposed to either an "acute" regimen of ibogaine administered at doses of 50, 100, or 150 mg/kg i.p. daily for 3 days or a "chronic" regimen of daily oral administration of 25, 75, or 150 mg/kg for 14 days. The acute i.p. regimen produced elevations of GFAP in animals of either gender that were not restricted to the cerebellum, and were observed in the cerebellum and hippocampus at the 50 mg/kg dosage level, and in the cortex, hippocampus, olfactory bulb, brain stem, and striatum at the 100 mg/kg level. The effect of the acute ibogaine regimen on GFAP was no longer evident at 14 days with either dosage in male rats, and was restricted to the cerebellum with the 100 mg/kg dose in female rats. GFAP levels were examined at 17 days after the completion of the chronic dosing regimen. No elevations of GFAP were found in any of the brain regions examined at any of the dosages administered utilizing the chronic regimen in males, and elevations of GFAP were found only in females, which were restricted to the hippocampus with the 25 mg/kg dosage regimen and were present in the hippocampus, olfactory bulb, striatum, and brain stem with the 150 mg/kg dosage regimen.

O'Hearn *et al.* (*159*) found GFAP elevations in the cerebellum only, and not the forebrain of male rats administered 100 mg/kg doses i.p on up to 3 consecutive days. Elevations of GFAP are relatively sensitive, but not specific to, neuronal degeneration (*162*). Using a silver degeneration-selective stain as a histologic marker of neurodegeneration, Scallet *et al.* (*165*) examined diverse brain regions in rats and mice treated with single 100 mg/kg doses of ibogaine administered i.p. and found evidence of neurodegeneration only in the cerebellum in rats, whereas mice showed no evidence of neurodegeneration. In rats that received a dose of ibogaine of 100 mg/kg i.p., neuronal degeneration was confined to the cerebellum as revealed by staining with Fluoro-Jade, a recently developed sensitive and definitive marker of neuronal degeneration (*166,167*).

Sensitivity to ibogaine neurotoxicity appears to vary significantly between species. The monkey appears to be less sensitive to potential ibogaine neurotoxicity than the rat (*36*). Mash *et al.* observed no evidence of neurotoxicity in monkeys treated for 5 days with repeated oral doses of ibogaine of 5 to 25 mg/kg, or subcutaneously administered doses of 100 mg/kg (36). Another species difference in sensitivity is the mouse, which unlike the rat shows no evidence of cerebellar degeneration at a 100 mg/kg i.p. dose of ibogaine (*165*).

2. *Mechanisms of Neurotoxicity*

Ibogaine's cerebellar toxicity could be related to excitatory effects mediated by $sigma_2$ receptors in the olivocerebellar projection, which sends glutaminergic excitatory input to cerebellar Purkinje cells, whose synaptic redundancy makes them particularly vulnerable to excitotoxic injury (*160*). $Sigma_2$ agonists are reported to potentiate the neuronal response to NMDA (*95*), and potentiation of glutamatergic responses at Purkinje cells might lead to the observed neurotoxicity. $Sigma_2$ agonists have also been shown to induce apoptosis, and activation of $sigma_2$ receptors by ibogaine results in direct neurotoxicity via induction of apoptosis in *in vitro* cell culture systems (*168,169*). Elsewhere in this volume, Bowen discusses the effects of *iboga* alkaloids at $sigma_2$ receptors. It is possible therefore that ibogaine's neurotoxic effect on the highly sensitive Purkinje neurons is the result of combined direct neurotoxicity and excitotoxicity due to the enhancement of glutamatergic activity, both effects being mediated by $sigma_2$ receptors. The agonist activity of ibogaine at the $sigma_2$ receptor might explain the apparent paradox of ibogaine-induced excitotoxicity, despite its properties as an NMDA antagonist (*42*). The neurotoxic effects of *iboga* alkaloids can apparently be dissociated from their putative effects on addiction, since $sigma_2$ receptors appear not to be involved in the suppression of drug self-administration. 18-MC, an ibogaine congener with relatively much less $sigma_2$ affinity, reportedly produces effects similar to ibogaine on morphine and cocaine administration in rats, but has shown no evidence of neurotoxicity, even at high dosages (*42,75,120*).

Ibogaine's NMDA antagonist activity has been cited as a rationale for a patent for its use as a neuroprotective agent to minimize excitotoxic damage in stroke and anoxic brain injury (*170*). In methamphetamine-treated mice, ibogaine is reported to protect against hyperthermia and the induction of heat shock protein, which are possible mediators of methamphetamine neurotoxicity (*171*). Binienda *et al.* in this volume report an accentuation of delta amplitude in ibogaine pretreated animals given cocaine, and they suggest a "paradoxical" proconvulsant effect resulting from the interaction of cocaine and ibogaine, similar to interactions reported between cocaine and other noncompetitive NMDA antagonists. However, ibogaine is reported to protect against convulsions produced by electroshock (*61*), or the administration of NMDA (*55*). Luciano *et al.* (*148*) did not observe EEG abnormalities in five human subjects during treatment with ibogaine in the dosage range of 20 to 25 mg/kg. There is apparently no reported human data on possible differences between the pre- and post-ibogaine treatment EEG, or effects persisting into extended periods of time after treatment.

3. *Tremor*

Ibogaine has been noted to produce tremor at dosages of 10 mg/kg i.p. in rats (*172*) and 12 mg/kg s.c. in mice (*173*). Glick *et al.* (*119*) evaluated ibogaine and several other *iboga* alkaloids, and found that their effects on drug self-administration and tendency to produce tremor were independent from one another. Studies of structure-activity relationships of the *iboga* alkaloids indicate that the tendency to cause tremor is enhanced by the presence of a methoxy group at position 10 or 11 and is diminished or eliminated by the presence of a carbomethoxy group at position 16 (*173,174*). Accordingly, tremors were not produced in rats administered noribogaine, which differs from ibogaine with respect to the absence of a methoxy group at position 10, at a dosage of 40 mg/kg i.p. (*124*). Likewise, tremors were not observed in rats administered a dosage of 18-MC as high as 100 mg/kg. 18-MC differs from ibogaine with respect to the absence of a methoxy group at position 10 and the presence of a carbomethoxy group at position 16 (*120*).

4. *Observations in Humans*

Concern over possible neurotoxicity led Mash *et al.* to quantitatively investigate ibogaine's effects on postural stability, body tremor, and appendicular tremor in humans (*36*). In U.S. FDA safety trials, nine subjects receiving 1 and 2 mg/kg of ibogaine showed only a statistically insignificant increase in body sway 6 hours after taking ibogaine. Ten patients evaluated 5 to 7 days after receiving doses of ibogaine ranging from 10 to 30 mg/kg showed no evidence of abnormality on quantitative measures of static or dynamic posturography or hand accelometry, or on clinical neurologic exam.

A woman died in the United States in 1994 who had been previously treated with ibogaine 25 days earlier (*36*). This woman had undergone four separate treatments with ibogaine in dosages ranging from 10 to 30 mg/kg in the 15 months prior to her death. The cause of death was concluded to have been a mesenteric arterial thrombosis related to chronic cellulitis, and a role for ibogaine in causing the fatality was not suspected. Of interest with regard to concerns over potential neurotoxicity, was the absence of any neuropathological abnormality not associated with chronic IV drug use. Neuropathological examination revealed only slight medullary neuroaxonal dystrophy and an old focal meningeal fibrosis, which were explainable on the basis of chronic IV drug use (*36*). There was no evidence of cytopathology or neurodegenerative changes in the cerebellum or any other brain area, nor was there evidence of astrocytosis or microglial activation.

B. Cardiovascular Effects

Glick *et al.* (*45*) found no changes in resting heart rate or blood pressure at a dose of ibogaine of 40 mg/kg i.p., which has been used in that laboratory in drug withdrawal or self-administration studies. Higher doses of ibogaine (100 and 200 mg/kg) decreased the heart rate without an effect on blood pressure, and 18-MC had no apparent effect on heart rate or blood pressure at any of the above doses. Binieda *et al.* (*90*) found a significantly decreased heart rate in rats given ibogaine 50 mg/kg i.p.

Mash *et al.* (*175*) reported on intensive cardiac monitoring in 39 human subjects dependent on cocaine and/or heroin who received fixed p.o. doses of ibogaine of 500, 600, 800, or 1000 mg. Six subjects exhibited some significant decrease of resting pulse rate relative to baseline, one of whom evidenced a significant decrease in blood pressure, which was attributed to a transient vasovagal response. Monitoring revealed no evidence of EKG abnormalities appearing or intensifying during ibogaine treatment. No significant adverse events were seen under the study conditions, and it was concluded that the single dose of ibogaine was apparently well tolerated. In their chapter in this volume, Mash *et al.* comment further that random regression of vital signs showed no changes across time or by dosage in opiate-dependent subjects. They did however observe the occurrence of a hypotensive response to ibogaine in some cocaine-dependent subjects, which was responsive to volume repletion.

C. Fatalities

The LD50 of ibogaine is reportedly 145 mg/kg i.p. and 327 mg/kg intragastrically in the rat, and 175 mg/kg i.p. in the mouse (*158*).

In June 1990, a 44 year-old woman died in France approximately 4 hours after receiving a dose of ibogaine of about 4.5 mg/kg p.o. The cause of death was concluded to have been acute heart failure in an autopsy carried out at the Forensic-Medical Institute in Zurich (*176*). Autopsy revealed evidence of a prior myocardial infarction of the left ventricle, severe atherosclerotic changes, and 70 to 80% stenosis of all three major coronary artery branches. This patient had a history of hypertension, and inverted T waves were noted on EKG three months prior to the patient's death. The autopsy report concluded that the patients preexisting heart disease was likely to have caused the patient's death, and it specifically excluded the possibility of a direct toxic effect of ibogaine. The report acknowledged the possibility that an interaction between ibogaine and the patient's preexisting heart condition could have been a contributing factor in the fatal outcome.

The autopsy report, which included information obtained from the patient's family physician, and the psychiatrist who administered ibogaine, makes reference to the possibility that the patient might have taken other drugs. The autopsy report noted the presence of amphetamine in the enzyme immunocytochemical (EMIT) assay of a dialysate of the kidney tissue (urine was reported not to be obtainable). This finding, however, was regarded as artifactual and possibly attributable to a false positive EMIT result due to the presence of phenylethylamine.

A fatality occurred during a heroin detoxification treatment of a 24-year-old female in the Netherlands in June 1993. This incident was a significant factor in the NIDA decision not to fund a clinical trial of ibogaine in 1995. The patient received a total ibogaine dose of 29 mg/kg p.o. and suffered a respiratory arrest and died 19 hours after the start of the treatment. Forensic pathological examination revealed no definitive conclusion regarding the probable cause of death (*177*) and cited the general lack of information correlating ibogaine concentrations with possible toxic effects in humans. The high levels of noribogaine found in the deceased patient were possibly consistent with saturation of elimination kinetics. However, the higher levels of noribogaine in heart, relative to femoral blood, also suggested significant postmortem redistribution of noribogaine. The potential artifact associated with a high volume of distribution and postmortem release of drug previously sequestered in tissue (*51,139,158*) limits the interpretability of postmortem levels of noribogaine.

Some evidence suggested the possibility of surreptitious opioid use in this case, which was noted in the Dutch inquiry (*178*) and which is another source of uncertainty in this fatality. There is evidence suggesting that the interaction of opioids and ibogaine potentiates opioid toxicity (*68,179*). Analysis of gastric contents for heroin or morphine, which might have confirmed recent heroin smoking, and analysis of blood for 6-monoacetyl morphine, a heroin metabolite whose presence indicates recent use (*180*), were not performed. This incident underscores the need for the security and medical supervision available in a conventional medical setting, and for completion of dose escalation studies to allow systematic collection of pharmacokinetic and safety data.

In London, in January 2000, a 40-year-old heroin addict died after having allegedly taken 5 g of *iboga* alkaloid extract 40 hours prior to his death (*38*, see the chapter by Alper *et al.* in this volume). The extract was said to have contained approximately five times the alkaloid content of the dried rootbark. The official British inquest regarding this matter is still in progress as of the time of the writing of this book.

D. Abuse Liability

The available evidence does not appear to suggest that ibogaine has significant potential for abuse. The 5-HT_{2A} receptor, the primary mediator of responding for LSD and other commonly abused drugs classified as "hallucinogenic" or "psychedelic," does not appear to be essential to discriminability of the ibogaine stimulus (*84,96*). Ibogaine is reportedly neither rewarding or aversive in the conditioned place preference paradigm (*132*). Rats given either 10 or 40 mg/kg ibogaine daily for 6 consecutive days did not show withdrawal signs (*129*). Animals do not self-administer 18-MC, an ibogaine analog, in paradigms in which they self-administer drugs of abuse (*45*). None of the consultants to NIDA in the 1995 Ibogaine Review Meeting identified the possible abuse of ibogaine as a potential safety concern.

VII. Learning, Memory, and Neurophysiology

A. Learning, Memory, and Addiction

Drug abusers may be viewed as having a disorder involving excess attribution of salience to drugs and drug-related stimuli (*181*), which suggests the possibility of a role of processes subserving learning and memory in the acquisition of the pathological motivational focus in addiction (*182-185*). Learning, in the most general sense, can be viewed as the modification of future brain activity, of which thought, motivation, consciousness, or sensory experience are emergent properties, on the basis of prior experience. This broad definition subsumes everything from social behavior to learning to read, to the neuroadaptations of drug tolerance and dependence.

Addiction can be argued to involve the pathological acquisition or "learning" of associations of drug related stimuli with motivational states corresponding to valuation and importance (*181,183,184*). The pathological learning of addiction differs from that of normal learning in at least two important respects. First, the acquisition of drug salience in addiction does not involve learned associations between drug-related external cues or internal representations, and the experience of external events as they actually occur. Instead, the "imprinting" or "stamping in" of drug incentives appears to involve alterations of neural plasticity in processes that subserve motivation, memory and learning, resulting in neural behavior that to a significant extent has escaped the constraint of validation by experience with external reality (*183-186*). Dopamine and glutamate transmission are thought to be involved in the modulation of neural plasticity of both normal learning and the neuroadaptations of drug salience (*184*). Second, drug-related "learning" does not apparently habituate (*184*). Unlike normal learning, the drug stimulus appears to be experienced as perpetually novel and continues to command attention and be attributed with salience unattenuated by habituation (*53,182*).

B. Effects of Ibogaine on Learning and Memory

Ibogaine appears to have significant effects on brain events involved in learning and the encoding of drug salience. Ibogaine interacts significantly with the NMDA receptor (*39,58,179*), which is involved in long term potentiation (LTP), a process thought to be important in neural plasticity, memory, and learning (*182,184,187*). Experiences apparently involving memory, such as panoramic recall, are prominent in descriptions by individuals who have taken ibogaine (*14*).

The observation of an effect of ibogaine on the expression of behavioral sensitization to amphetamine, but not a conditioned place preference (*188*), raises the interesting possibility of a relatively selective effect of ibogaine on the pathological encoding of drug salience, distinguished from learning involving non-drug incentives. Ibogaine reportedly attenuates the acquisition of place preference for morphine or amphetamine (*131,132*). A general effect of interference with learning has been suggested (*189*), but studies on spatial learning show an actual enhancement by ibogaine (*102,190*). Consistent with a selective effect on neuroadaptations acquired from drug exposure are ibogaine's effects on locomotor activity and dopamine efflux in the NAc, which are relatively more evident in animals with prior experience with morphine (*112,113*) or cocaine (*111*).

C. Ibogaine and the EEG

Studies of animals treated acutely with ibogaine report a desynchronized EEG with fast low amplitude activity, a state described as "activated" or "aroused"

(*89,90,191*). Binienda *et al.* (*90*) noted a decline in delta amplitude and interpreted this as consistent with activation of dopaminergic receptors. However, observations on the interaction of atropine and ibogaine with respect to the EEG suggest the involvement of ascending cholinergic input. Depoortere (*191*) found that ibogaine enhanced an atropine-sensitive theta frequency EEG rhythm in rats. Schneider and Sigg (*89*) observed a shift toward high-frequency low-voltage EEG activity following the administration of ibogaine to cats, and they noted that this effect was blocked by the administration of atropine. Luciano *et al.* (*148*) observed no changes in the visually evaluated EEG in humans administered 20 to 25 mg/kg ibogaine.

D. Goutarel's Hypothesis

The French chemist Robert Goutarel (*14*) hypothesized that ibogaine treatment involves a state with functional aspects shared by the brain states of REM sleep, with important effects on learning and memory. During the REM state, there is believed to be reconsolidation of learned information in a state of heightened neural plasticity, with the reprocessing of previously learned information and the formation of new associations (*192,193*). Goutarel suggested that a REM-like state may be induced by ibogaine, which corresponds to a window of heightened neural plasticity, during which there may be weakening of the pathological linkages between cues and representations of the drug incentive and the motivational states with which they have become paired (*14*). Analogous to the reconsolidation of learned information that is thought to occur during the REM state (*192,193*), Goutarel theorized that the pathological learning of addiction was modified during ibogaine treatment. He appears to have based his theoretical formulation mainly on reports of the phenomenological experiences of awake ibogaine-treated subjects that share features in common with dreams. Goutarel's hypothesis is speculative, but nonetheless has an interesting apparent consistency with the literature on the relationship of learning and addiction and the physiologic function of the REM EEG state with regard to the consolidation of learned information.

There is some evidence that may be viewed as consistent with Goutarel's hypothesis. Goutarel's belief in a relationship of the ibogaine-treated EEG state to that of REM is supported by studies in animals treated with ibogaine that report an apparently activated or desynchronized EEG state consistent with arousal, vigilance, or REM sleep (*90,191*). The observation that ibogaine enhanced an atropine-sensitive theta frequency rhythm (*191*) suggests the possible involvement of ascending cholinergic input, which is an essential determinant of EEG desynchronization during REM sleep (*192*). The possible reconsolidation of learned information due to heightened plasticity during both the REM and ibogaine-induced desynchronized EEG states is suggested by the observation that

EEG dyssynchrony is associated with an increased facilitation of Hebbian covariance (*194*), which is believed to be an important determinant of the neural plasticity involved in consolidation of learning and memory. Also, with regard to a possible analogy of the REM and ibogaine induced brain states, some ibogaine treatment guides have anecdotally mentioned that they have observed REM-like eye movements in awake patients during treatments (*195,196*).

VIII. Anthropological and Sociological Perspectives

As discussed in various aspects by this volume by the Fernandezes, Frenken, and Lotsof and Alexander, ibogaine's use appears to involve distinctive interactions of psychopharmacologic effects with set and setting in both the subcultures of the United States and Europe, and the centuries older, sacramental context of the use of iboga in Bwiti, the religious movement in West Central Africa. In the Bwiti religious subculture, and arguably to some extent in the European ibogaine subculture, there is the common attribute of a group of initiates that seek to facilitate healing through the affiliation of the collective with the individual. In both the African and U.S./European contexts, the ibogaine experience has been attributed to serving the objective of facilitating personal growth and change. Use of ibogaine in both contexts has been criticized as involving the use of an "addictive" or "hallucinogenic" agent, and it appears to some extent to involve the formation of a subculture among individuals confronted with marginalizing social circumstances such as colonialism, or the state of addiction (*197-199*, see also Fernandez and Fernandez in this volume).

Galanter (*200*) identifies three important psychological features that he regards as descriptive of the process of charismatic groups or zealous self-help movements such as 12-step programs that appear to also be relevant to Bwiti. These three processes are group cohesiveness, shared belief, and altered consciousness, such as that of religious ecstasy or insight to which the group can attribute a new construction of reality in their life. An understanding of these powerful behavioral influences could be useful in optimizing the clinical milieu and interpersonal dynamics of present conventional treatment settings, or of future treatment settings, if ibogaine or a congener should receive official approval.

The application of ethnographic techniques to the analysis of the phenomenological features of the acute treatment experience could be informative from a neuropsychiatric, as well as from a cultural perspective. For example, similar subjective phenomena are frequently described in both ibogaine treatment and near death experiences (NDEs) (*14,152,199,201*) such as panoramic memory;

calm, detached emotional tone; specific experiences, such as passage along a long path or floating; "visions" or "waking dream" states featuring archetypal experiences such as contact with transcendent beings; and the frequent attribution of transcendent significance to the experience. Such shared features between ibogaine and NDEs suggest a common transcultural phenomenology of transcendent or religious experience or, alternatively, the possibility of a similar subjective experience due to the influence of a common underlying neurobiological mechanism such as NMDA transmission (*202*).

IX. Economic and Political Perspectives

A. Economic Incentives and the Development of Ibogaine

The academic research community working in the public sector has a crucial role in studying ibogaine as a paradigm for the development of new treatment approaches. The strategy of relying on the pharmaceutical industry to underwrite the cost of drug development works extremely well in many instances, but appears to present some limitations with regard to the development of pharmacotherapy for addiction in general, and specifically ibogaine.

In the public sector, the major economic incentives for the development of addiction treatment are the saved costs associated with preventing lost economic productivity, medical morbidity, or crime. In the private sector, decisions are based on weighing the expense of development against the expected profit, and not the magnitude of saved economic or social costs. Owing to limited financial incentives in the form of insurance reimbursements and a perceived lack of "breakthrough" compounds, the U.S. pharmaceutical industry has not generally viewed addiction as an attractive area for development (*203*), and expenditures for the development of medications for addiction are small relative to those to develop drugs for other indications. Ibogaine is particularly unattractive to industry for several reasons: its mechanism of action is apparently complex and incompletely understood, it may present significant safety issues, it is a naturally occurring alkaloid whose structure cannot itself be patented, and some of its use patents are close to expiration.

There is arguably an important role for academic/public-sector development in the case of a theoretically interesting drug with a limited profit potential and significant developmental expense such as ibogaine. However, the entire annual expenditures for medications development in NIDA, which accounts to about 90% of U.S. public sector spending on developing addiction pharmacotherapy, is on the order of approximately $60 million, a fraction of the average cost of successfully developing a drug to market, which is estimated to exceed $300 million (*204*). Opportunities to fund research on ibogaine are limited by factors that generally affect the development of other drugs to treat addiction: a limited public sector budget in the presence of disproportionately low private-sector expenditures on the development of pharmacotherapies for addiction relative to other indications (*203*).

B. Political Issues

The chapter by Alper *et al.* in this volume describes the medical subculture of the informal ibogaine treatment scene and the political subculture of advocacy for the development and availability of ibogaine. These scenes are a distinctive and significant aspect of ibogaine's history, which arguably have impacted on decisions regarding its development. From a clinical standpoint, the informal treatment subculture has been an important source of information on human experience with ibogaine (*32*).

From a political or historical standpoint, the informal treatment subculture has viewed itself as a form of activism or civil disobedience on the part of its participants seeking a treatment, despite a lack of official approval (*34*). Ibogaine has been associated with a vocal activist subculture, which views its mission as making controversial treatments available to a stigmatized minority group of patients suffering from a life-threatening illness, and has utilized tactics intended to engage the attention of the press (*34*). These confrontational media-oriented tactics may well have provoked negative reactions at times, but may also have influenced Curtis Wright, the former FDA ibogaine project officer, to write in 1995 that ". . . a significant portion of the public we serve believes the drug merits investigation" (*205*).

X. Conclusions

Evidence that supports the possible efficacy of ibogaine as a treatment for addiction includes case reports in humans, and effects in preclinical models of drug dependence. The case report evidence has mainly involved the indication of acute opioid withdrawal, and there appears to be consistency between earlier observations derived from informal treatment contexts (*32-36,146-150*) and more recent work from a setting that appears to conform to a conventional medical model (*123*, Mash *et al.* in this volume). The continued existence of informal treatment scenes parallels case report evidence indicating possible efficacy. Animal work has provided observations of attenuation of opiate withdrawal signs and reductions of self-administration of a variety of drugs including morphine, cocaine, alcohol, and nicotine. Preclinical models have also yielded evidence that with respect to certain abused drugs, ibogaine may dampen responses that may be associated with dependence, such as dopamine efflux in the NAc or locomotor activation.

Ibogaine's pharmacologic profile includes interactions with multiple neurotransmitter systems that could plausibly be related to addiction, including NMDA, nicotinic, mu- and kappa-opioid, and serotonergic systems. The putative efficacy of ibogaine does not appear fully explainable on the basis of interactions with any single neurotransmitter system, or on the basis of currently utilized pharmacologic strategies such as substitution therapies, or monoamine reuptake inhibition. Ibogaine's effects may result from interactions between multiple neurotransmitter systems, and might not be attributable to actions at any single type of receptor. The apparently persistent effect of ibogaine has been suggested to involve a long-lived metabolite. Some evidence suggests effects on second messenger signal transduction, an interesting possibility that could conceivably result from interactions between multiple neurotransmitter systems and produce persistent effects lasting beyond the duration of occupancy at receptor sites. Work with ibogaine congeners suggests that other *iboga* alkaloids can be developed that might minimize unwanted toxic, or possibly behavioral effects, while retaining apparent efficacy in drug dependence. In summary, the available evidence suggests that ibogaine and the *iboga* alkaloids may have efficacy in addiction on the basis of mechanisms that are not yet known and which can possibly be dissociated from toxic effects, and may present significant promise as a paradigm for the study and development of pharmacotherapy for addiction.

References

1. H. Pope, *Econ. Bot.* **23,** 174 (1969).
2. S. Budvari and M.J. O'Neil, *Merck Index* (S. Budvari and M.J. O'Neil, eds.), Chapman and Hall, New York, NY, 1996.
3. G. Arai, J. Coppola and C.A. Jeffrey, *Acta Cryst.* **13,** 553 (1960).
4. K. Biemann and Friedmann-Spiteller, *J. Am. Chem. Soc.* **83,** 4805 (1961).
5. G.V. Binst, C. Danhoux, C. Hootele, J. Pecher and R.H. Martin, *Tetrahedron Lett. 973* (1964).
6. J. Pecher, R.H. Martin, N. Defay, M. Kaisin, J. Peeters, G.V. Binst, N. Verzele, and F. Alderweireldt, *Tetrahedron Lett. 270* (1961).
7. M. Damak, C. Poupat, and A. Ahond, *Tetrahedron Lett.* **39,** 3531 (1976).
8. F. Ladhar, N. Gorbel, and M. Damak, *J. de la Societé Chimique de Tunisie* **5,** 43 (1981).
9. E. Wenkert, D.W. Cochran, H.E. Gottlieb, and E.W. Hagaman, *Helv. Chim. Acta.* **59,** 2437 (1976).
10. M.E. Alburges, R.L. Foltz, and D.E. Moody, *J Anal. Toxicol.* **19,** 381 (1995).
11. C.A. Gallagher, L.B. Hough, S.M. Keefner, A. Syed-Mozaffari, S. Archer, and S.D. Glick,

Biochem. Pharmacol. **49,** 73 (1995).
12. W.L. Hearn, J. Pablo, G.W. Hime, and D.C. Mash, *J Anal. Toxicol.* **19,** 427 (1995).
13. F.R. Ley, A.R. Jeffcoat, and B.F. Thomas, *J. Chromatogr.* **723,** 101 (1996).
14. R. Goutarel, O. Gollnhofer, and R. Sillans, *Psyched. Mono. Essays* **6,** 70 (1993).
15. J. Dybowsky and E. Landrin, *Compt. Rend. Acad. Sci.* **133,** 748 (1901).
16. A. Haller and E. Heckel, *Compt. Rend. Acad. Sci.* **133,** 850 (1901).
17. M. Lambert and E. Heckel, *Compt. Rend. Acad. Sci.* **133,** 1236 (1901).
18. H. Isbell, "Ciba Ibogaine File Document No. AB0491-492 410," 1955.
19. W.I. Taylor, *J. Am. Chem. Soc.* **79,** 3298 (1957).
20. D.F. Dickel, C.I. Holden, R.C. Maxfield, L.E. Paszak, and W.I. Taylor, *J. Am. Chem. Soc.* **80,** 123 (1958).
21. G. Büchi, D.L. Coffen, K. Kocsis, P.E. Sonnet, and F.E. Ziegler, *J. Am. Chem. Soc.* **88,** 3099 (1966).
22. H.S. Lotsof, *U.S. Patent* **4,499,096;** *Chem. Abstr.* **102,**160426w (1985).
23. H.S. Lotsof, *U.S. Patent* **4,587,243;** *Chem. Abstr.* **106,**12967r (1986).
24. D.P. Bocher and C. Naranjo, *French Special Drug Patent No.* 138.081;081713m, Inst. Class A61k (1969).
25. H.S. Lotsof, *U.S. Patent* **4,857,523;** *Chem. Abstr.* **112,**32041m (1989).
26. H.S. Lotsof, *U.S. Patent* **5,026,697;** *Chem. Abstr.* **116,**17031x (1991).
27. H.S. Lotsof, *U.S. Patent* **5,152,994;** *Chem. Abstr.* **116,**100980b (1992).
28. E.D. Dzoljic, C.D. Kaplan, and M.R. Dzoljic, *Arch. Int. Pharmacodyn. Ther.* **294,** 64 (1988).
29. S.D. Glick, K. Rossman, S. Steindorf, I.M. Maisonneuve, and J.N. Carlson, *Eur. J. Pharmacol.* **195,** 341 (1991).
30. S.D. Glick, C.A. Gallagher, L.B. Hough, K.L. Rossman, and I.M. Maisonneuve, *Brain Res.* **588,** 173 (1992).
31. T.S.L. Cappendijk and M.R. Dzoljic, *Eur. J. Pharmacol.* **14,** 261 (1993).
32. K.R. Alper, H.S. Lotsof, G.M.N. Frenken, D.J. Luciano, and J. Bastiaans, *Am. J. Addict.* **8,** 234 (1999).
33. H.S. Lotsof, *Multidiscip. Assoc. Psyched. Stud.* **5,** 16 (1995).
34. P. DeRienzo and D. Beal, *The Ibogaine Story*, Autonomedia, Brooklyn, New York, 1997.
35. B. Sisko, *Multidiscip. Assoc. Psyched. Stud.* **4,** 15 (1993).
36. D.C. Mash, C.A. Kovera, B.E. Buck, M.D. Norenberg, P. Shapshak, W.L. Hearn, and J. Sanchez-Ramos, *Ann. N. Y. Acad. Sci.* **844,** 274 (1998).
37. MDD-NIDA, *Draft Protocol: Rising Dose Tolerance Study using Single Administration to Assess the Safety and Preliminary Efficacy of Ibogaine for the Treatment of Cocaine and/or Heroin Dependence*, NIDA, Rockville, MD, 1995.
38. egroups, *http://www.egroups.com/group/ibogaine*, 2000.
39. P. Popik and P. Skolnick, *in* "The Alkaloids" (G.A. Cordell, ed.), p. 197. Academic Press, New York, 1999.
40. P.M. Sweetnam, J. Lancaster, A. Snowman, J.L. Collins, S. Perschke, C. Bauer, and J. Ferkany, *Psychopharmacology* **118,** 369 (1995).
41. H. Sershen, A. Hashim, and A. Lajtha, *Brain Res. Bull.* **42,** 161 (1997).
42. S.D. Glick and I.M. Maisonneuve, *Ann. N. Y. Acad. Sci.* **844,** 214 (1998).
43. D.C. Deecher, M. Teitler, D.M. Soderlund, W.G. Bornmann, M.E. Kuehne, and S.D. Glick, *Brain Res.* **571,** 242 (1992).
44. J.K. Staley, Q. Ouyang, J. Pablo, W.L. Hearn, D.D. Flynn, R.B. Rothman, K.C. Rice, and D.C. Mash, *Psychopharmacology* **127,** 10 (1996).
45. S.D. Glick, I.M. Maisonneuve, M.E. Kuehne and U.K. Bandarage, *CNS Drug Rev.* **5,** 27 (1999).
46. M.E.M. Benwell, P.E. Holton, R.J. Moran, and D.J.K. Balfour, *Br. J. Pharmacol.* **117,** 743 (1996).
47. A.S. Schneider, J.E. Nagel, and S.J. Mah, *Eur. J. Pharmacol.* **317,** R1 (1996).
48. B. Badio, W.L. Padgett and J.W. Daly, *Mol. Pharmacol.* **51,** 1 (1997).

49. S.J. Mah, Y. Tang, P.E. Liauw, J.E. Nagel, and A.S. Schneider, *Brain Res.* **797,** 173 (1998).
50. J.D. Fryer and R.J. Lukas, *J. Pharmacol. Exp. Ther.* **288,** 88 (1999).
51. L.B. Hough, S.M. Pearl, and S.D. Glick, *Life Sci.* **58,** L119 (1996).
52. A. Bisaga and P. Popik, *Drug & Alcohol Depend.* **59,** 1 (2000).
53. B.H. Herman, F. Vocci, and P. Bridge, *Neuropsychopharmacology* **13,** 269 (1995).
54. B.H. Herman and C.P. O'Brien, *Seminars Neurosci.* **9,** 158 (1997).
55. K. Chen, T.G. Kokate, S.D. Donevan, F.I. Carroll, and M.A. Rogawski, *Neuropharmacology* **35,** 423 (1996).
56. R.T. Layer, P. Skolnick, C.M. Bertha, U.K. Bandarage, M.E. Kuehne, and P. Popik, *Eur. J. Pharmacol.* **309,** 159 (1996).
57. P. Popik, R.T. Layer, and P. Skolnick, *Psychopharmacology* **114,** 672 (1994).
58. P. Popik, R.T. Layer, L.H. Fossom, M. Benveniste, B. Geter-Douglass, J.M. Witkin, and P. Skolnick, *J. Pharmacol. Exp. Ther.* **275,** 753 (1995).
59. D.C. Mash, J.K. Staley, J. Pablo, A.M. Holohean, J.C. Hackman, and R.A. Davidoff, *Neurosci. Lett.* **192,** 53 (1995).
60. Y. Itzhak and S.F. Ali, *Ann. N. Y. Acad. Sci.* **844,** 245 (1998).
61. B. Geter-Douglass and J.M. Witkin, *Psychopharmacology* **146,** 280 (1999).
62. S.F. Ali, G.D. Newport, W. Slikker, Jr., R.B. Rothman, and M.H. Baumann, *Brain Res.* **737,** 215 (1996).
63. M.H. Baumann, R.B. Rothman, and S.F. Ali, *Ann. N. Y. Acad. Sci.* **844,** 252 (1998).
64. E.E. Codd, *Pharmacol. Lett.* **57,** 315 (1995).
65. J.P. Pablo and D.C. Mash, *Neuroreport* **9,** 109 (1998).
66. S.M. Pearl, K. Herrick-Davis, M. Teitler, and S.D. Glick, *Brain Res.* **675,** 342 (1995).
67. M.K. Mundey, N.A. Blaylock, R. Mason, S.D. Glick, I.M. Maisonneuve, and V.G. Wilson, *Br. J. Pharmacol.* **129,** 1561 (2000).
68. J.A. Schneider and M. McArthur, *Experientia* **12,** 323 (1956).
69. B. Frances, R. Gout, J. Cros, and J.M. Zajac, *Fundament. Clin. Pharmacol.* **6,** 327 (1992).
70. H.N. Bhargava, Y.J. Cao, and G.M. Zhao, *Brain Res.* **752,** 234 (1997).
71. S.D. Glick, *Personal Communication*, October 19, 2000.
72. R.A. Rabin and J.C. Winter, *Eur. J. Pharmacol.* **316,** 343 (1996).
73. Y.J. Cao and H.N. Bhargava, *Brain Res.* **752,** 250 (1997).
74. S.D. Glick, I.M. Maisonneuve, and S.M. Pearl, *Brain Res.* **749,** 340 (1997).
75. S.D. Glick, I.M. Maisonneuve, K.E. Visker, K.A. Fritz, U.K. Bandarage, and M.E. Kuehne, *Psychopharmacology* **139,** 274 (1998).
76. S.D. Glick, I.M. Maisonneuve, J. Raucci, and S. Archer, *Brain Res* **681,** 147 (1995).
77. S.M. Pearl and S.D. Glick, *Neurosci Lett.* **213,** 5 (1996).
78. H. Sershen, A. Hashim, and A. Lajtha, *Brain Res. Bull.* **36,** 587 (1995).
79. D.C. Mash, J.K. Staley, M.H. Baumann, R.B. Rothman, and W.L. Hearn, *Life Sci.* **57,** L45 (1995).
80. D. Wei, I.M. Maisonneuve, M.E. Kuehne, and S.D. Glick, *Brain Res.* **800,** 260 (1998).
81. G.B. Wells, M.C. Lopez, and J.C. Tanaka, *Brain Res. Bull.* **48,** 641 (1999).
82. R. Glennon, *Neuropsychopharmacology* **3,** 509 (1990).
83. D. Repke, D. Artis, and E.H.Z. Wong, *J. Org. Chem.* **59,** 2164 (1994).
84. S. Helsley, D. Fiorella, R.A. Rabin, and J.C. Winter, *Pharmacol. Biochem. Behav.* **59,** 419 (1998).
85. S. Helsley, R.A. Rabin, and J.C. Winter, *Prog. Neuro-Psychopharmacol. Biol. Psychiat.* **23,** 317 (1999).
86. E.D. French, K. Dillon, and S. Ali, *Life Sci.* **59,** PL 199 (1996).
87. H. Sershen, A. Hashim, L. Harsing, and A. Lajtha, *Life Sci.* **50,** 1079 (1992).
88. M.H. Baumann, R.B. Rothman, and S.F. Ali, *Drug & Alcohol Dependence* **59,** 143 (2000).
89. J.A. Schneider and E.B. Sigg, *Ann NY Acad Sci* **66,** 765 (1957).
90. Z. Binienda, M.A. Beaudoin, B.T. Thorn, D.R. Papurna, R.A. Johnson, C.M. Fogle, W. Jr. Slikker, and S.F. Ali, *Ann. N. Y. Acad. Sci.* **844,** 265 (1998).

91. E. Vincent and M. Serro, *Compt. Rend Soc. Biol.* **136,** 612 (1942).
92. I.M. Maisonneuve, G.L. Mann, C.R. Deibel, and S.D. Glick, *Psychopharmacology* **129,** 249 (1997).
93. R.H. Mach, C.R. Smith, and S.R. Childers, *Life Sci.* **57,** L57 (1995).
94. W.D. Bowen, B.J. Vilner, W. Williams, C.M. Bertha, M.E. Kuehne, and A.E. Jacobson, *Eur. J. Pharmacol.* **279,** R1 (1995).
95. S. Courtre and G. Debonnel, *Synapse* **29,** 62 (1998).
96. S. Helsley, D. Fiorella, R.A. Rabin, and J.C. Winter, *Prog. Neuro-Psychopharmacol. Biol. Psychiat.* **22,** 649 (1998).
97. M.D. Schecter, *Life Sci* **60,** 83 (1997).
98. H.E. Jones, H. Li, and R.L. Balster, *Pharmacol. Biochem. Behav.* **59,** 413 (1998).
99. S. Helsley, R.A. Filipink, W.D. Bowen, R.A. Rabin, and J.C. Winter, *Pharmacol. Biochem. Behav.* **59,** 495 (1998).
100. C. Zubaran, M. Shoaib, I.P. Stolerman, J. Pablo, and D.C. Mash, *Neuropsychopharmacology* **21,** 119 (1999).
101. S. Helsley, R.A. Rabin and J.C. Winter, *Eur. J. Pharmacol.* **345,** 139 (1998).
102. S. Helsley, C.A. Dlugos, R.J. Pentney, R.A. Rabin, and J.C. Winter, *Brain Res.* **759,** 306 (1997).
103. B. Grella, M. Dukat, R. Young, M. Teitler, K. Herrick-Davis, C.B. Gauthier, G and R.A. Glennon, *Drug & Alcohol Depend.* **50,** 99 (1998).
104. S. Helsley, R.A. Rabin, and J.C. Winter, *Life Sci.* **60,** L147 (1997).
105. M.E. Alburges and G.R. Hanson, *Brain Res.* **844,** 96 (1999).
106. M.E. Alburges, B.P. Ramos, L. Bush, and G.R. Hanson, *Eur. J. Pharmacol.* **390,** 119 (2000).
107. M.E. Alburges and G.R. Hanson, *Brain Res.* **847,** 139 (1999).
108. S.D. Glick, K.E. Visker, and I.M. Maisonneuve, *Eur. J. Pharmacol.* **357,** 9 (1998).
109. S. Schenk, B. Partridge, and T.S. Shippenberg, *Psychopharmacol. (Berl.)* **144,** 339 (1999).
110. K.K. Szumlinski and I.M. Maisonneuve, *Toxicon* **39,** 75 (2001).
111. K.K. Szumlinski, I.M. Maisonneuve, and S.D. Glick, *Eur. J. Pharmacol.* **398,** 259 (2000).
112. S.M. Pearl, D.W. Johnson, and S.D. Glick, *Psychopharmacology* **121,** 470 (1995).
113. S.M. Pearl, I.M. Maisonneuve, and S.D. Glick, *Neuropharmacology* **35,** 1779 (1996).
114. S.S. Sharma and H.N. Bhargava, *Pharmacol.* **57,** 1079 (1998).
115. I.M. Maisonneuve, R.W.J. Keller, and S.D. Glick, *Eur. J. Pharmacol.* **199,** 35 (1991).
116. R.A. Rabin and J.C. Winter, *Brain Res.* **731,** 226 (1996).
117. F.J. White and P.W. Kalivas, *Drug & Alcohol Dependence* **51,** 141 (1998).
118. S.I. Dworkin, S. Gleeson, D. Meloni, T.R. Koves, and T.J. Martin, *Psychopharmacology* **117,** 257 (1995).
119. S.D. Glick, M.E. Kuehne, J. Raucci, T.E. Wilson, D. Larson, J.R.W. Keller, and J.N. Carlson, *Brain Res.* **657,** 14 (1994).
120. S.D. Glick, M.E. Kuehne, I.M. Maisonneuve, U.K. Bandarage, and H.H. Molinari, *Brain Res.* **719,** 29 (1996).
121. H. Sershen, A. Hashim, and A. Lajtha, *Pharmacol. Biochem. Behav.* **47,** 13 (1994).
122. A.H. Rezvani, D.H. Overstreet, Y. Yang, I.M. Maisonneuve, U.K. Bandarage, M.E. Kuehne, and S.D. Glick, *Pharmacol. Biochem. Behav.* **58,** 615 (1995).
123. D.C. Mash and C.A. Kovera, *Abstracts, College of Problems of Drug Dependence, Sixty-Second Annual Scientific Meeting,* p. 121. San Juan, Puerto Rico, 2000.
124. S.D. Glick, S.M. Pearl, J. Cai, and I.M. Maisonneuve, *Brain Res.* **713,** 294 (1996).
125. A.H. Rezvani, D.H. Overstreet, Y. Yang, I.M. Maisonneuve, U.K. Bandarage, M.E. Kuehne, and S.D. Glick, *Pharmacol. Biochem. Behav.* **58,** 615 (1997).
126. T.S.L. Cappendijk, D. Fekkes, and M.R. Dzoljic, *Behav. Brain Res.* **65,** 117 (1994).
127. S.D. Glick, K. Rossman, N.C. Rao, I.M. Maisonneuve, and J.N. Carlson, *Neuropharmacology* **31,** 497 (1992).
128. B. Rho and S.D. Glick, *Neuroreport* **9,** 1283 (1998).
129. M.D. Aceto, E.R. Bowman, and L.S. Harris, *NIDA Research Monograph* **95,** 576 (1990).

130. L.G. Sharpe and J.H. Jaffe, *Neuroreport* **1,** 17 (1990).
131. I. Moroz, L.A. Parker, and S. Siegel, *Exp. Clin. Psychopharmacol.* **5,** 119 (1997).
132. L.A. Parker, S. Siegel, and T. Luxton, *Learning & Memory* **3,** 344 (1995).
133. T. Luxton, L.A. Parker, and S. Siegel, *Prog. Neuro-Psychopharmacol. Biol. Psychiat.* **20,** 857 (1996).
134. I.M. Maisonneuve and S.D. Glick, *Eur. J. Pharmacol.* **212,** 263 (1992).
135. S.M. Pearl, L.B. Hough, D.L. Boyd, and S.D. Glick, *Pharmacol. Biochem. Behav.* **57,** 809 (1997).
136. I.M. Maisonneuve, K.L. Rossman, R.W.J. Keller, and S.D. Glick, *Brain Res.* **575,** 69 (1992).
137. H. Sershen, L.G.J. Harsing, A. Hashim, and A. Lajtha, *Life Sci.* **51,** 1003 (1992).
138. H. Sershen, A. Hashim, and A. Lajtha, *Pharmacol. Biochem. Behav.* **53,** 863 (1996).
139. P.A. Broderick, F.T. Phelan, F. Eng, and R.T. Wechsler, *Pharmacol. Biochem. Behav.* **49,** 771 (1994).
140. K.K. Szumlinski, I.M. Maisonneuve, and S.D. Glick, *Pharmacol. Biochem. Behav.* **63,** 457 (1999).
141. K.K. Szumlinski, I.M. Maisonneuve, and S.D. Glick, *Psychopharmacology* **145,** 227 (1999).
142. K.K. Szumlinski, M.Y. Balogun, I.M. Maisonneuve, and S.D. Glick, *Psychopharmacol.* **151,** 234 (2000).
143. I.M. Maisonneuve and S.D. Glick, *Eur. J. Pharmacol.* **383,** 15 (1999).
144. M.S. Reid, K.J. Hsu, K.H. Souza, P.A. Broderick, and S.P. Berger, *J. Neural Transm. (Budapest)* **103,** 967 (1996).
145. S. Ikemoto and J. Panksepp, *Brain Res. Rev.* **31,** 6 (1999).
146. M. Cantor, *The Truth Seeker* **117,** 23 (1990).
147. D. Luciano, *Am. J. Addict.* **7,** 89 (1998).
148. D. Luciano, E.A. Della Sera, and E.G. Jethmal, *Multidiscip. Assoc. Psyched. Stud.* **9,** 27 (2000).
149. S.G. Sheppard, *J. Subst. Abuse Treat.* **11,** 379 (1994).
150. C.D. Kaplan, E. Ketzer, J. DeJong, and M. DeVires, *Social Neurosci. Bull.* **6,** 6 (1993).
151. H.S. Lotsof, *Presented at the NIDA Ibogaine Review Meeting*, Rockville, MD, 1995.
152. G. Roberts and J. Owen, *Br. J. Psychiat.* **153,** 607 (1988).
153. A.R. Jeffcoat, C.E. Cook, J.M. Hill, D.P. Coleman, and G.M. Pollack, *NIDA Research Monograph* **141,** 309 (1994).
154. R.A. Upton, *Presented at the NIDA Ibogaine Review Meeting*, Rockville, MD, 1995.
155. R.S. Obach, J. Pablo, and D.C. Mash, *Drug Metab. Disp.* **26,** 764 (1998).
156. C.R. Wolf and G. Smith, *Br. Med. Bull.* **55,** 366 (1999).
157. J. Pablo, R. Tyndale, S. Obach, W.L. Hearn, and D.C. Mash, *Abstracts, College of Problems of Drug Dependence, Sixtieth Annual Scientific Meeting*, p. 108. Scottsdale, AZ, 1998.
158. H.I. Dhahir, *A Comparative Study of the Toxicity of Ibogaine and Serotonin. Doctoral Thesis,* **71-25-341,** University Microfilm International, Ann Arbor, MI, 1971.
159. E. O'Hearn and M.E. Molliver, *Neuroscience* **55,** 303 (1993).
160. E. O'Hearn and M.E. Molliver, *J. Neurosci.* **17,** 8828 (1997).
161. H.H. Molinari, I.M. Maisonneuve, and S.D. Glick, *Brain Res.* **737,** 255 (1996).
162. Z. Xu, L.W. Chang, Jr., W. Slikker, S.F. Ali, R.L. Rountree, and A.C. Scallet, *Toxicol. Sci.* **57,** 95 (2000).
163. E. O'Hearn and M.E. Molliver, *in* "Cell Death and Diseases of the Nervous System" (V.E. Koliatos and R.R. Ratan, eds.), p. 221. Humana Press, Totowa, NJ, 1998.
164. J.P. O'Callaghan, T.S. Rogers, L.E. Rodman, and J.G. Page, *Ann. NY Acad. Sci.* **801,** 205 (1996).
165. A.C. Scallet, X. Ye, R. Rountree, P. Nony, and S.F. Ali, *Ann. NY Acad. Sci.* **801,** 227 (1996).
166. L.C. Schmued, C. Albertson, and W. Slikker, Jr., *Brain Res.* **751,** 37 (1997).
167. L.C. Schmued and K.J. Hopkins, *Toxicol. Pathol.* **28,** 91 (2000).
168. B.J. Vilner, U.K. Bandarage, M.E. Kuehne, C.M. Bertha, and W.D. Bowen, *NIDA Research Monograph* **178,** 235, (1998).

169. W.D. Bowen, B.J. Vilner, W. Williams, U.K. Bandarage, and M.E. Kuehne, *Soc. Neurosci. Abstr.* **23,** 2319, #905.7 (1997).
170. J.W. Olney, *U.S. Patent* 5,629,307; *Chem. Abstr.* **127,** 819q (1997).
171. X. Yu, S.Z. Imam, G.D. Newport, W. Slikker, Jr., and S.F. Ali, *Brain Res.* **823,** 213 (1999).
172. S. Helsley, D. Fiorella, R.A. Rabin, and J.C. Winter, *Pharmacol., Biochem. Behav.* **58,** 37 (1997).
173. G. Zetler, G. Singbartl, and L. Schlosser, *Pharmacology* **7,** 237 (1972).
174. G. Singbartl, G. Zetler, and L. Schlosser, *Neuropharmacology* **12,** 239 (1973).
175. D.C. Mash, K. Allen-Ferdinand, M. Mayor, C.A. Kovera, J.F. Ayafor, I.C. Williams, and F.R. Ervin, *NIDA Research Monograph* **179,** 294 (1999).
176. W. Baer, Forensic Subsequent Autopsy/Report Case # N-138 1991. University of Zurich, Switzerland.
177. G. van Ingen, *Pro Justitia No. 93221/I057*, Dept. Justice, The Netherlands, Lab. Forensic Pathol., 1994.
178. Court of Appeal at the Hague, *Order of the Court. Cause List Number: 997179K09*, 1999.
179. P. Popik, R.T. Layer, and P. Skolnick, *Pharmacol. Rev.* **47,** 235 (1995).
180. P. Kintz, P. Mangin, A.A. Lugnier, and A.J. Chaumont, *Human Toxicol.* **8,** 487 (1989).
181. T.E. Robinson and K.C. Berridge, *Brain Res. Rev.* **18,** 247 (1993).
182. I. Wickelgren, *Science* **280,** 2045 (1998).
183. S. Siegel, *Addiction* **94,** 1113 (1999).
184. G. DiChiara, *Eur. J. Pharmacol.* **375,** 13 (1999).
185. V. Deroche, M. Le Moal, and P.V. Piazza, *Eur. J. Neurosci.* **11,** 2731 (1999).
186. N.M. White, *Addiction* **91,** 921 (1996).
187. X. Nogues, *Prog. Neuro-Psychopharmacol. Biol. Psychiat.* **21,** 507 (1997).
188. J.R. Blackburn and K.K. Szumlinski, *Behav. Brain. Res.* **89,** 99 (1997).
189. R.P. Kesner, P.Jackson-Smith, C. Henry, and K. Amann, *Pharmacol., Biochem. Behav.* **51,** 103 (1995).
190. P. Popik, *Life Sci.* **59,** L379 (1996).
191. H. Depoortere, *Neuropsychobiology* **18,** 160 (1987).
192. E. Hennevin, B. Hars, C. Maho, and V. Bloch, *Behav. Brain. Res.* **69,** 125 (1995).
193. M. Jouvet, *J. Sleep Res.* **7,** 1 (1998).
194. S.J. Cruikshank and N.M. Weinberger, *J. Neurosci.* **16,** 861 (1996).
195. G.M. Frenken, *Personal Communication*, 2000.
196. E. Taub, *Personal Communication*, August 23, 2000
197. S.J. Blatt, B. Rounsaville, S.L. Eyre, and C. Wilber, *J. Nerv. Ment. Dis.* **172,** 342 (1984).
198. C.J. Acker, *J. Psychoactive Drugs* **25,** 193 (1993).
199. J.W. Fernandez, *in* "Flesh of the Gods: The Ritual Use of Hallucinogens" (P.T. Furst, ed.), p. 237, Waveland Press, Prospect Heights, IL, 1990.
200. M. Galanter, *Am. J. Psychiat.* **147,** 543 (1990).
201. J.W. Fernandez, "Bwiti: An Ethnography of Religious Imagination in Africa," Princeton University Press, Princeton, NJ, 1982.
202. K.L. Jansen, *Med. Hypoth.* **31,** 25 (1990).
203. W.K. Schmidt, C.W. Gorodetzky, N.L. Teti, F. Vocci, S.A. Grossman, and R.S. Mansbach, *in* "Proceedings of the 60th Annual Meeting, The College of Problems on Drug Dependence," p. 20. U.S. Department of Health & Human Services, Bethesda, MD, 1999.
204. H. Grabowski, *Pharmacoeconomics* **11,** 389 (1997).
205. C.R. Wright, "Memorandum to Franck Vocci: Ibogaine." March 10, 1995.

—CHAPTER 2—

MECHANISMS OF ACTION OF IBOGAINE: RELEVANCE TO PUTATIVE THERAPEUTIC EFFECTS AND DEVELOPMENT OF A SAFER *IBOGA* ALKALOID CONGENER

STANLEY D. GLICK, ISABELLE M. MAISONNEUVE, AND KAREN K. SZUMLINSKI

Center for Neuropharmacology and Neuroscience
Albany Medical College
Albany, NY 12208

I. Introduction

Ibogaine, an alkaloid extracted from *Tabernanthe iboga* (Apocynaceae), is being used in uncontrolled clinical trials as a long-acting treatment for opioid and stimulant abuse, alcoholism, and smoking. In this laboratory, animal models have been used to study ibogaine's interactions with drugs of abuse, to investigate its mechanisms of action, and to help develop an ibogaine derivative that will have an improved safety profile. An outline illustrating the kinds of studies we have conducted is shown in Table I. In this review, we will describe, in parallel, the results of these studies with ibogaine and with 18-methoxycoronaridine (18-MC), a novel *iboga* alkaloid congener.

The structures of ibogaine and 18-MC are shown in Figure 1. It is of interest

0099-9598/01 $35.00

TABLE I. PRECLINICAL EVALUATION OF IBOGAINE AND 18-MC

Strategies	Methods
Efficacy • Addiction (drug-self-administration) • Dependence (opioid withdrawal) Safety • Side effects • Pharmacokinetics Mechanisms of action • Receptor interactions • Effects on neurotransmitters	Animals • Rats (Sprague-Dawley and Long-Evans) • Ns = 4-9 (usually 6) Route of administration • i.p. and p.o. routes Techniques • Intravenous and oral drug self-administration, morphine withdrawal signs • Cerebellar morphology, heart rate and blood pressure, GCMS • Radioligand binding, in vivo microdialysis, locomotor activity

that the structure-activity relationships elucidated in the 1970s suggested that 18-MC might have fewer side effects than ibogaine. Singbartl *et al.* (*1*) found that, when injected intracerebrally, several *iboga* alkaloids caused tremors in mice. Tremorigenic activity was increased by the addition of a methoxy group at position 10 or 11 and was reduced or abolished by the addition of a carbomethoxy group at position 16 (note that an alternative numbering scheme refers to these positions as 12, 13, and 18, respectively). 18-MC has both of these non-tremorigenic features (i.e., lacking ibogaine's 10-methoxy group and having a 16-carbomethoxy group) and thus, in at least one respect, should be safer than ibogaine.

Ibogaine has an active metabolite, noribogaine (*2,3*), and both ibogaine and noribogaine appear to have multiple mechanisms of action in the nervous system. 18-MC also appears to have multiple targets. Table II shows the reported affinities of ibogaine and noribogaine for several binding sites, as well as the affinities of 18-MC for these same sites. The evidence to date suggests that actions at several of these sites may together mediate the putative antiaddictive effects of these

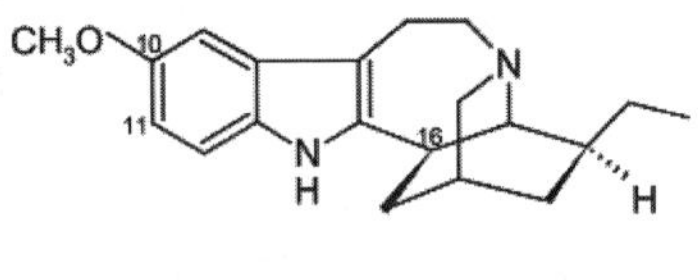

Ibogaine

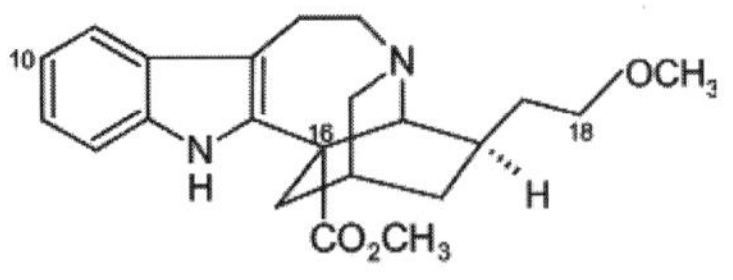

(±)-18-Methoxycoronaridine

FIGURE 1. Structures of ibogaine and 18-methoxycoronaridine.

TABLE II. INTERACTIONS OF 18-MC, IBOGAINE, AND NORIBOGAINE WITH THE INDICATED TARGET SITES (VALUES ARE μM K_i)

Site	Ligand	Tissue	18-MC	Ibogaine	Noribogaine
κ opioid	[^{3}H]-U69593	Calf cortex	5.1 ± 0.50	2.2 ± 0.10	0.61 ± 0.015
μ opioid	[^{3}H]-DAGO	Calf cortex	1.1 ± 0.30	2.0 ± 0.15	0.68 ± 0.016
δ opioid	[^{3}H]-DPDPE	Calf caudate	3.5 ± 0.05	>10	5.2 ± 0.64
Nociceptin	[^{3}H]-nociceptin	Bovine cortex	>100	>100	>100
NMDA	[^{3}H]-MK801	Bovine cortex	>100	3.1 ± 0.30	15 ± 2.0
D1	[^{3}H]-SCH23390	Calf caudate	>100	>10	>10
D2	[^{3}H]-N-methyl-spiperone	Calf caudate	16 ± 0.60	>10	>10
D3	[^{3}H]-7-OH-DPAT	Calf caudate	25 ± 2.5	70 ± 1.7	>100
M1	[^{3}H]-pirenzepine	Calf cortex	32 ± 3	16 ± 1.0	15 ± 1.0
M2	[^{3}H]-QNB	Calf cortex	>100	31± 3.4	36 ± 3.7
5-HT1A	[^{3}H]-8-OH-DPAT	Rat hippocampus	46 ± 4.9	>100	>100
5-HT1B	[^{3}H]-serotonin	Calf caudate	>100	>100	>100
5-HT1C	[^{3}H]-mesulergine	Calf cortex	>100	>100	>100
5-HT1D	[^{3}H]-serotonin	Calf caudate	>10	>100	>100
5-HT2A	[^{3}H]-ketanserin	Gf-6 cells	40 ± 3.4	16	>100
5-HT2C	[^{3}H]-mesulergine	J-1 cells	>100	>10	>10
5-HT3	[^{3}H]-GR-65,630	NG-108 cells	3.8 ± 0.067	2.6 ± 0.23	>100
Sodium channel	[^{3}H]-BTX-B	Bovine cortex	6.4 ± 0.68	3.6 ± 0.35	17 ± 0.6
Sigma 1	[^{3}H]-(+)-pentazocine	Calf caudate	>100	2.5 ± 0.6	11 ± 1.7
Sigma 2	[^{3}H]-DTG	Calf hippocampus	13 ± 1.2	0.4 ± 0.036	19 ± 1.3
GABA B	[^{3}H]-GABA	Calf cortex	>100	>100	>100
NE uptake	[^{3}H]-nisoxetine	Bovine cortex	>10	>100	39 ± 1.5
5-HT uptake	[^{3}H]-paroxetine	Bovine brain stem	>10	4.1 ± 0.83	0.57 ± 0.083

drugs. While most pharmaceutical development efforts focus on single mechanisms of action, a drug capable of treating diverse addictions may, of necessity, have to have multiple actions. Hence, as reviewed below, the peculiarly broad efficacy of ibogaine and 18-MC may be precisely attributable to their peculiarly complex pharmacology.

II. Behavioral and Neurochemical Methods

All subjects were naïve female Sprague-Dawley (Taconic) or Long-Evans (Charles River) rats, approximately 3 months old and weighing 230-250 g at the beginning of an experiment. Rats were maintained on a normal light/dark cycle (lights on/off at 0700 h/1900 h).

The intravenous self-administration procedure was described previously (*4-6*).

The intravenous self-administration system consisted of polyethylene-silicone cannulas constructed according to the design of Weeks (*7*), Instech harnesses and commutators, and Harvard Apparatus infusion pumps (#55-2222). Responses on either of two levers produced a 10 or 50 μl infusion of drug solution, 0.01 mg (0.04 mg/kg) morphine sulfate, or 0.1 mg (0.4 mg/kg) cocaine hydrochloride, respectively, in 0.2 to 1.0 second.

Nicotine was self-administered via the oral route using an operant procedure previously described (*8*). Rats received nicotine (1.4 μg/μl of the base; 0.1 ml per response) by pressing one lever and water by pressing another lever.

Locomotor activity was assessed using cylindrical photocell activity cages (60 cm, three crossing beams) interfaced to an IBM compatible computer (*9*).

The microdialysis procedures used to assess the effects of drug treatments on extracellular levels of dopamine and its metabolites have been used extensively in this laboratory (*3,5,6,10-14*). Rats were implanted stereotaxically with guide cannulae so that, when inserted, the tips of the dialysis probes would be located in the intended brain areas (e.g., nucleus accumbens, striatum, medial prefrontal cortex). All microdialysis experiments were carried out in freely moving animals. Perfusate samples were analyzed by HPLC with electrochemical detection.

III. Opioid Interactions

The acute intraperitoneal (i.p.) administration of either ibogaine or 18-MC, 15 minutes prior to testing, dose-dependently decreased the self-administration of morphine (*4,6*) in rats. As shown in Figure 2, although ibogaine and 18-MC were about equally potent, 18-MC was more selective in that ibogaine, but not 18-MC, acutely (on the day of treatment) depressed responding for a nondrug reinforcer (water). The effects of ibogaine and 18-MC on morphine self-administration were protracted; pretreatment with 40 mg/kg ibogaine or 18-MC had significant effects for 24 and 48 hours, respectively (*4-6*). Although the acute effects of ibogaine on morphine self-administration can be attributed to the induction of whole body tremors, the protracted effects of ibogaine occur at times when the drug is eliminated from the body and tremors are absent (*4*). 18-MC does not induce tremors, but its effects on morphine self-administration also persist long after 18-MC itself is eliminated (*6,14*).

Comparable effects of 18-MC on morphine self-administration were also observed following oral treatment (*14*), suggesting that 18-MC, like ibogaine, will be pharmacologically active when given orally to humans. Furthermore, in a recent study, oral 18-MC treatment (40 mg/kg) was found to produce a downward shift in the entire dose-response relationship for self-administered morphine (*13*).

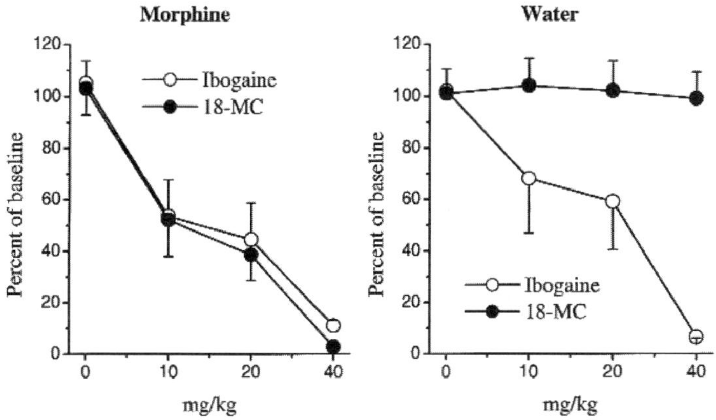

FIGURE 2. Comparison of acute effects of ibogaine and 18-MC on morphine self-administration and on responding for water.

This indicated that 18-MC, and probably ibogaine as well (*15*), decreases the reinforcing efficacy of morphine. Prolonged access to opioids (*16*), as well as to stimulants (*17*), has been shown to result in an "escalation" of drug intake such that the drug becomes more reinforcing and the dose-response relationship is shifted higher. Our results (*13*) suggest that 18-MC may reverse this trend, counteracting the neural adaptations produced by chronic drug administration.

Consistent with human anecdotal reports, the antiaddictive efficacies of both ibogaine and 18-MC seem to increase with repeated treatment. Weekly or biweekly injections of ibogaine (3-4 injections of 40 mg/kg, i.p.) can increasingly suppress morphine intake for up to a week in some rats (*4*), and repeated daily administration of low doses of 18-MC (e.g., 5 injections of 20 mg/kg, p.o.), while having little or no effect on self-administration on day 1 of treatment, decreased morphine intake by day 4 (*14*). This suggests that rather than giving people ibogaine or 18-MC in a single large dose, as is currently done for ibogaine, it might be advisable, at least for reasons of safety, to give smaller doses repeatedly.

Ibogaine has been claimed to "suppress the multiple symptoms and physical discomfort of narcotic withdrawal" (ENDABUSE™ product information). Accordingly, we assessed the effects of ibogaine (*18*) and 18-MC (*19*) treatment in an animal model of morphine withdrawal, in which signs of withdrawal were induced in morphine-dependent rats by the acute administration of a μ-opioid receptor antagonist (naltrexone). Both ibogaine and 18-MC reduced the intensity of several signs of morphine withdrawal. However, their effects were not identical, suggesting that ibogaine and 18-MC may act via somewhat different mechanisms.

Ibogaine and 18-MC also differ with regard to their acute effects on morphine-induced locomotion. Ibogaine decreased morphine's efficacy to induce locomotion, shifting morphine's dose-response curve downward (*9,15,20*), whereas 18-MC enhanced morphine's potency, shifting its dose-response curve to the left (*21*). Ibogaine (40 mg/kg, i.p., 19 hours beforehand) also produced a greater attenuation of morphine-induced (5 mg/kg, i.p.) locomotion in rats previously (2-4 times) administered morphine (30 mg/kg, i.p.) compared to acutely treated rats (*9*). Interestingly, however, 18-MC (40 mg/kg, i.p., 19 hours earlier) blocked the expression of locomotor sensitization following chronic morphine administration (*21*). The dose-effect curve of control rats sensitized by chronic morphine administration was shifted to the left of control rats that did not sensitize in response to chronic morphine, whereas the dose-effect curves of 18-MC-pretreated sensitized and nonsensitized rats were virtually identical. Thus, it appears that whereas ibogaine produces a greater effect on morphine-induced locomotion in drug-experienced animals, compared to naïve animals, 18-MC masks, or possibly reverses, the alterations in behavior produced by chronic morphine experience, apparently returning the animal to its initial nonsensitized state.

Some of the effects of ibogaine appear to be at least partially mediated by a combination of κ-opioid agonist and NMDA antagonist actions. Thus, a combination of a κ-opioid antagonist (nor-binaltorphimine; norBNI) and an NMDA agonist (NMDA) significantly antagonized the effect of ibogaine on morphine self-administration, while neither norBNI nor NMDA alone had this

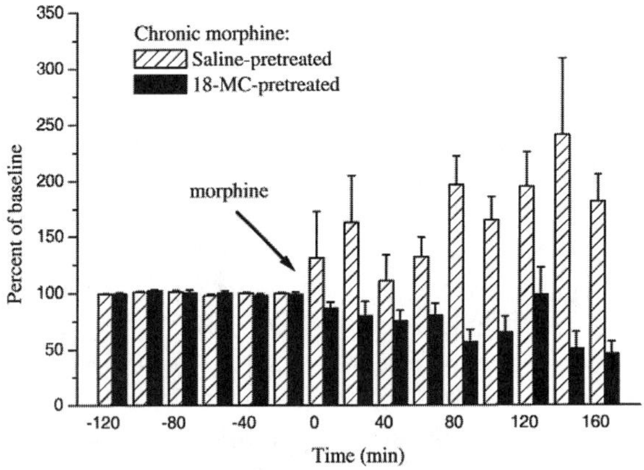

FIGURE 3. Effects of 18-MC (40 mg/kg, 19 hours beforehand) on the sensitized dopamine response to morphine (20 mg/kg, i.p.); morphine was administered daily for 5 consecutive days, and again (test for sensitization), after a 3-day withdrawal period.

effect (*22*). Other effects of ibogaine were also blocked by a combination of norBNI and NMDA (*22*). These included ibogaine (40 mg/kg, i.p., administered 19 hours beforehand) inhibition of morphine-induced (5 mg/kg, i.p.) locomotor stimulation and ibogaine inhibition of dopamine release in the striatum. Comparable studies with 18-MC have not been conducted.

All addictive drugs (including opioids, stimulants, ethanol, and nicotine) examined to date share an ability to enhance dopamine transmission in the nucleus accumbens (*23,24*), a critical mediator of the "rewarding" or "incentive motivational" effects of drugs (*25,26*). Consistent with their putative antiaddictive actions, ibogaine and 18-MC (40 mg/kg, i.p.) were both found to decrease accumbal dopamine release during the first 3 hours after their administration (*6,10*). Both compounds, administered 19 hours earlier, also blocked acute morphine-induced (*6,10*) increases in extracellular levels of dopamine in the nucleus accumbens.

Like its effects on sensitized locomotor behavior, we have recently found that 18-MC (40 mg/kg, i.p., 19 hours earlier) similarly abolishes the sensitized dopamine response to morphine in rats chronically administered morphine (Figure 3). Again, the data suggest that 18-MC (and probably ibogaine) counteracts or reverses the homeostatic disturbances that are a consequence of repetitive opioid use.

IV. Stimulant Interactions

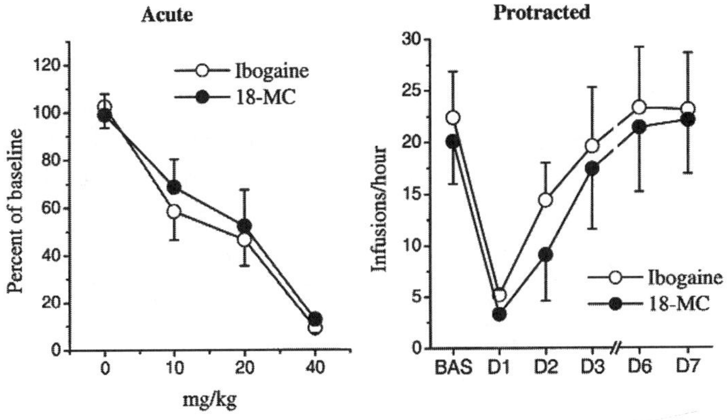

FIGURE 4. Comparison of acute and protracted effects of ibogaine and 18-MC on cocaine self-administration.

The acute intraperitoneal (i.p.) administration of either ibogaine or 18-MC, 15 minutes prior to testing, dose-dependently decreased the self-administration of cocaine (*5,6*) in rats. As shown in Figure 4, ibogaine and 18-MC were again about equally potent; and similar to previous results with morphine, their effects on cocaine self-administration were protracted, lasting for approximately 24 hours. In contrast, 18-MC seemed to be approximately twice as potent as ibogaine in decreasing oral nicotine preferences (*12*), and recent work with an intravenous nicotine self-administration paradigm suggests that 18-MC is at least twice as potent in decreasing nicotine intake as in decreasing either morphine or cocaine intake.

With respect to stimulant-induced locomotion, both ibogaine and 18-MC augmented the expression of locomotor behavior in response to cocaine (*27-30*) and amphetamines (*31,32*). Ibogaine and 18-MC both shifted the dose-response curve of acute cocaine-treated animals to the left of controls (*30*), indicating that pretreatment with these agents renders an animal more sensitive to cocaine's acute locomotor effects (Figure 5). In an early study, ibogaine (40 mg/kg, i.p., 24 hours beforehand) attenuated the locomotor response to *d*-amphetamine (1.5 mg/kg, i.p.) in rats repeatedly administered *d*-amphetamine (4 x 1.5 mg/kg, every other day) (*34*). More recently, pretreatment with ibogaine or 18-MC (40 mg/kg, i.p., 19 hours earlier) was found to shift the inverted U-shaped dose-response curves for locomotion in chronic cocaine-treated rats to the left of controls such that *iboga*-pretreated rats displayed augmented locomotor activation at lower cocaine doses (e.g., 5 and 10 mg/kg) (*29,30*) and lower levels of locomotor

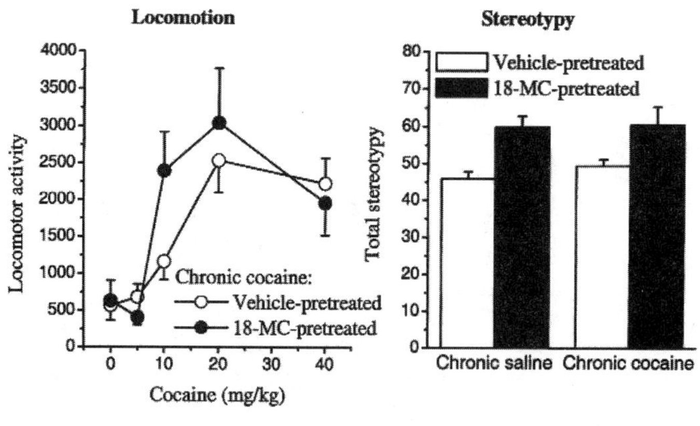

FIGURE 5. 18-MC (40 mg/kg, 19 hours before test doses of cocaine) shifted the chronic cocaine locomotor dose-response curve to the left and enhanced the stereotypic response to cocaine (40 mg/kg); similar findings occurred with ibogaine.

activation at higher cocaine doses (e.g., 20 and 40 mg/kg) (*30*), compared to control animals (Figure 5). The locomotor-attenuating effects of *iboga* pretreatment at higher cocaine doses can be attributed to the induction of repetitive, species-specific behaviors (stereotypy), which can be physically incompatible with locomotion (e.g., focused sniffing, grooming, gnawing). Ibogaine and 18-MC (40 mg/kg, i.p., 19 hours earlier) promoted the expression of high levels of cocaine-induced stereotypic behavior in both acute and chronic cocaine-treated rats, compared to controls (Figure 5).

Virtually identical effects of ibogaine and 18-MC pretreatment were observed for methamphetamine-induced stereotypy. This latter finding may possibly account for the previously reported (*32*) attenuating effect of ibogaine on *d*-amphetamine-induced locomotion in chronic d-amphetamine-treated rats. Combined, these findings indicate that pretreatment with either ibogaine or 18-MC will enhance rats' sensitivity to the behavioral-activating effects of stimulant drugs, and that this increase can be above and beyond the sensitization produced by chronic stimulant administration alone.

Distinctions between ibogaine and 18-MC have been reported with respect to some of their neurochemical effects. For one, acute ibogaine, as well as noribogaine, increase extracellular levels of serotonin in the nucleus accumbens, whereas 18-MC has no effect (*33*). Secondly, ibogaine pretreatment (19 hours earlier) augments (*27*), whereas 18-MC has no effect on (*14*), acute cocaine-induced increases in extracellular levels of dopamine in the nucleus accumbens. Curiously, however, both agents block acute nicotine-induced dopamine release

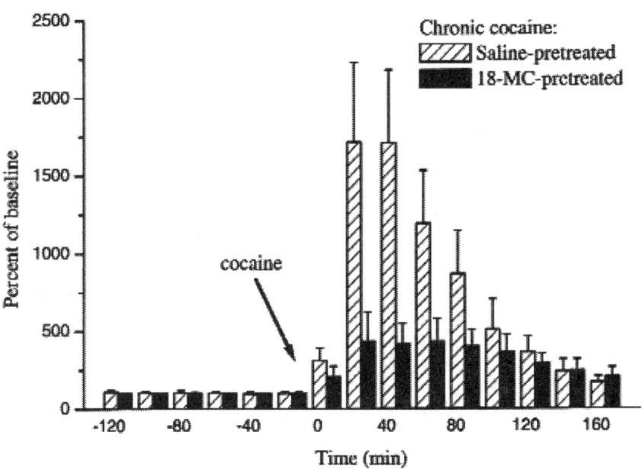

FIGURE 6. Effects of 18-MC (40 mg/kg, 19 hours beforehand) on the sensitized dopamine response to cocaine (15 mg/kg, i.p.); cocaine was administered daily for 5 consecutive days and, again (test for sensitization), after a two-week withdrawal period.

in the nucleus accumbens (*11,12*). The studies with 18-MC have recently been extended to animals sensitized by chronic cocaine and, interestingly, 18-MC pretreatment blocked the sensitized dopamine response to chronic cocaine (Figure 6). Although the effect of ibogaine pretreatment on cocaine-sensitized levels of dopamine in the nucleus accumbens has not yet been assessed, the 18-MC results suggest that different mechanisms may mediate the interactions of ibogaine and related agents with the effects of acute versus chronic cocaine. The data also suggest that the changes in nucleus accumbens dopamine are more directly related to cocaine's reinforcing or addictive property than to its locomotor stimulant effects, since, as reviewed earlier, 18-MC decreased cocaine self-administration, but enhanced both acute and chronic cocaine-induced locomotor behavior.

V. Metabolism and Distribution of Ibogaine and 18-MC

Plasma and tissue levels of both ibogaine and 18-MC have been determined using gas chromatography-mass spectrometry (*14,34*). Both compounds have short initial half-lives of 5 to 10 minutes and terminal half-lives of slightly over 100 minutes. Consistent with a two-compartment model of their elimination, both ibogaine and 18-MC are highly sequestered in fat (*14,34*). In absolute terms, however, the fat levels of either ibogaine or 18-MC account for only a small fraction of the administered dose (approximately 10%), suggesting that both compounds are rapidly metabolized. Indeed, an active metabolite of ibogaine, noribogaine, has already been well characterized both *in vivo* (e.g., *2,3*) and *in vitro* (e.g., 35,36). Although some investigators (*37*) consider noribogaine to be the major determinant of ibogaine's pharmacology *in vivo*, studies in this laboratory (*20*) indicated that the elimination of noribogaine was also too fast for it to be responsible for all of ibogaine's prolonged effects. Recent work in this laboratory has provided evidence that 18-MC also has metabolites, but it remains to be determined whether they are active, and whether they contribute to the protracted behavioral effects of 18-MC.

VI. Toxicity

Ibogaine induces whole body tremors at moderate doses (20-40 mg/kg) and Purkinje cell loss in the cerebellum at high doses (≥100 mg/kg) (*38-40*).

However, 18-MC (40 mg/kg) is non-tremorigenic, and even multiple, high-dose (100 mg/kg) injections of 18-MC fail to produce damage to cerebellar Purkinje cells (*6*). The neurotoxic effect of ibogaine appears to be mediated by an agonist action at sigma-2 receptors (*41*). Consistent with this, 18-MC has a much lower affinity than ibogaine for sigma-2 sites (Table II and ref. *42*).

Anecdotal reports in humans indicate that ibogaine can slow heart rate. Consistent with these reports, recent work in this laboratory showed that, in awake and freely moving rats, high doses (100 and 200 mg/kg, i.p.) of ibogaine decreased heart rate, without altering blood pressure. In contrast, even at 200 mg/kg (i.p.), 18-MC had no effect on either heart rate or blood pressure (*14*).

Noribogaine has about a 10-fold higher affinity for the serotonin transporter than ibogaine and, consistent with this, noribogaine is more potent than ibogaine in raising extracellular levels of serotonin in the nucleus accumbens (*2*). However, the efficacy of ibogaine to increase serotonin levels appears to be substantially greater than that of noribogaine (*33*). Ibogaine may directly release serotonin. Compared to its effects on the dopamine systems, these serotonergic effects of ibogaine and noribogaine appear to be relatively short lasting, dissipating within 3 hours. Similarly, while effects of ibogaine on tissue levels of dopamine metabolites are still apparent on the day after administration (*15,43*), there are no effects on tissue levels of serotonin's metabolite (*43*). Serotonin would thus seem to have a role in mediating only the acute behavioral effects of ibogaine. These might include its acute discriminative stimulus effect in rats (*44,45*) and possibly its acute hallucinogenic effect in humans. 18-MC neither inhibits the reuptake of (Table II), nor releases, serotonin (*33*) and, to the extent that these actions are involved in ibogaine-induced hallucinations, it is predicted that 18-MC will not be hallucinogenic.

VII. Mechanisms

Table II shows the results of a receptor screen comparing the binding affinities of 18-MC, ibogaine, and its active metabolite, noribogaine. The binding profiles for 18-MC are somewhat different from that of its parent compound. Similar to results reported by others (cf. *46*), our studies show that ibogaine and noribogaine have low micromolar affinities for the κ- and μ-opioid receptors, the NMDA-subtype of glutamate receptor, 5-HT3 receptors, sigma-2 sites, sodium channels, and the serotonin transporter. In contrast, 18-MC has low micromolar affinities at all three opioid receptors (κ, μ, and δ) and at the 5-HT3 receptor, and no affinity at NMDA receptors, or the serotonin transporter (*14*). For both ibogaine and 18-MC, all of these receptor affinities are in the low micromolar range and therefore are not likely to be responsible for effects lasting 24 to 48 hours. However, some of the differences between ibogaine and 18-MC might account for a potentially higher therapeutic index for 18-MC. Ibogaine's affinities at muscarinic (M1 and M2) receptors and at sodium channels, which are two to three times greater than those of 18-MC, may mediate its tendency to lower heart rate. Ibogaine's action at sigma-2 sites has been linked to its neurotoxicity (*42*), and 18-MC has a 30-fold lower affinity for this site. In addition, as suggested previously, the hallucinogenic effect of ibogaine may be mediated via serotonin release, an effect not produced by 18-MC (*33*). Lastly, in functional assays, ibogaine was reported to be a noncompetitive antagonist at nicotinic receptors, possibly acting as an open channel blocker (*47,48*). The latter, as well as preliminary data from this laboratory, suggest that both ibogaine and 18-MC might have nanomolar affinities for nicotinic channels—and this action could well contribute to prolonged antiaddictive effects.

VIII. Discussion

Most drug development programs focus on single mechanisms of action, and the development of pharmacotherapies for drug addiction has been no exception to this practice. The use of methadone to treat heroin addiction and the use of nicotine formulations (e.g., gum, patch, nasal spray) to treat smoking are representative of a pharmacokinetic approach in which long-acting replacement therapies are used to dampen both the "highs" and "lows" associated with the short-acting addictive substances. This approach has limitations in that replacement therapies maintain physical dependence and often have other significant side effects as well. Newer, and still mostly experimental, approaches to this problem have attempted to develop agents that should modulate or directly interfere with the action of the abused drug. Representatives of such potential therapies are dopamine transporter inhibitors, dopamine receptor agonists and antagonists, GABA B receptor agonists, and partial μ-opioid receptor agonists. In general, treatments have been sought that are site specific, usually acting selectively at a particular receptor or receptor subtype; and most often, treatments are targeted to one particular addictive disorder.

Viewed in relationship to a "normal" pharmaceutical development program, the proposed use of ibogaine, 18-MC, and possibly their metabolites, to treat several varieties of drug addiction may appear, depending on one's bias, to be extraordinarily innovative or outrageously foolish. However, if the many anecdotal reports of efficacy are ever substantiated in well-controlled clinical trials, ibogaine will have taught us at least one important if not obvious truth—

namely, that addiction is a multifaceted brain disorder, and that to be effective, a treatment or treatments having multiple actions may be required. Certainly science, rather than politics, should determine whether or not ibogaine will have any clinical utility. Moreover, the apparent advantages of 18-MC, and perhaps other congeners yet to be tested, have already highlighted the significance of ibogaine's discovery. If only because it is the prototype, ibogaine would still merit a great deal of investigation.

The data reviewed here indicate that there are several ways in which ibogaine and 18-MC could exert antiaddictive effects. Both compounds have affinities for 5-HT3 receptors, the manipulation of which has been reported to alter amphetamine-induced euphoria in humans (*49*) and cocaine-induced locomotion, cocaine discrimination, alcohol consumption, and morphine withdrawal signs in rodents (*50-54*). These alkaloids also have similar affinities for μ-and κ-opioid receptors, and other data (*55-57*) have indicated that μ-antagonists and κ-agonists can modulate the self-administration of cocaine and morphine. However, as noted earlier, the protracted antiaddictive effects of ibogaine and 18-MC are hard to reconcile with their micromolar affinities for these receptors. In addition, both ibogaine (*18*) and 18-MC (*19*) attenuate naltrexone-precipitated withdrawal symptoms in morphine-dependent rats, findings that are inconsistent with μ-antagonist activity; and both ibogaine (*58*) and 18-MC have little or no analgesic activity, findings that are inconsistent with μ-agonist activity. Although NMDA antagonist (*59*) and serotonergic (*2*) actions of ibogaine have been invoked to explain ibogaine's effects, it is noteworthy that 18-MC appears to have neither action. The possibility that ibogaine and 18-MC have important actions at nicotinic receptors requires further investigation.

The short-half lives of ibogaine and 18-MC strongly suggest that the pharmacological actions of both alkaloids are attributable to one or more active metabolites; although noribogaine has been proposed (*2,37*) as the mediator of ibogaine's prolonged action, it would appear that noribogaine alone cannot account for ibogaine's effects since brain levels of noribogaine also decline rapidly after ibogaine administration to rats (*20*). As both ibogaine and 18-MC are deposited in fat (*14,34*), it is possible that the slow release of these compounds, or perhaps their metabolites, may contribute to their protracted effects.

In summary, although the pharmacology of ibogaine and 18-MC is complex, the study of their pharmacology represents an entirely novel approach to the development of pharmacotherapies for drug addiction. This approach will indeed have proven its worth if 18-MC, or another structural congener, is eventually found to be a safe and effective treatment for multiple forms of drug abuse. At the very least, continued investigation of ibogaine and 18-MC should help us further understand the neurobiology of addiction; and this, in the long term, may be a prerequisite for developing optimal antiaddictive agents.

Acknowledgments

The authors would like to thank Drs. M.E. Kuehne, L. Hough, and U. Bandarage for their valuable help and collaboration is various phases of this research, and D. Boyd, H. Dickinson, R. Murdoch, K. Visker, and L. Warner for their excellent technical assistance. This research was supported by NIDA grant DA 03817.

References

1. G. Singbartl, G. Zetler, and L. Schlosser, *Neuropharmacology* **12,** 239 (1973).
2. D.C. Mash, J.K. Staley, M.H. Baumann, R.B. Rothman, and W.L. Hearn, *Life Sci.* **57,** PL 45 (1995).
3. S.D. Glick, S.M., Pearl, Z. Cai, and I.M. Maisonneuve, *Brain Res.* **713,** 294 (1996).
4. S.D. Glick, K. Rossman, S. Steindorf, and J.N. Carlson, *Eur. J. Pharmacol.* **195,** 341 (1991).
5. S.D. Glick, M.E. Kuehne, J. Raucci, T.E. Wilson, T.D. Larson, R.W. Keller, and J.N. Carlson, *Brain Res.* **657,** 14 (1994).
6. S.D. Glick, M.E. Kuehne, I.M. Maisonneuve, U.K. Bandarage, and H.H. Molinari, *Brain Res.* **719,** 29 (1996).
7. J.R. Weeks, *in* "Methods in Psychobiology" (R.D. Myers, ed), Vol. 2, p. 155, Academic Press, Orlando, 1972.
8. S.D. Glick, K.E. Visker, and I.M. Maisonneuve, *Psychopharmacology* **128,** 426 (1996).
9. S.M. Pearl, D.W. Johnson, and S.D. Glick, *Psychopharmacology* **121,** 470 (1995).
10. I.M. Maisonneuve, R.W. Keller, and S.D. Glick, *Eur. J. Pharmacol.* **199,** 35 (1991).
11. I.M. Maisonneuve, G.L. Mann, C.R. Deibel, and S.D. Glick, *Psychopharmacology* **129,** 206 (1997).
12. S.D.Glick, I.M. Maisonneuve, K.E. Visker, K.A. Fritz, U.K. Bandarage, and M.E. Kuehne, *Psychopharmacology* **139,** 274 (1998).
13. I.M. Maisonneuve and S.D. Glick, *Eur. J. Pharmacol.* **383,** 15 (1999).
14. S.D. Glick, I.M. Maisonneuve, L.B. Hough, M.E. Kuehne, and U.K. Bandarage, *CNS Drug Rev.* **5,** 27 (1999).
15. I.M. Maisonneuve, K.L. Rossman, R.W. Keller, and S.D. Glick, *Brain Res.* **575,** 69 (1992).
16. S.H. Ahmed, J.R. Walker, and G.F. Koob, *Soc. Neurosci. Abstr.* **25,** 1324 (1999).
17. S. H. Ahmed and G.F. Koob, *Science* **282,** 298 (1998).
18. S.D. Glick, K. Rossman, N.C. Rao, I.M. Maisonneuve, and J.N. Carlson, *Neuropharmacology* **31,** 497 (1992).
19. B. Rho and S.D. Glick, *NeuroRep.* **9,** 1283 (1998).
20. S.M. Pearl, L.B. Hough, D.L. Boyd, and S.D. Glick, *Pharmacol. Biochem. Behav.* **57,** 809 (1997).
21. S.D. Glick, I.M. Maisonneuve, and K.K. Szumlinski, *Soc. Neurosci. Abstr.* **25,** 559 (1999).
22. S.D. Glick, I.M. Maisonneuve, and S.M. Pearl, *Brain Res.* **749,** 340 (1997).
23. R.A. Wise and M.A. Bozarth, *Psychol. Rev.* **94,** 469 (1987).
24. G. DiChiara and A. Imperato, *Proc. Natl. Acad. Sci USA* **85,** 5274 (1988).
25. C. Kornetsky and L.J. Porrino, *in* "Addictive States" (C.P. O'Brien and J.H. Jaffe, eds.), p. 59, Raven Press., New York, 1992.
26. K.C. Berridge and T.E. Robinson, *Brain Res.Rev.* **28,** 309 (1999).
27. I.M. Maisonneuve and S.D. Glick, *Eur. J. Pharmacol.* **212,** 263 (1992).
28. I.M. Maisonneuve, K.E. Visker, G.L. Mann, U.K. Bandarage, M.E. Kuehne, and S.D. Glick,

Eur. J. Pharmacol. **336,** 123 (1997).
29. K.K. Szumlinski, K.K., I.M. Maisonneuve, and S.D. Glick, *Pharmacol. Biochem Behav.* **63,** 457 (1999).
30. K.K. Szumlinski, I.M. Maisonneuve, and S.D.Glick, *Psychopharmacology* **145,** 227 (1999).
31. I.M. Maisonneuve, R.W. Keller, and S.D. Glick, *Brain Res.* **579,** 87 (1992).
32. J.R. Blackburn and K.K. Szumlinski, *Behav Brain Res.* **89,** 99 (1977).
33. D. Wei, I.M. Maisonneuve, M.E. Kuehne, and S.D. Glick, *Brain Res.* **800,** 260 (1998).
34. L.B. Hough, S.M. Pearl, and S.D. Glick, *Life Sci.* **58,** PL 119 (1996).
35. S.M. Pearl, K. Herrick-Davis, M. Teitler, and S.D. Glick, *Brain Res.* **675,** 342 (1995).
36. J.K. Staley, Q. Ouyang, J. Pablo, W.L. Hearn, D.D. Flynn, R.B. Rothman, K.C. Rice, and D.C. Mash, *Psychopharmacology* **127,** 10 (1996).
37. D.C. Mash, C.A. Kovera, B.E. Buck, M.D. Norenberg, P. Shapshak, W.L. Hearn, and J. Sanchez-Ramos, *Ann. N.Y. Acad. Sci.* **844,** 274 (1998).
38. E. O'Hearn, D.B. Long, and M.E. Molliver, *NeuroRep.* **4,** 299 (1993).
39. E. O'Hearn and M.E. Molliver, *Neuroscience* **55,** 303 (1993).
40. H.H. Molinari, I.M. Maisonneuve, and S.D. Glick, *Brain Res.* **737,** 255 (1996).
41. W.D. Bowen, B.J. Vilner, W. Williams, C.M. Bertha, M.E. Kuehne, and A.E. Jacobson, *Eur. J. Phamacol.* **279,** R1 (1995).
42. B.J. Vilner, U.K. Bandarage, M.E. Kuehne, C.M. Bertha, and W.D. Bowen, *NIDA Res. Monogr.* **178,** 235 (1998).
43. S.F. Ali, G.D. Newport, W. Slikker, Jr., R.B. Rothman, and M.H. Baumann, *Brain Res.* **737,** 215 (1996).
44. P.A. Palumbo and J.C. Winter, *Pharmacol. Biochem. Behav.* **43,** 1221 (1992).
45. M.D. Schechter and T.L. Gordon, *Eur. J. Pharmacol.* **249,** 79 (1993).
46. S.D. Glick and I.M. Maisonneuve, *Ann. N.Y. Acad. Sci.* **844,** 214 (1998).
47. B. Badio, W.L. Padgett, and J.W. Daly, *Mol. Pharmacol.* **51,** 1 (1997).
48. A.S. Schneider, J.E. Nagel, and S.J. Mah, *Eur. J. Pharmacol.* **317,** R1 (1996).
49. T.A. Grady, A. Broocks, S.K. Canter, T.A. Pigot, B. Dubbert, J.L. Hill, and D.L. Murphy, *Psychiatry Res.* **64,** 1 (1996).
50. A.L. Svingos and R. Hitzemann, *Pharmacol. Biochem. Behav.* **43,** 871 (1992).
51. G.R. King, Z. Xiong, and E.H. Ellinwood, Jr., *Psychopharmacology* **135,** 263 (1998).
52. M.D. Schechter, Pharmacol. *Biochem. Behav.* **44,** 661 (1993).
53. E.M. Sellers, G.A. Higgins, D.M. Tomkins, M.K. Romach, and T. Toneatto, *J. Clin. Psychi.* **52,** 49 (1991).
54. A. Pinelli, S. Trivulzio, and L. Tomasoni, *Eur. J. Pharmacol.* **340,** 111 (1997).
55. T.R. Kosten, T.R., H.D. Kleber, and C. Morgan, *Life Sci.* **44,** 887 (1989).
56. S.S. Negus, S.J. Henriksen, A. Mattox, G.W. Pasternak, P.S. Portoghese, A.E. Takemori, M.B. Weinger, and G.F. Koob, *J. Pharmacol. Exp. Ther.* **265,** 1245 (1993).
57. .D. Glick, I.M. Maisonneuve, J. Raucci, and S. Archer, *Brain Res.* **681,** 147 (1995).
58. A.A. Bagal, L.B. Hough, J.W. Nalwalk, and S.D. Glick, *Brain Res.* **741,** 258 (1996).
59. P. Popik, R.T. Layer, L.H. Fossom, M. Benveniste, B. Geter-Douglass, J.M. Witkin, and P. Skolnick, *J. Pharmacol. Exp. Ther.* **275,** 753 (1995).

——CHAPTER 3——

IBOGAINE AS A GLUTAMATE ANTAGONIST: RELEVANCE TO ITS PUTATIVE ANTIADDICTIVE PROPERTIES

PHIL SKOLNICK

Dov Pharmaceutical, Inc.
Hackensack, NJ 07601

I. Overview

Beginning in the mid 1980s, Howard Lotsof (*1-4*) filed a series of patents claiming that ibogaine, an alkaloid derived from *Tabernanthe iboga*, possessed antiaddictive qualities. At the time, the concept that a single molecule could treat dependence across classes of abused drugs (e.g., cocaine, nicotine, ethanol, opiates) was viewed as radical, if not revolutionary. In the absence of rigorously controlled, double-blind clinical trials, these claims have engendered skepticism and controversy. During the past 5 to 7 years, the biomedical research community has made a concerted effort to characterize the neurochemical actions of ibogaine with the implicit understanding that such studies may provide insight into the putative antiaddictive actions of this compound. In parallel with these neurochemical studies, preclinical behavioral studies have established that ibogaine can interfere with tolerance and dependence phenomena (reviewed in reference *5*).

If the dictum, "Clinical data trumps preclinical data" has merit, then the pragmatist may legitimately question the value of such preclinical studies in the face of anecdotal reports that ibogaine does possess antiaddictive properties.

THE ALKALOIDS, Vol.56
0099-9598/01 $35.00

Certainly, there are many examples where safe and effective drugs have been used for decades (e.g., benzodiazepines, NSAIDs) before a molecular mechanism of action was evinced. However, in view of the safety concerns raised by both preclinical and clinical reports (e.g., reference *6*), and in the absence of controlled clinical studies, basic research on ibogaine is clearly mandated.

Pharmacologically relevant concentrations (doses) of ibogaine can affect several neurotransmitter systems (reviewed in reference *5*). These multiple actions pose the challenge of separating "wheat from chaff"—that is, discriminating those effects relevant to the putative antiaddictive properties of ibogaine from epiphenomena. The majority of these "mechanism of action" investigations, including work from our studies at the NIH, have focused on "traditional" targets, such as ion channels, transporters, and the seven transmembrane superfamily of transmitter receptors. Such studies have largely neglected a host of potential intracellular targets that may either act independently, or in concert with, extracellular targets to produce the antiaddictive properties described anecdotally in the clinic and documented in preclinical studies. Absent these studies, there remain sufficient *in vitro* and *in vivo* data to both formulate testable hypotheses and create a diversity of opinion (clearly evident at the First International Congress on Ibogaine) about the neurochemical processes responsible for these antiaddictive actions. This contribution will overview data demonstrating that pharmacologically relevant concentrations of ibogaine produce a blockade of *N*-methyl-D-aspartate (NMDA) receptors, and relate the relevance of these findings to its antiaddictive properties.

II. NMDA Antagonist Properties of Ibogaine

There is a striking similarity between the claims that have been made for ibogaine and an emerging body of preclinical evidence that NMDA antagonists interfere with tolerance and dependence phenomena to a wide variety of abused drugs. This prompted us to determine if the basis for the apparent mimicry between ibogaine and NMDA antagonists could be due to an identical locus of action. In our initial studies, we examined the ability of ibogaine to inhibit radioligand binding to native NMDA receptors from rat brain (*7*). Ibogaine inhibited the binding of [^{3}H]dizocilpine (MK-801) in a concentration-dependent manner with a K_i of ~1 μM. This inhibition by ibogaine reflected an increase in the K_d of [^{3}H]MK-801 without striking changes in B_{max}, characteristic of two ligands acting at the same site (i.e., a competitive interaction). Subsequent neurochemical studies from our laboratory and others confirmed that the apparent affinity of ibogaine is in the low μM range using other radioligands acting at the same locus as MK-801 (e.g., [^{3}H]TCP) and NMDA receptors derived from a variety of sources, including human brain (*8-11*). In contrast, ibogaine does not remarkably affect radioligand binding to other members of the ionotropic glutamate receptor family (i.e., kainate and AMPA receptors), nor does it inhibit radioligand binding to the glutamate recognition site on NMDA receptors (*7*). While the affinity of ibogaine for NMDA receptors is low relative to MK-801 (and other NMDA antagonists belonging to this same class such as TCP and PCP), brain concentrations of the parent alkaloid are in the range of 1 to 10 μM after administering pharmacologically relevant doses (i.e., doses capable of interfering with tolerance and/or dependence phenomena) to rodents (*12*).

Such neurochemical studies are valuable because they provide a mechanistic link between ibogaine and a class of uncompetitive NMDA antagonists (including MK-801, PCP, memantine, and ketamine) that has been extensively characterized both *in vitro* and *in vivo*. Uncompetitive NMDA antagonists can be envisioned as channel "plugs" (analogous to placing a cork in one end of a tube) and exhibit a number of characteristic features including use (i.e., the channel lumen must be open in order for such compounds to enter and bind) and voltage (the "block" is relatively more efficient at hyperpolarized membrane potentials) dependence. Because of the potential therapeutic applications of uncompetitive NMDA antagonists, this class of compound has been extensively studied at all levels of cellular organization (ranging from effects on single channel activity to behavior). This "prior art" allows us to make predictions about the pharmacological actions of ibogaine that may be NMDA receptor-mediated, and provides strategies to isolate and assess the contribution of this effect relative to its putative antiaddictive actions.

While radioligand binding studies indicate that ibogaine acts as an uncompetitive NMDA antagonist (i.e., acting at the same locus and by the same mechanism as, for example, dizocilpine and phencyclidine), several independent lines of investigation have provided compelling evidence that supports this hypothesis. Thus, in electrophysiological studies, the inhibition of NMDA responses by ibogaine exhibits the voltage and use dependence characteristic of this class of compounds (*8,10,11*). Further, there is very good agreement between the estimated potencies of ibogaine obtained in neurochemical and electrophysiological studies. For example, analysis of the NMDA receptor block using the Woodhull equation permits a calculation of the K_d of ibogaine as a function of membrane potential. In cultured hippocampal neurons, the K_d of ibogaine ranged from ~8.6 μM at 0 mV to ~2.3 μM at –60 mV (8).

The neuroprotective effects of NMDA antagonists are perhaps the best described pharmacological actions produced by this class of compounds (*13*). These neuroprotective actions can readily be demonstrated in both simple systems and whole animals using a variety of insults, ranging from glutamate-induced cell death in primary neuron culture to animal models of focal ischemia.

If the neurochemical and electrophysiological studies with ibogaine are pharmacologically meaningful, then like other NMDA antagonists, ibogaine should protect against NMDA receptor-mediated neurotoxicity. To test this hypothesis, we examined (*8*) the ability of ibogaine to prevent glutamate-induced death of cerebellar granule neurons in primary culture. Many studies have shown that activation of NMDA receptors is a necessary condition for glutamate-induced death of these neurons, and as such, NMDA antagonists (including uncompetitive antagonists such as MK-801) are effective in blocking this "excitotoxic" process. In our hands, ibogaine decreased glutamate-induced neurotoxicity in a concentration dependent manner with an IC_{50} of ~4.9 μM; this value closely approximates the potency of ibogaine as an NMDA antagonist estimated by neurochemical and electrophysiological techniques. By comparison, MK-801 was ~500-fold more potent, with an IC_{50} value of ~9.6 nM. At face value, a neuroprotective action of ibogaine appears at variance with reports that this alkaloid produces degeneration of cerebellar Purkinje neurons (*14,15*). However, it is unlikely that this latter action is a consequence of NMDA receptor blockade since the prototypic uncompetitive NMDA antagonist, MK-801, does not produce a similar effect (*16*). Based on its side effect profile, it is unlikely that the neuroprotective properties of ibogaine will be reduced to clinical practice. Nonetheless, Olney (*17*) has patented the use of ibogaine as a neuroprotective agent!

III. Are the NMDA Antagonist Actions of Ibogaine Relevant to Its Putative Antiaddictive Properties?

These *in vitro* data provide compelling evidence that ibogaine can act as an NMDA antagonist. Further, ibogaine concentrations that are required to produce this action are well within the range found in the rodent central nervous system (*12*) at doses that affect both tolerance and dependence phenomena. This same dose range of ibogaine can substitute as a discriminate stimulus in mice trained to recognize the prototypic uncompetitive NMDA antagonist, MK-801 (*8*). These findings, coupled with an emerging preclinical literature (*18-20*) demonstrating that NMDA antagonists interfere with tolerance and dependence phenomena to a variety of abused drugs (*7,8*), indicate that it is this NMDA antagonist action that is responsible, either wholly or in part, for the antiaddictive properties of ibogaine. If ibogaine produces its antiaddictive actions via a voltage-dependent block of NMDA receptors, then reversal of this block should reduce or abolish these actions. One strategy that has been employed to relieve this block relies on increasing brain concentrations of glycine (or a glycine-mimetic such as D-serine) at strychnine-insensitive glycine receptors. Glycine is a coagonist at NMDA receptors. Due to the presence of specific transporters that appear colocalized with NMDA receptors (*21*), it is unlikely that these strychnine-insensitive glycine sites are saturated under physiological conditions (*22*). Thus, raising glycine concentrations increases the probability of NMDA receptor-coupled channel opening, which in turn increases the likelihood that ibogaine (and other channel blockers) will dissociate from the binding site. This "unblocking" strategy has been shown to reduce some of the pharmacological effects of dizocilpine and phenyclidine (*23-25*).

It was demonstrated that like other NMDA antagonists, memantine (a low-affinity, uncompetitive NMDA antagonist) blocks the expression of morphine withdrawal in mice (*18*). This is evidenced by a dose-dependent reduction in naloxone-precipitated jumping in morphine-dependent animals. Parenteral administration of glycine (at doses that significantly elevate brain glycine levels [*26*]) blocked this action of memantine, but did not remarkably affect naloxone-precipitated jumping when administered alone (*18*). Similarly, this regimen of glycine abolished the ability of ibogaine to reduce naloxone-precipitated jumping (*8*). Clearly, it is not possible to extrapolate the importance of this single measure of morphine withdrawal in mice to the complex phenomena associated with opiate dependence in humans. Nonetheless, these data indicate that the NMDA antagonist properties of ibogaine are responsible for its "antiaddictive actions" in this measure.

This "unblocking" paradigm may be useful as a means of examining the relative contribution of NMDA receptor blockade to a particular "antiaddictive" property of ibogaine (or an ibogaine derivative). This issue transcends academic minutiae because there are a number of NMDA antagonists that are in clinical use with an established safety and side effect profile. For example, memantine has been used in Europe to treat neurodegenerative disorders such as senile dementia (*27*). Thus, if the putative antiaddictive properties of ibogaine are due to its NMDA antagonist action, then there are established therapeutic alternatives. In support of this hypothesis, the ability of a low affinity NMDA antagonist (dextromethorphan) to attenuate opiate withdrawal and craving has already been examined in a small, open clinical trial. In this study (*28*), six patients addicted to heroin were detoxified using dextromethorphan. Two patients requested methadone on the first day of the study, but the four patients completing the study: "had a rapid and complete attenuation of signs, symptoms, and craving by the fourth day of treatment." Particular improvement in the alleviation of craving was noted during the first 2 days (*28*). This report, while preliminary, is consistent with preclinical data demonstrating that NMDA antagonists block the expression of opiate withdrawal (*18,29*). However, in view of the number of targets that can be affected by pharmacologically relevant concentrations of ibogaine (*5,9,30*), it may be argued that NMDA antagonists may only be effective in treating a subset of abused drugs (or a subset of signs and symptoms), despite the striking similarities between this class of compounds and ibogaine in preclinical studies.

Several ibogaine derivatives (*31*) were synthesized in an attempt to relate the potency of these compounds to the expression of morphine withdrawal in mice (i.e., blockade of naloxone-precipitated jumping) and to NMDA receptor affinity. All of these derivatives (including a number of coronaridine derivatives) were less potent than ibogaine as NMDA antagonists *in vitro*. Notably, the K_i values of noribogaine, (±)-ibogamine, and (±)-coronaridine were ~5-fold lower than ibogaine (i.e., 5 to 6 μM). At the highest "nontoxic" doses tested (80 mg/kg), none of these compounds significantly reduced naloxone-precipitated jumping in morphine-dependent mice. Limiting side effects, such as profound ataxia and convulsions, prevented testing higher doses (i.e., 120 mg/kg) of several of these alkaloids (e.g., noribogaine). At face value, it may be argued that this study supports the hypothesis that the NMDA antagonist properties of ibogaine are essential to its "antiaddictive" actions. However, in the absence of pharmacokinetic data (e.g., brain levels of these alkaloids), these data may be considered inconclusive. The affinity of noribogaine (also known as desmethylibogaine and 10-hydroxy-ibogamine) at NMDA receptors (K_I of 5 to 6 μM) is noteworthy since this compound appears to be the primary metabolite of ibogaine (*30*). If this metabolite enters the central nervous system as readily as its parent, then the NMDA antagonist action of noribogaine could also contribute to its pharmacological properties.

Glick and coworkers (*32*) have reported that addition of a methoxy moiety to coronaridine results in a compound that lacks the tremorigenic properties of ibogaine, but retains many of its putative antiaddictive properties in animals. Thus, like ibogaine, 18-methoxycoronaridine has been reported to reduce morphine and cocaine self-administration in rats (*32*), attenuate alcohol consumption in alcohol-preferring rats (*33*), and reduce nicotine intake (*34*). It has been reported that neither racemic 18-methoxycoronaridine nor its optically active isomers (i.e., (+)- and (-)-18-methoxycoronaridine) possess NMDA antagonist properties, but retain μM affinities for opioid (κ, μ, and δ) receptors, sodium channels, 5HT-3 receptors, and $sigma_2$ sites (*35*). Because it seems unlikely that ibogaine and 18-methoxycoronardine produce their antiaddictive actions through different mechanisms, it may be concluded that one or more of the neurochemical properties common to these closely related compounds are necessary for these effects. However, following intravenous administration, 18-methoxycoronaridine has a very short half-life (~5 to 10 minutes) (*35*). This raises the possibility that it is not the parent alkaloid, but rather a metabolite of 18-methoxycoronaridine that is responsible for the observed antiaddictive actions. Short of identifying an active metabolite(s), there are several experiments that could be done to determine if administration of the parent compound produces an NMDA antagonist. One simple experiment would be to determine if the ability of 18-methoxycoronaridine to interfere with morphine withdrawal can be attenuated by glycine administration. This experimental strategy was successfully employed to link the antiaddictive properties of ibogaine and memantine to an NMDA antagonist action. Second, if rodents trained to recognize MK-801 as a discriminative stimulus (*8*) also recognize 18-methoxycoronaridine at doses that interfere with tolerance/dependence phenomena, then it is likely that a metabolite with NMDA antagonist properties is formed in vivo. Such experiments are necessary to critically assess the contribution of NMDA receptor blockade in the putative antiaddictive actions of 18-methoxycoronaridine. This compound appears to lack the tremorigenic actions of ibogaine (*32*). However, in the absence of basic toxicological studies, the claim that 18-methoxycoronaridine is nontoxic (*32*) must be viewed as premature.

Acknowledgments

Many of the preclinical studies described in this contribution were generated by Drs. Morris Benveniste, Linda Fossom, Richard Layer, and Piotr Popik during their tenure as Postdoctoral Fellows in the Laboratory of Neuroscience, National Institutes of Health. I thank my former Fellows for their insights and dedication to this challenging research area.

References

1. H. S. Lotsof, *U.S. Patent* **4,499,096;** *Chem. Abstr.* **102** 160426w (1985).
2. H. S. Lotsof, *U.S. Patent* **4,587,243;** *Chem. Abstr.* **106** 12967r (1986).
3. H. S. Lotsof, *U.S. Patent* **4,857,523;** *Chem. Abstr.* **112** 32041m (1989).
4. H. S. Lotsof, *U.S. Patent* **5,026,697;** *Chem. Abstr.* **116** 17031x (1991).
5. P. Popik and P. Skolnick, in *"The Alkaloids: Chemistry and Biology,"* (G.A. Cordell, ed.), Vol. 52, p. 197. Academic Press, San Diego, 1999.
6. N. Touchette, *J. NIH Res.* **5,** 50 (1993).
7. P. Popik, R.T. Layer, and P. Skolnick, *Psychopharmacology* **114,** 672 (1994).
8. P. Popik, R.T. Layer, L. Fossom, M. Benveniste, B. Getter-Douglass, J.M. Witkin, and P. Skolnick, *J. Pharmacol. Exp. Ther.* **275,** 753 (1995).
9. P.M. Sweetnam, J. Lancaster, A. Snowman, J.L. Collins, S. Pershke, C. Bauer, and J. Ferkany, *Psychopharmacology* **118,** 369 (1995).
10. D.C. Mash, J.K. Staley, J.P. Pablo, A.M. Holohean, J.C. Hackman, and R.A. Davidoff, *Neurosci. Lett.* **192,** 53 (1995).
11. T. Chen, S. Kokate, F. Donevan, I. Carroll, and M.A. Rogawski, *Neuropharmacology* **35,** 423 (1996).
12. S.M. Pearl, L.B. Hough, D.L. Boyd, and S.D. Glick, *Pharmacol. Biochem. Behav.* **57,** 809 (1997).
13. D.W. Choi, *J. Neurobiol.* **23,** 1261 (1992).

14. E. O'Hearn and M.E. Molliver, *Neurosci.* **55,** 303 (1993).
15. E. O'Hearn and M.E. Molliver, *J. Neurosci.* **15,** 8828 (1997).
16. R. Nakki, J. Koistinaho, F.R. Sharp, and S.M. Sagar, *J. Neurosci.* **15,** 2097 (1995).
17. J.W. Olney *U.S. Patent* **5**,629,307; *Chem. Abstr.* **127,** 819q (1997).
18. P. Popik and P. Skolnick, *Pharmacol. Biochem. Behav.* **53,** 791 (1996).
19. K.A. Trujillo and H. Akil, *Drug Alc. Depend.* **38,** 139 (1995).
20. J. Witkin, *Life Sci.* **53,** PL405-PL410 (1993).
21. K.E. Smith, L.A. Borden, P.R. Hartig, T. Branchek, and R.L. Weinshank, *Neuron* **8,** 927 (1992).
22. R. Bergeron, T.M. Meyer, J.T. Coyle, and R.W. Greene, *Proc. Natl. Acad. Sci. USA* **95,** 15730 (1998).
23. E. Toth and A. Lajtha, *Neurochem. Res.* **11**, 393 (1986).
24. G.E. Evoniuk, R.P. Hertzman, and P. Skolnick, *Psychopharmacology* **105,** 125 (1991).
25. P. Contreras, *Neuropharmacology* **29,** 291 (1990).
26. R. Trullas and P. Skolnick, *Eur. J. Pharmacol.* **185,** 1 (1990).
27. C.G. Parsons, W. Danysz, and G. Quack, *Neuropharmacology* **38,** 85 (1999).
28. A. Bisaga, P. Gianelli, and P. Popik, *Am. J. Psychiat.* **154,** 584 (1997).
29. K.A. Trujillo and H. Akil, *Science* **251,** 85 (1991).
30. J.K. Staley, Q. Ouyang, J. Pablo, W.L. Hearn, D.D. Flynn, R.B. Rothman, K.C. Rice, and D.C. Mash, *Psychopharmacology* **127,** 10 (1996).
31. R.T. Layer, P. Skolnick, C.M. Bertha, U.K. Bandarage, M.E. Kuehne, and P. Popik, *Eur. J. Pharmacol.* **309,** 159 (1996).
32. S.D. Glick, M.E. Kuehne, I.M. Maisonneuve, U.K. Bandarge, and H.H. Molinari, *Brain Res.* **719,** 29 (1996).
33. A.H. Rezvani, D.H. Overstreet, Y. Yang, I.M. Maisonneuve, U.K. Bandarage, M.E. Kuehne, and S.D. Glick, *Pharmacol. Biochem. Behav.* **58,** 615 (1997).
34. S.D. Glick, I.M. Maisonneuve, K.E. Visker, K.A. Fritz, U.K. Bandarage, and M.E. Kuehne, *Psychopharmacology* **139,** 274 (1998).
35. S.D. Glick, I.M. Maisonneuve, L.B. Hough, M.E. Kuehne, and U.K. Bandarage, *CNS Drug Rev.* **5,** 27 (1999).

—CHAPTER 4—

DRUG DISCRIMINATION STUDIES WITH IBOGAINE

SCOTT HELSLEY[1], RICHARD A. RABIN,[2] AND J. C. WINTER[2]

[1]*Department of Anesthesiology*
Duke University Medical Center
Durham, NC 27710

[2]*Department of Pharmacology and Toxicology*
SUNY Buffalo
Buffalo, New York 14214

I. Introduction

Drug discrimination is a useful technique in the search for the mechanism of action of psychotropic drugs. Perhaps its utility has been best demonstrated in efforts to elucidate those neurotransmitter receptors involved in the mediation of the perceptual effects of phenethylamine (DOM) and indoleamine (LSD) hallucinogens (*1*).

As far as biological assays of pharmacological activity are concerned, drug discrimination is relatively easy to understand. Specifically, the goal of such studies is to produce drug-induced stimulus control. This condition is said to exist when, in the presence of a stimulus, an animal subject emits a conditioned response (*2*). An example of a conditioned response might be pressing a specific lever when presented with a two-lever choice, or entering a specific arm of a two-arm maze (T-maze). The stimulus (or discriminative stimulus) is the interoceptive state produced by the drug used to establish stimulus control. To illustrate this by way of example, assume that one wishes to establish stimulus control with drug X. The discriminative stimulus will be the interoceptive cue, which is simply the effect experienced by the subject after receiving drug X. In this case the subject is a rat. However, pigeons, other rodents, and primates can also be used. The subject, over several weeks, has been taught to press levers in a two-lever operant chamber for a food reward in a process known as shaping. During this process, the ratio of response to reward is gradually increased from 1 to 10 or more. Since a certain fixed number of responses is required for a reward, this is known as a fixed ratio schedule of reinforcement. If 10 responses are required for a reward, then a fixed ratio of 10 (FR 10) is the schedule. Once the rat has mastered this, it is time to establish stimulus control.

In order to accomplish this, the rat is trained with once daily sessions, which alternate between drug and vehicle treatment. Thus, using a specific pretreatment time (e.g., 15 minutes prior to the session), the training drug (drug X) or its vehicle are given on alternating days. On days when drug X is given, only responses on one lever (left or right) are rewarded. On vehicle days, only responses on the other lever are rewarded. After a number of sessions the subject will learn to press the drug appropriate lever only on those days when that drug is administered. On vehicle days, the subject should only respond on the vehicle-appropriate lever. Once the subject is responding reliably in a treatment-appropriate manner, stimulus control is said to be present. Having established stimulus control in a given subject investigations can begin regarding the receptor interactions mediating the discriminative cue of drug X. Test drugs (drug Y) that are known agonists or antagonists at certain receptors can be given during test sessions. If these produce responses on the drug X-appropriate lever, then the test drug is said to substitute for the training drug, alternatively, by convention, the training drug stimulus is said to generalize to that of the test drug. This terminology is often reversed by some authors resulting in confusion on the part of the readers. If responding is on the vehicle-appropriate lever then no substitution or generalization is present. Often there is an intermediate degree of substitution or generalization (partial generalization) suggesting that the interoceptive cues are similar, but not identical. Another test is that of antagonism. In these tests, a drug with known receptor binding properties is given with the training drug (e.g., drugs X and Y together). If responding is seen only on the vehicle appropriate-lever then drug Y has antagonized the interoceptive cue produced by drug X. Alternatively, responding on the drug X-appropriate lever would indicate absence of antagonism. During test sessions, rewards should not be given and the session should be terminated after a fixed number of responses. This is to minimize learning during the test session, which might confuse the subjects. Furthermore, the subjects should only be tested every third or fourth day (after they have demonstrated reliable treatment appropriate responding in thc prior two or three training sessions). Training doses, test doses, and pretreatment times are usually based on reports of pharmacological effects of the drug in question found in the literature.

Drug discrimination is a remarkably simple, yet elegant, technique. Using this paradigm, investigators can often gain insight into the mechanism of action of a given agent. The results of these studies can then be correlated with other studies, such as receptor binding studies and second messenger assays.

Like other techniques, drug discrimination does have its shortcomings. Because the interoceptive cue is what is being evaluated, other drug effects and their specific mechanisms may not be accounted for. No technique is perfect. Nonetheless, for studying drugs with psychotropic effects, drug discrimination remains a powerful weapon in the arsenal of the behavioral pharmacologist.

II. Ibogaine in Drug Discrimination Studies

A. Ibogaine as a Discriminative Stimulus

Before there was any evidence supporting the antiaddictive effects of ibogaine, this agent was known first and foremost as a hallucinogen. Because of this, drug discrimination seems well suited to the study of ibogaine. Knowledge of those receptors involved in the ibogaine discriminative cue could contribute significantly to our understanding of its mechanism of action, both as a hallucinogen and possibly as a therapy for substance abuse conditions.

Figure 1a shows the dose-response effects ibogaine (●), harmaline (▲), and noribogaine (10-hydroxyibogamine) (■) in rats trained with 10.0 mg/kg ibogaine as a discriminative stimulus. All drugs were administered i.p. 60 minutes before testing. Each point represents one determination in each of 10 subjects unless otherwise noted by the number of subjects completing the test over the number of subjects tested. The ED_50 for ibogaine was 4.6 mg/kg.

Figure 1b shows the time course for the ibogaine-trained stimulus. Maximal substitution was observed at a pretreatment time of 60 minutes (94%). Following this, a time-dependent decrease in ibogaine-appropriate responding was observed. (*Modified from reference 3 with permission.*)

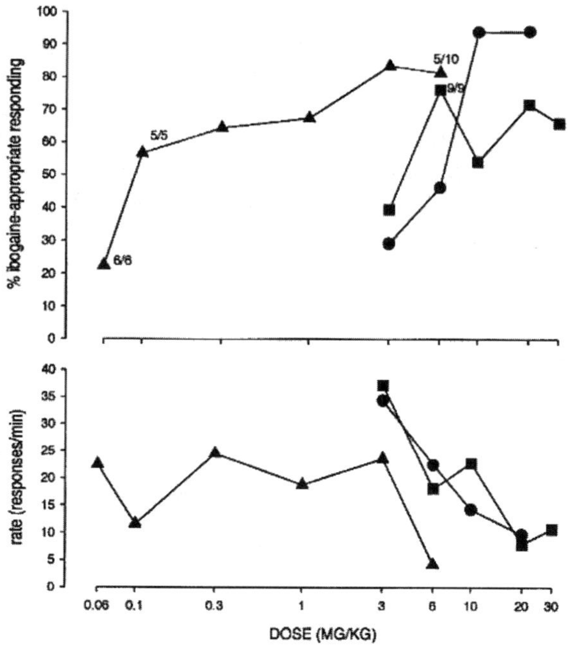

FIGURE 1A. Dose response for ibogaine ●, harmaline ▲, and noribogaine (10-hydroxyibogamine) ■.

B. SEROTONERGIC AGENTS IN IBOGAINE-TRAINED RATS

The structural similarity between ibogaine and serotonin taken together with the fact that the 5-HT_{2A} receptor is the primary mediator of the discriminative stimulus effects of the classical hallucinogens lysergic acid diethylamide (LSD) and (-)-2,5-dimethoxy-4-methyl-amphetamine (DOM) (*4-6*) makes serotonergic agents a natural starting point in the study of ibogaine.

Palumbo and Winter (*7*) were the first to look at ibogaine in drug discrimination studies. They found that ibogaine produced an intermediate level of substitution in both LSD and DOM-trained subjects. This effect was blocked by the 5-HT_2 antagonist pizotylene. The first report in which ibogaine was trained as a discriminative stimulus was by Schecter and Gordon (*8*). These authors observed an intermediate level of substitution by the 5-HT releasing agent fenfluramine. This evidence suggested a possible role for serotonergic receptors in the stimulus effects of ibogaine. Further investigations have revealed that the 5-HT_{2A} receptor plays a role, although this does not appear to be essential to the ibogaine-induced discriminative stimulus. This is evidenced by the observation that both

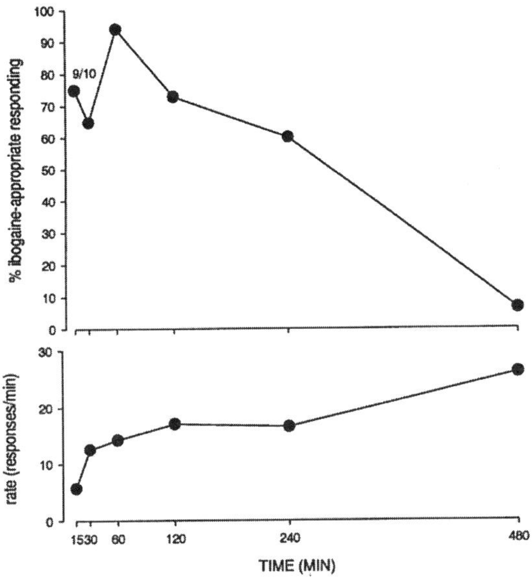

FIGURE 1B. Ibogaine time course.

DOM and LSD produced intermediate levels of substitution for ibogaine that was blocked by the 5-HT_{2A} antagonist pirenpirone (*9*). The conclusion that this component is nonessential stems from the fact that while pirenpirone blocked the ibogaine-appropriate responding produced by LSD and DOM, it did not affect the ibogaine-appropriate responding produced by ibogaine itself (*9*) (Figure 3). For a detailed discussion of nonessential stimulus components, see reference *10*.

A possible explanation for the differences in ibogaine-appropriate responding produced by LSD illustrated above, compared to the work of Schecter and Gordon (34.5% substitution) (*8*), could be accounted for by rat strain differences and/or pretreatment time differences for ibogaine training (60 minutes vs. 30

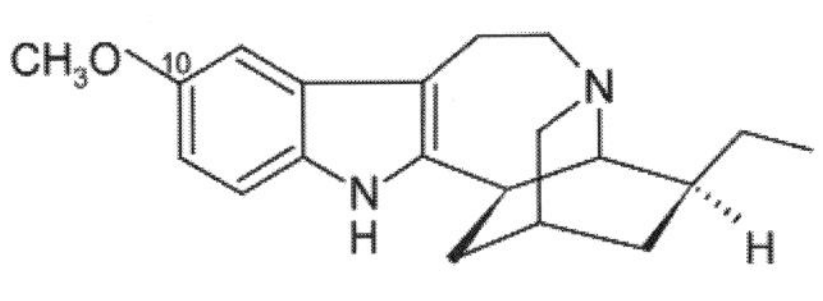

Ibogaine

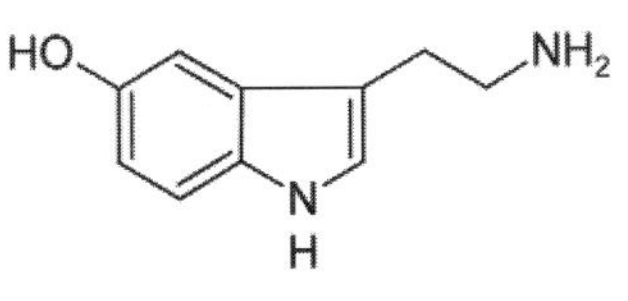

Serotonin

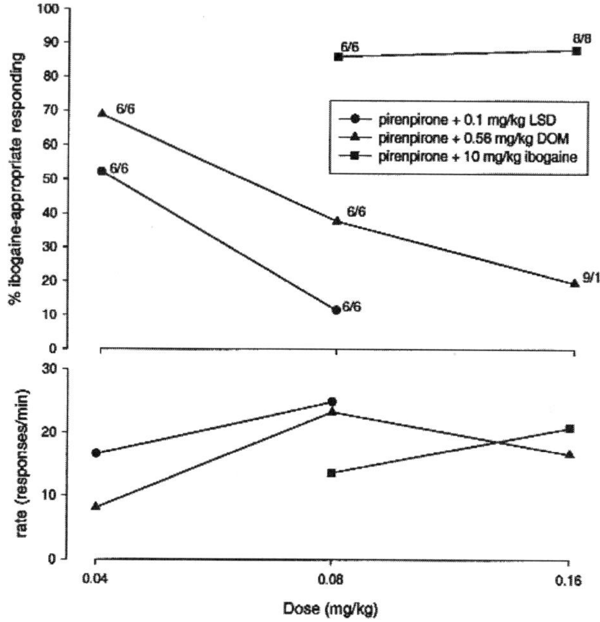

FIGURE 2. The dose response relationships for pirenpirone (75 minutes presession) in the presence of LSD (●), DOM (▲), and ibogaine (■) in rats trained to discriminate ibogaine (10.0 mg/kg, i.p., 60 minutes presession) from vehicle. The ratio adjacent to each of the points represents the number of subjects completing the test session over the number of subjects participating in each test session. *From reference 9 with permission.*

minutes, respectively).

In addition to 5-HT_{2A} receptors, there is evidence for the involvement of the 5-HT_{2C} receptor. In contrast, the 5-HT_{1A} and 5-HT_3 subtypes do not appear to play a major role in the ibogaine-mediated discriminative stimulus (*11*). Interestingly, the phenomenon observed with the 5-HT_{2A} component is also seen with the 5-HT_{2C} component of the ibogaine cue. That is, involvement of the 5-HT_{2C} receptor in the ibogaine-trained discriminative stimulus appears to be nonessential. As illustrated in Table I, the 5-HT_{2C} agonists MK 212 and mCPP both produced intermediate levels of substitution, which were blocked by metergoline, an agent that has antagonist properties at 5-HT_{2C} receptors. In contrast, the ibogaine-appropriate responding produced by ibogaine itself was not affected by mesulergine (*10*) or metergoline (*11*).

These studies have demonstrated a role for 5-HT_{2A} and 5-HT_{2C} receptors in the ibogaine discriminative cue. A role for the 5-HT_{2A} receptor is further supported by biochemical studies, which provided evidence for *in vivo* occupancy of these

TABLE I.

Drug treatment	% ibogaine-appropriate Responses	Rate (Responses/min)	n/N
Ibogaine	94.0	14.3	10/10
Ibogaine (10 mg/kg) + Metergoline (1.0 mg/kg)	100.0	36.4	4/4
MK 212 (0.3 mg/kg) + Metergoline (1.0 mg/kg)**	10.0 [79.6]*	16.1	7/8
mCPP (0.8 mg/kg) Metergoline (1.0 mg/kg)**	23.8 [76.4]*	13.0	6/8

The ratio n/N represents the number of animals responding (n) out of the number of animals tested (N). The % ibogaine-appropriate responding produced by both mCPP and MK-212 alone is enclosed in brackets. Treatment sessions were compared to immediately preceding ibogaine training sessions using Wilcoxon's signed ranks test.

*Reflects significant differences from the ibogaine-treatment condition ($p<0.05$).

**Reflects significant differences between the drug alone and the drug + antagonist conditions as determined by the Mann-Whitney Rank Sum test. *Modified from reference 11 with permission.*

receptors by ibogaine (*10*). Although these receptor interactions are not essential to the ibogaine stimulus, they provide a link between ibogaine and classical hallucinogens such as LSD and DOM. This is further supported by a recent study investigating the effects of monoamine reuptake inhibitors on the stimulus effects of hallucinogens. Specifically, the DOM, LSD, and ibogaine discriminative cues were all potentiated by the monoamine reuptake inhibitors fluoxetine, venlafaxine, and fluvoxamine (*12*). The exact mechanism for this is unknown at present. Certainly, further insights into this area of study will enhance our knowledge both of hallucinogens and of antidepressant medications, which are commonplace in psychiatric practice.

C. Beta-Carboline Agents In Ibogaine-Trained Rats

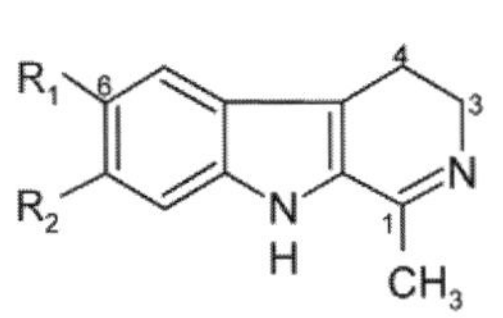

6-Methoxyharmalan	$R_1 = OCH_3, R_2 = H$
Harmaline	$R_1 = H, R_2 = OCH_3$
Harmine	$R_1 = H, R_2 = OCH_3, \Delta^{3,4}$
THBC	$R_1 = H, R_2 = H, 1,2\text{-}H_2$

One group of hallucinogens that has received little attention is the beta-carboline (or Harmala) alkaloids group. Interestingly, these agents bear a strong structural resemblance to ibogaine. Anecdotal reports suggest that the tremorigenic and subjective effects of agents, such as harmaline and harmine, are not unlike those of ibogaine (*13*). Several of these alkaloids were tested in ibogaine-trained rats (*10*). The results are shown in Figure 3. Full generalization was observed with 6-methoxyharmalan and harmaline, while partial generalization was seen with harmine, harmane, harmalol, and THBC (tetrahydro-beta-carboline). No generalization was seen to 6,7-dimethoxy-4-ethyl-β-carboline-3-carboxylate (DMCM) or norharmane.

Unfortunately, the mechanism of action of the harmala alkaloids remains unknown. However, this is not the case for other beta-carbolines like DMCM. This agent has inverse agonist properties at benzodiazepine sites (*14*). We found

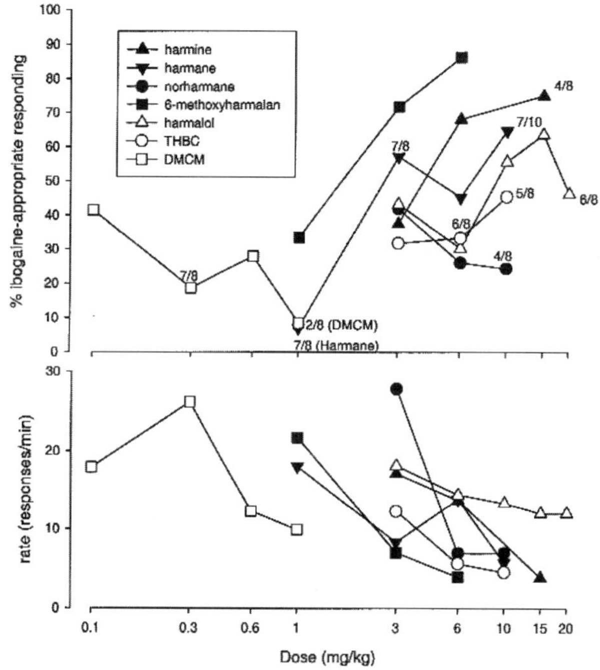

FIGURE 3. Dose-response relationships for beta-carbolines in rats trained with 10.0 mg/kg ibogaine as a discriminative stimulus. All agents were administered i.p., 60 minutes presession. The ratio adjacent to each of the points is the number of subjects completing the test session over the number of subjects participating in each test session. Where no ratio is shown, a ratio of 8/8 is implied. *Modified from reference 19 with permission.*

it interesting that the *iboga* alkaloid, tabernanthine, is reported to have benzodiazepine inverse agonist effects in rats (*15*). Nonetheless, as shown above, ibogaine itself did not generalize to DMCM. Likewise, *in vitro* work by Deecher and colleagues (*16*) showed ibogaine, as well as harmaline, to be without effect on GABA-stimulated chloride uptake in the mouse brain.

The results of these studies with beta carbolines have implications regarding the potential antiaddictive effects of ibogaine. In morphine self-administration studies, Glick *et al.* (*17*) showed that, unlike ibogaine, harmaline did not produce a sustained decrease in morphine consumption by rats. If the self-administration paradigm used by these researchers is an accurate model of substance abuse in humans, then it appears that mimicry of the ibogaine discriminative stimulus is not effective in predicting antiaddictive activity in light of the fact that harmaline fully mimics ibogaine (*3*). Furthermore, norharmane, an agent that did not substitute for ibogaine in the present study, attenuates naloxone-precipitated withdrawal from morphine, as does ibogaine (*18*). For further information on the effects of beta carbolines in ibogaine-trained subjects, see reference *19*.

Interestingly, recent work by Grella and colleagues (*20*) shows significant harmaline-appropriate responding by the hallucinogen DOM (76%). These observations, taken together with our findings (*10*), support a role for the 5-HT_{2A} receptor subtype in the stimulus effects of ibogaine and harmaline.

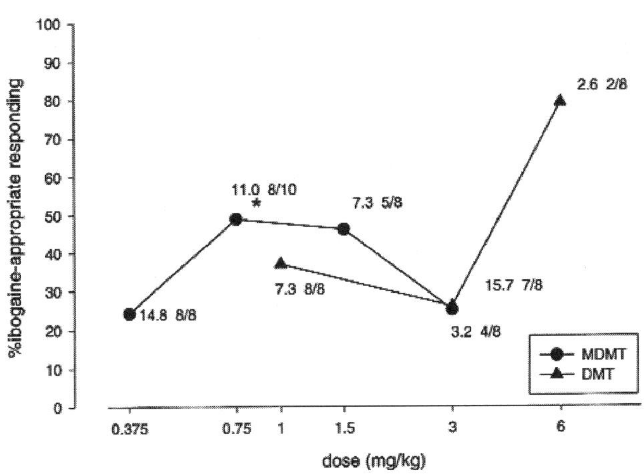

FIGURE 4. Dose response relationships for DMT and MDMT in rats trained with ibogaine. Note that * refers to statistically significant difference from vehicle condition ($p<0.05$) (Helsley, *et al.*, previously unpublished data).

D. Tryptaminergic Hallucinogens in Ibogaine-Trained Rats

The tryptamines form another subgroup of indoleamine hallucinogens. As shown in Figure 4, these agents appear to be less like ibogaine in terms of the stimulus cues compared to the beta-carboline, indoleamine, and phenethylamine hallucinogens, although weak partial generalization is seen with MDMT. A possible explanation for this finding lies in the observation that the stimulus effects of the tryptaminergic hallucinogen MDMT are mediated chiefly through interactions with 5-HT_{1A} receptors (*21, 22*). These receptors do not appear to be involved in the ibogaine stimulus (*11*).

Figure 4 also shows dose response relationships for DMT and MDMT in rats trained with ibogaine (10 mg/kg) as a discriminative stimulus. All agents were administered i.p., 15 minutes presession. The ratio adjacent to each of the points is the number of subjects completing the test session over the number of subjects participating in each test session. Also next to each point is the response rate in responses per minute.

E. NMDA Antagonists in Ibogaine-Trained Subjects

NMDA antagonists are often classified as hallucinogens because of their psychotomimetic effects. These agents occupy a binding site within a calcium channel; in so doing they occlude the channel. This calcium channel is normally gated by the excitatory neurotransmitter glutamate (for review of NMDA antagonists, see reference *23*). The best known of these agents is phencyclidine (angel dust, PCP). This agent was originally marketed as an anesthetic agent, but it was quickly removed from the market because of extreme psychotic reactions. Interestingly, ketamine, another NMDA antagonist, is still used today as an anesthetic. Emergence reactions are significantly milder than those seen with phencyclidine and are usually easily controlled with benzodiazepines.

A possible role for ibogaine acting at NMDA receptors is supported by the observation that ibogaine has appreciable affinity for this binding site (*24,25*). Popik and associates investigated the effects of ibogaine on MK 801 (dizocilpine)-trained mice in a T-maze. These authors observed approximately 70% dizocilpine-appropriate responding at an ibogaine dose of 100 mg/kg (*26*). Subsequent studies suggest that this interaction does not play a major role in the stimulus effects of ibogaine at lower doses, as neither phencyclidine nor MK 801 produced significant substitution in ibogaine-trained subjects (*27*). Correspondingly, ibogaine failed to substitute for phencyclidine in phencylidine-trained rats and monkeys (*28*) and MK 801-trained rats (*29*). The most plausible explanation for the contrast between these results and those of Popik, *et al.* would be dose and species differences.

F. Opioids in Ibogaine-Trained Subjects

Ibogaine's effects on addiction and withdrawal phenomena naturally lead to questions regarding its effects on opiate systems. The results of drug discrimination studies with opioids in ibogaine-trained subjects are quite interesting. Specifically neither mu- nor kappa-agonists substituted for ibogaine. No substitution, but weak partial antagonism, was seen with the pure antagonist naloxone. An intermediate level of substitution, but no significant antagonism, was observed with naltrexone (55.6%) (*27*). Also, in rats trained with the kappa-agonist U50,488, no substitution was seen with ibogaine (*29*). However, significant generalization (60-70%) was observed with the mixed agonist/antagonist compounds pentazocine, diprenorphine, and nalorphine (*27*).

Interestingly, the intermediate substitution produced by diprenorphine and nalorphine was antagonized by naloxone (*27*). Although the implications of these

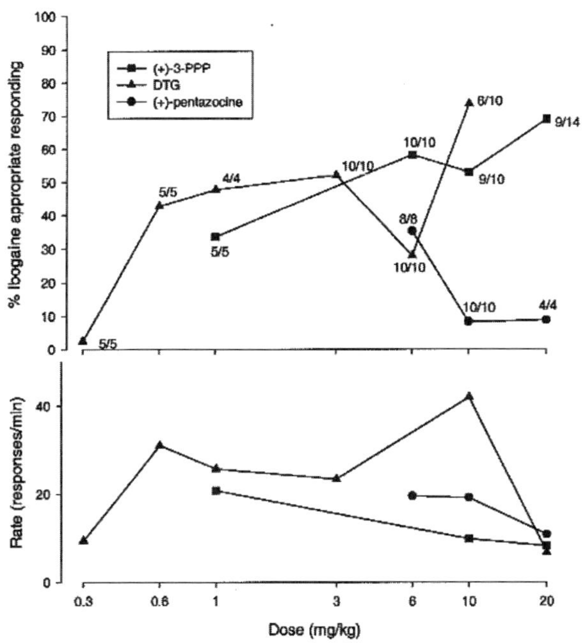

FIGURE 5. The dose response relationships for s ligands in rats trained to discriminate ibogaine (10.0 mg/kg, i.p., 60 minutes presession) from water. All agents were administered i.p., 30 minutes presession. The ratio adjacent to each of the points represents the number of subjects completing the test session over the number of subjects participating in each test session. *Taken from reference 27 with permission.*

results are not clear, they do suggest a role for opiate receptors in the mechanism of action of ibogaine. Further studies will be necessary to clarify this.

G. Sigma Ligands in Ibogaine-Trained Rats

In keeping with its unusual pharmacological profile, ibogaine has appreciable affinity for sigma (σ) receptors of the σ_2 subtype (*30,31*). Sigma receptors are a relatively new discovery, and hence, compared to other receptors, little is known about them. Thus, we are only beginning to characterize pharmacological agents as agonists or antagonists at these receptors. Nonetheless, ibogaine appears relatively selective for the σ_2 subtype. Several sigma ligands were tested in ibogaine-trained rats, and it was observed that nonselective sigma ligands (DTG, (+)-3-PPP) produced intermediate levels of substitution (Figure 5), while the σ_1 selective agents (+)-SKF 10,047 and (+)-pentazocine failed to substitute (*27*).

Unfortunately no σ_2-selective agents were available at the time of these studies. Because we are still in the early stages in the study of the pharmacology of these receptors, all that can be concluded from these studies is that sigma receptors of the σ_2 subtype play a role in the ibogaine discriminative stimulus. As more selective sigma ligands are discovered, the exact role of these receptors in ibogaine's mechanism of action will hopefully become more clear.

H. Noribogaine

In radioligand binding assays, ibogaine shows remarkably low affinity for most known receptors (micromolar vs. nanomolar) (*32*). This, taken together with the observation that ibogaine's pharmacological activity is relatively long lived, suggests the possibility that a long-acting metabolite may mediate many of ibogaine's pharmacological effects; noribogaine (10-hydroxyibogamine) is thought to be such an agent (*33*).

10-Hydroxyibogamine appears to be similar to ibogaine in its stimulus properties, but it does not substitute completely for the parent compound, as illustrated in Figure 1 (*3*). Zubaran and colleagues (*29*) observed similar results. These authors also looked at brain levels of ibogaine and its metabolite. Interestingly, they found the metabolite to be more potent than the parent compound in eliciting ibogaine-appropriate responding with ED_{50} values of 1.98 and 4.51 mg/kg, respectively. Brain levels of noribogaine were similar at behaviorally equi-effective doses of both agents (1.11 μg/g after administration of the ED_{50} of 10-hydroxyibogamine and 1.23 μg/g after the ED_{50} of ibogaine). These results suggest that the stimulus effects of ibogaine are mediated mainly by 10-hydroxyibogamine (*29*). Further studies will be necessary to confirm this.

III. Summary

The results of the studies described here support the hypothesis that ibogaine produces its effects via selective interactions with multiple receptors. It appears that 5-HT_{2A}, 5-HT_{2C}, and σ_2 receptors are involved in mediating the stimulus effects of ibogaine. In addition, opiate receptors may also be involved. In contrast, σ_1, PCP/MK-801, 5-HT_3, and 5-HT_{1A} receptors do not appear to play a major role.

Ibogaine's hallucinogenic effects may be explained by its interactions with 5-HT_{2A} and 5-HT_{2C} receptors, while its putative antiaddictive properties may result from its interactions with σ_2 and opiate receptors. Alternatively, the possibility that ibogaine's hallucinogenic properties underlie its antiaddictive effects, as previously suggested (*34*), would support a role for 5-HT_2 receptors in mediating the reported therapeutic effects of ibogaine.

Certainly many questions remain regarding ibogaine's mechanism of action. Although drug discrimination will be useful for answering some of those questions, the true potential of this technique is realized when it is combined with other techniques. The next few years promise to be fruitful with respect to our understanding of this agent. Reasons supporting this belief include advances in the study of sigma receptors, interest in ibogaine's effects on second messenger systems, and the development of ibogaine congeners such as 18-methoxycoronaridine (*35*).

In conclusion, the aforementioned studies should serve to guide further endeavors. Pertinent questions have been generated: What is the role of sigma receptors in the effects of ibogaine, especially with regard to addiction? How does ibogaine affect opiate neurotransmission? What effects, if any, do the Harmala alkaloids have on addiction phenomena? What is the mechanism of action of harmaline? Can 10-hydroxyibogamine serve as a discriminative stimulus and, if so, what receptor interactions mediate its stimulus effects? Does the ibogaine-trained stimulus generalize to novel agents, including 18-methoxycoronaridine?

Acknowledgments

These studies were supported, in part, by U.S. Public Health service grant DA 03385 [JCW, RAR], by National Research Service Award DA 05735 [SH], and by a grant from Schering-Plough Research Institute [SH]. We thank Ms. Deborah Timineri for her technical assistance.

References

1. J.C. Winter, *NIDA Res. Monograph* **146,** 157 (1994).
2. R.T. Kelleher and L.H. Morse, *in* "Reviews of Physiology, Biochemistry, and Behavior," Volume 60, pp. 2-56. Springer-Verlag, New York, 1968.
3. S. Helsley, R.A.Rabin, and J.C. Winter, *Life Sci.* **60,** PL147 (1997).
4. D. Fiorella, R.A. Rabin, and J.C. Winter, *Psychopharmacology* **121,** 347 (1995).
5. D. Fiorella, R.A. Rabin, and J.C. Winter, *Psychopharmacology* **121,** 357 (1995).
6. D. Fiorella, S. Helsley, D.S. Lorrain, R.A. Rabin, and J.C. Winter, *Psychopharmacology* **121,** 364 (1995).
7. P.A. Palumbo and J.C. Winter, *Pharmacol. Biochem. Behav.* **43,** 1221 (1992).
8. M.D. Schechter and T.L. Gordon, *Eur. J. Pharmacol.* **249,** 79 (1993).
9. S. Helsley, D. Fiorella, R.A. Rabin, and J.C. Winter, *Pharmacol. Biochem. Behav.* **59,** 419 (1998).
10. J.C. Winter, *Pharmacol. Biochem. Behav.* **20,** 201 (1984).
11. S. Helsley, R.A. Rabin, and J.C. Winter, *Prog. Neuro-Psychopharmacol. Biol. Psych.* **23,** 317 (1999).
12. J.C. Winter, S. Helsley, D. Fiorella, and R.A. Rabin, *Pharmacol. Biochem. Behav.* **63,** 507 (1999).
13. C. Naranjo, *Clin. Toxicol.* **2,** 209 (1969).
14. E.N. Petersen, *Eur. J. Pharmacol.* **94,** 117 (1983).
15. J.-H. Trouvin, P. Jacqmin, C. Rouch, M. Lesne, and C. Jacquot, *Eur. J. Pharmacol.* **140,** 303 (1987).
16. D.C. Deecher, M. Teitler, D.M. Soderlund, W.G. Bornmann, M.E. Kuehne, and S.D. Glick, *Brain Res.* **571,** 242 (1992).
17. S.D. Glick, M.E. Kuehne, J. Raucci, T.E. Wilson, D. Larson, R.W. Keller, Jr., and J.N. Carlson, *Brain Res.* **657,** 14 (1994).
18. S.L.T. Cappendijk, D. Fekkes, and M.R. Dzoljic, *Behav. Brain Res.* **65,** 117 (1994).
19. S. Helsley, R.A. Rabin, and J.C. Winter, *Eur. J. Pharmacol.* **345,**139 (1998).
20. B. Grella, M. Dukat, R. Young, M. Teitler, K. Herrick-Davis, C.B. Gauthier, and R.A. Glennon, *Drug Alc. Dep.* **50,** 99 (1998).
21. D.G. Spencer, T. Glaser, and J. Traber, *Psychopharmacology* **93,** 158 (1987).
22. J.C. Winter, R.A. Fillipink, D.M. Timineri, S. Helsley, and R.A. Rabin, *Pharmacol. Biochem. Behav.* **65,** 75 (2000).
23. J. Willetts, R.L. Balster, and J.D. Leander, *Trends Pharmacol. Sci.* **11,** 423 (1990).
24. P. Popik, R.T. Layer, and P. Skolnick, *Psychopharmacology* **114,** 672 (1994).
25. R.T. Layer, P. Skolnick, C.M. Bertha, U.K. Bandarage, M.E. Kuehne, and P. Popik, *Eur. J. Pharmacol.* **309,** 159 (1996).
26. P. Popik, R.T. Layer, L.H. Fossom, M. Benveniste, B. Geter-Douglass, J.M. Witkin, and P. Skolnick, *J. Pharmacol. Exp. Ther.* **275,** 753 (1995).
27. S. Helsley, R. Filipink, W.D. Bowen, R.A. Rabin, and J.C. Winter, *Pharmacol. Biochem. Behav.* **59,** 495 (1998).
28. H.E. Jones, H. Li, and R. Balster, *Pharmacol. Biochem. Behav.* **59,** 413 (1998).
29. Zubaran, M. Shoaib, I.P. Stolerman, J. Pablo, and D.C. Mash, *Neuropsychopharmacol.* **21,** 119 (1999).
30. R.H. Mach, C.R. Smith, and S.R. Childers, *Life Sci.* **57,** PL 57 (1995).
31. W.D. Bowen, B.J. Vilner, W. Williams, C.M. Bertha, M.E. Kuehne, and A.E. Jacobson, *Eur. J. Pharmacol.* **279,** R1 (1995).
32. P.M. Sweetnam, J. Lancaster, A. Snowman, J.L. Collins, S. Perschke, C. Bauer, and J. Ferkany, *Psychopharmacology* **118,** 369 (1995).
33. D.C. Mash, J.K. Staley, M.H. Baumann, R.B. Rothman, and W.L. Hearn, *Life Sci.* **57,** PL 45

(1995).
34. J.H. Halpern, *Addiction Res.* **4,**177 (1996).
35. S.D. Glick, M.E. Kuehne, I.M. Maisonneuve, U.K. Bandarage, and H.H. Molinari, *Brain Res.* **719,** 29 (1996).

—CHAPTER 5—

COMPARATIVE NEUROPHARMACOLOGY OF IBOGAINE AND ITS *O*-DESMETHYL METABOLITE, NORIBOGAINE

MICHAEL H. BAUMANN[1], JOHN PABLO[2], SYED F. ALI[3], RICHARD B. ROTHMAN[1], AND DEBORAH C. MASH[2]

[1]*Clinical Psychopharmacology Section*
Intramural Research Program
National Institute on Drug Abuse
National Institutes of Health
Baltimore, MD 21224

[2]*Department of Neurology*
University of Miami School of Medicine
Miami, FL 33136

[3]*Division of Neurotoxicology*
National Center for Toxicological Research
Food and Drug Administration
Jefferson, AR 72079

THE ALKALOIDS, Vol.56
0099-9598/01 $35.00

I. Introduction

Drug addiction is a disease that affects millions of people worldwide (*1*). The severity of the drug addiction problem, coupled with a lack of effective medications, has prompted investigators to explore the plant kingdom as a source of novel therapeutics. One example of a plant-derived compound with potential utility in treating drug addiction is the indole alkaloid, ibogaine (*2,3*). Ibogaine is found in the roots of the African shrub, *Tabernanthe iboga*. Historically, native peoples of West Central Africa have used the root bark of this plant as a sacrament in their rituals of initiation into adulthood (*4*). More recently, ibogaine has gained a reputation as an "addiction interrupter," based on findings in animals and humans (reviewed in *5,6*). In rats, acute administration of ibogaine (40 mg/kg, i.p.) produces long-lasting decreases in the self-administration of cocaine and morphine (*7-9*, see Glick *et al.* in this volume). Ibogaine also alleviates symptoms of opioid withdrawal in morphine-dependent rats (*10,11*) and heroin-dependent human addicts (*12,13*, see Alper *et al.* this volume). These promising findings support the development of ibogaine as a pharmacological adjunct in the treatment of substance use disorders.

Despite extensive investigation, the mechanisms underlying the antiaddictive properties of ibogaine are not fully understood (*14,15*). Radioligand binding studies show that ibogaine binds with low micromolar (μM) affinity to a number of molecular targets in nervous tissue, resulting in a complex pharmacology (*16-27*). Some of these ibogaine binding sites include sigma-2 receptors (*16,17*), serotonin (5-HT) and dopamine (DA) transporters (*18-21*), mu- and kappa-opioid receptors (*21-24*), and NMDA-coupled ion channels (*25-27*). Biodistribution studies in rats demonstrate that brain concentrations of ibogaine range from 10 to

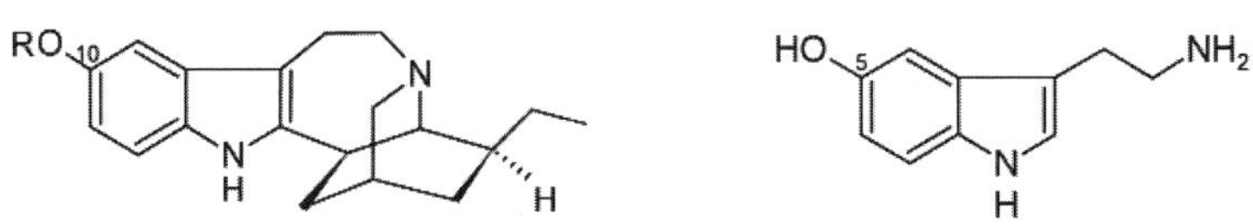

Ibogaine, R = CH_3 (Le Men-Taylor numbering)
Noribogaine (10-Hydroxyibogamine)*, R = H

Serotonin (5-HT)

FIGURE 1. Chemical structures of ibogaine, its *O*-demethylated metabolite (noribogaine), and the neurotransmitter serotonin (5-HT).

* Noribogaine (10-hydroxyibogamine) has frequently been referred to as 12-hydroxyibogamine in the biological and medical literature based on the Chemical Abstracts numbering system for the ibogamine alkaloid skeleton.

20 μM when measured 1 hour after acute administration of 50 mg/kg p.o. (*21*) or 40 mg/kg i.p. (*28,29*). Thus, the interaction of ibogaine with μM-affinity binding sites may be functionally relevant *in vivo*. Few studies have been able to attribute *in vivo* pharmacological effects of ibogaine to activation of specific binding sites. In fact, there is speculation that the key to ibogaine's antiaddictive potential is related to the simultaneous activation of multiple neurotransmitter systems in the brain (*14,15*).

An intriguing aspect of ibogaine pharmacology is the long-lasting action of the drug. In rats, a single administration of ibogaine elicits behavioral and neurochemical effects that can last for days (*7-9,18,30,31*), even though the biological half-life of the drug is only a few hours (*32,33*). Such observations suggest the possibility that ibogaine is converted to a long-acting metabolite (*7-9*). Mash *et al.* (*19*) and Hearn *et al.* (*34*) provided the first direct evidence for the formation of a major ibogaine metabolite *in vivo*. These investigators used sensitive analytical methods to identify an *O*-demethylated metabolite of ibogaine, 10-hydroxyibogamine (noribogaine), in the blood and urine from monkeys and humans treated with ibogaine. Figure 1 shows the chemical structures of ibogaine, noribogaine, and the neurotransmitter serotonin (5-hydroxytryptamine, 5-HT). The methoxy group at the 10-position of ibogaine is converted to a hydroxyl group to form noribogaine. Note the presence of an indole moiety in the structure of the *iboga* alkaloids and 5-HT. Subsequent pharmacokinetic studies have demonstrated that ibogaine is converted to noribogaine in rats (*21,29*).

Interestingly, as summarized in Table I, the *in vitro* pharmacology of noribogaine differs significantly from that of ibogaine. For example, noribogaine displays a higher affinity for 5-HT transporters (SERTs) and opioid receptor subtypes when compared to ibogaine. A growing body of preclinical evidence demonstrates that noribogaine is biologically active *in vivo* and undoubtedly

TABLE I.
AFFINITIES OF IBOGAINE AND NORIBOGAINE AT SELECTED SITES

Binding site (ref.)	Radioligand[a]	Ibogaine (K_i, μM)	Noribogaine (K_i, μM)
Sigma-2 receptors (*16*)	$[^3H]$-DTG	0.20	5.26
5-HT transporters (*19*)	$[^{125}I]$-RTI-55	0.55	0.04
DA transporters (*19*)	$[^{125}I]$-RTI-121	1.98	2.05
Mu-opioid receptors (*24*)	$[^3H]$-DAMGO	3.76	0.16
Kappa-opioid receptors (*23*)	$[^3H]$-U69,593	3.77	0.96
NMDA sites (*21*)	$[^3H]$-MK-801	5.20	31.41

[a] Specific details of the radioligand binding assays and the calculation of inhibitory constants, K_i, are described in the cited references in parentheses.

contributes to the spectrum of effects produced by systemically administered ibogaine (*19,35-41*). The primary aim of this chapter is to review the comparative neurobiology of ibogaine and noribogaine in rodent species. The chapter focuses on data collected from the laboratories of the authors and attempts to integrate these findings with the available literature on *iboga* alkaloids.

II. Pharmacokinetics

Pharmacokinetic studies carried out in the early 1970s showed that systemic administration of ibogaine to rats and mice is followed by rapid distribution of the drug into various organs, including brain, liver, and kidney (*32,33*). The same studies determined the elimination kinetics of ibogaine from the brain and reported a half-life of about 1 hour. More recently, a number of laboratories have developed sensitive analytical methods to detect ibogaine in blood and other tissues (*34, 42-44*). These methods generally involve organic extraction of tissue samples, a derivatization procedure, and subsequent quantitation of ibogaine via gas chromatography-mass spectroscopy (GC/MS). Using these GC/MS methods, Hough *et al.* (*28*) examined the tissue distribution of ibogaine in rats after administration of i.p. or s.c. injections. One hour after an i.p. injection of ibogaine (40 mg/kg), tissue concentrations of the drug varied widely, ranging from 100 ng/ml in plasma to 10 μg/g in fat. These drug concentrations decreased nearly 10-fold by 12 hours postinjection. In all instances, tissue levels of ibogaine were greater after s.c. administration when compared to i.p. administration.

Findings from the work of Hough *et al.* indicate two possible mechanisms whereby ibogaine could have long-lasting actions *in vivo*. First, the high concentration of ibogaine in fat tissue suggests that fat can serve as a storage depot for the drug. Under these circumstances, it seems feasible that small amounts of ibogaine could be released from fat tissue into the circulation for extended periods after a single dose. Second, the lower tissue concentrations of ibogaine observed after i.p. administration suggest that ibogaine is extensively metabolized when given by the i.p. route.

As mentioned previously, Mash and coworkers (*19,34*) identified noribogaine as the major metabolite of ibogaine in monkeys and humans. These investigators postulated that noribogaine is formed via first-pass metabolism of ibogaine in the liver. In agreement with this notion, Obach *et al.* (*45*) reported that ibogaine is *O*-demethylated by cytochrome P450 enzymes in human liver microsomes *in vitro*. In particular, cytochrome P450 2D6 appears to be the main isoform responsible for ibogaine *O*-demethylase activity in humans. While noribogaine has been identified in plasma and brain tissue from rats treated with ibogaine (*21,29,41*),

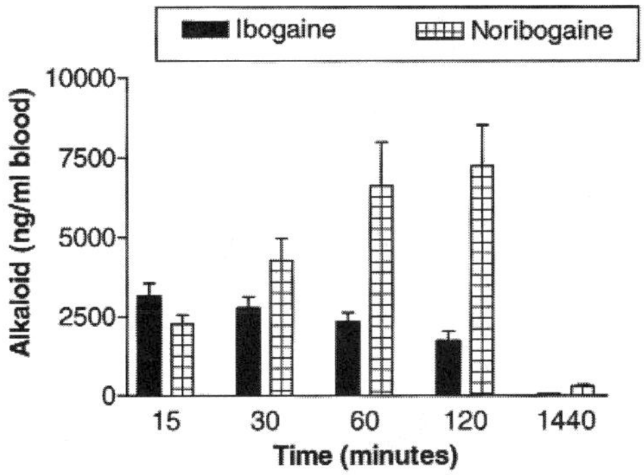

FIGURE 2. Time-concentration profiles for ibogaine and its *O*-demethylated metabolite, noribogaine, in rats. Male rats fitted with indwelling jugular catheters received i.p. ibogaine (40 mg/kg, i.p.) at time zero, and blood samples were drawn at various times thereafter. Whole blood levels of ibogaine and noribogaine were assayed using GC/MS. Data are mean ± SEM for N = 6/group.

no study has determined the specific cytochrome P450 isoform(s) responsible for formation of noribogaine in rats or other species.

In our laboratory, we have been interested in the pharmacokinetics and metabolism of ibogaine in rats, because this animal species is the principal model system used for evaluating the antiaddictive properties of ibogaine. With this in mind, we carried out investigations to examine the metabolic conversion of ibogaine to noribogaine in rats (*46*). Male rats were fitted with indwelling jugular catheters and allowed one week to recover. On the morning of an experiment, rats received an i.p. injection of ibogaine (40 mg/kg), and repeated blood samples were withdrawn via the catheters at various times thereafter for 24 hours. Whole blood samples were assayed for ibogaine and noribogaine using GC/MS methods (*34*).

Figure 2 shows that ibogaine is rapidly metabolized to noribogaine in rats, and the maximal blood concentration of noribogaine exceeds that of ibogaine by more than 2-fold. At 24 hours postinjection, blood levels of ibogaine are undetectable whereas blood levels of noribogaine are in the range of 300 ng/ml. Thus, noribogaine is present in the bloodstream at pharmacologically relevant concentrations for at least one day postinjection, long after ibogaine has been cleared. Biodistribution studies in rats have shown that noribogaine readily penetrates the blood-brain barrier and enters into the brain (*21,29,41*). In fact, brain concentrations of noribogaine are equal to, or greater than, brain concentrations of ibogaine after i.p. or oral administration of ibogaine (*21,29,41*). These data clearly show that noribogaine can contribute to the acute and long-lasting effects of ibogaine administered systemically in rats.

It should be mentioned that the concentrations of ibogaine and noribogaine in rat blood shown in Figure 2 are much higher than the concentrations of these alkaloids in rat plasma reported by others (*28,29*). This observation suggests that ibogaine and noribogaine are sequestered in some cellular fraction of whole blood. One possibility is that *iboga* alkaloids are taken up into blood platelets by a process involving SERT sites present on platelet cell membranes. While this hypothesis is speculative, Table I shows that both ibogaine and noribogaine have significant affinity for SERT. The nature of the interaction between *iboga* alkaloids and platelet SERTs has not been well characterized and deserves to be studied.

Pearl *et al.* (29) have reported gender differences in responsiveness to ibogaine, with females exhibiting a greater sensitivity to the effects of the drug. These investigators also showed that female rats have significantly higher levels of ibogaine and noribogaine in brain tissue after i.p. administration of ibogaine. Thus, the enhanced ibogaine sensitivity in females may be due to pharmacokinetic differences between sexes. In order to further assess the role of gender and gonadal steroids on the kinetics and metabolism of ibogaine, we carried out an investigation using groups of male and female rats with differing sex hormone status (*47*). Five groups of rats were used: (1) intact sham-operated males, (2) castrated males, (3) intact females prior to ovulation (i.e. in proestrus phase), (4) intact females after ovulation (i.e., in diestrus phase), and (5) ovariectomized females. All rats were fitted with indwelling jugular catheters at the time of sham surgery or gonadectomy, and allowed one week to recover. In the intact female groups, vaginal cytology was monitored to track the stage of the estrous cycle. Preovulatory females were subjected to experimental testing during the proestrus stage of the cycle, when circulating levels of endogenous estrogen are high. Postovulatory females were tested the day after the estrus stage of the cycle when levels of estrogen are lower. On the day of an experiment, rats received 40 mg/kg i.p. ibogaine, and repeated blood samples were withdrawn via the catheters at various times thereafter. Levels of *iboga* alkaloids in whole blood were assayed by GC/MS.

Table II summarizes the effects of gender and gonadectomy on blood levels of ibogaine and noribogaine after i.p. ibogaine injection. In all groups, ibogaine concentrations in blood reach maximum within 10 to 12 minutes, whereas noribogaine concentrations reach maximum between 1 and 3 hours postinjection. Importantly, preovulatory females with high circulating estrogen display nearly 2-fold greater blood levels of ibogaine when compared to all other groups. This observation supports the findings of Pearl *et al.*, who showed female rats have higher plasma and brain levels of ibogaine when compared to male rats (*29*). Thus, it appears that estrogen increases the bioavailability of ibogaine, and this effect may be mediated by enhanced absorption of the drug from the peritoneal cavity into the circulation. In the castrated males and ovariectomized females,

TABLE II.
EFFECTS OF GENDER AND GONADECTOMY ON THE KINETICS AND METABOLISM OF IBOGAINE AND NORIBOGAINE IN RATS

Condition/ Gender	Ibogaine AUC[a] (ng*h/ml)	Tmax (h)	Noribogaine AUC (ng*h/ml)	Tmax (h)
Control Male	8350 ± 243[b]	0.167 ± 0.010	27181 ± 2102	0.834 ± 0.074
Castrated Male	7983 ± 496	0.195 ± 0.028	11631 ± 1003*	1.667 ± 0.211*
Preovulatory Female	15898 ± 939*	0.195 ± 0.028	34811 ± 1948	3.000 ± 0.447*
Postovulatory Female	8477 ± 396	0.167 ± 0.010	30960 ± 2246	2.780 ± 0.211*
Ovariectomized Female	8933 ± 614	0.167 ± 0.010	15891 ± 1805*	3.330 ± 0.667*

[a] Area under the curve (AUC) and time to achieve maximal drug concentration (Tmax) were calculated using PCNONLIN computer software.
[b] Values are mean ± SEM for N = 6 rats/group.
* = $P < 0.05$ with respect to sham-operated control males.

noribogaine concentrations are significantly lower than in the other groups. These data suggest that gonadectomy decreases the metabolic conversion of ibogaine to noribogaine. Taken together, the data demonstrate that the sex steroid modulation of ibogaine kinetics and metabolism is complex. While elevations in estrogen can increase bioavailability of ibogaine, removal of gonadal steroids impairs the metabolism of ibogaine to noribogaine. Such gender differences in ibogaine pharmacokinetics have important implications. First, caution should be exercised when administering ibogaine to female animals or humans because females will be more sensitive to the effects of the drug. Second, when interpreting data from studies using female rats, it must be remembered that pharmacological effects of ibogaine may be greater in females as compared to males for the same dose of drug.

III. Neurochemical Mechanisms

A. Effects on Dopamine Systems

A large body of preclinical evidence shows that mesolimbic dopamine (DA) neurons are involved in drug-associated reward processes (*48,49*). The mesolimbic DA system in rodents consists of cell bodies residing in the ventral tegmental area (VTA) that send axonal projections to numerous limbic forebrain regions, most notably the nucleus accumbens (NAC) and prefrontal cortex (PFC). Acute administration of abused drugs, such as morphine and cocaine, causes elevations in the concentration of extracellular DA in rat NAC (*50,51*). Withdrawal from chronic exposure to these abused drugs, in contrast, results in significant reductions in extracellular DA in the NAC (*52,53*). Thus, increased synaptic DA is associated with the positive rewarding effects of drugs (i.e., euphoria), whereas decreased synaptic DA is associated with negative withdrawal effects (i.e., dysphoria). Current theories of addiction suggest that both the positive and negative effects of drugs are involved in the maintenance of a drug-dependent state (*54,55*).

Because of the prominent role of DA in drug addiction, it seems conceivable that the antiaddictive properties of ibogaine might be related to effects of the drug on DA systems in the brain. Radioligand binding studies demonstrate that ibogaine does not interact with DA receptor subtypes *in vitro* (*20-22*), thus ibogaine is not a direct DA agonist or antagonist. A number of investigators have shown that ibogaine binds with low μM affinity to DA transporter proteins (DATs) labeled with the cocaine analogs $[^3H]$WIN-35,428 and $[^{125}I]$RTI-121 (*18-21*). In contrast, Broderick *et al.* (*56*) reported that concentrations of ibogaine up to 100 μM do not affect DAT binding when transporters are labeled with the piperazine analog $[^3H]$GBR12935. These apparently discordant results may be explained by the findings of Vaughan (*57*), who showed that cocaine-like drugs (i.e. phenyltropanes) and GBR-like drugs (i.e. phenylpiperazines) bind to different regions of the DAT polypeptide. It seems plausible that ibogaine exhibits selective affinity for a cocaine-binding domain located on DAT proteins.

DAT sites are important regulatory elements in the brain. Under normal circumstances, DATs function to recapture released DA from the synapse and transport it back into the intraneuronal cytoplasm (*58,59*). The DA uptake activity of DAT is the principal mechanism for inactivating DA transmission. Mash *et al.* (*19, 21*) demonstrated that ibogaine and noribogaine display low μM affinity for

TABLE III.
INHIBITORY POTENCY OF IBOGAINE AND NORIBOGAINE IN ASSAYS OF TRANSPORTER BINDING AND MONOAMINE UPTAKE IN RAT BRAIN

Assay[a]	Ibogaine IC_{50} (μM)	Noribogaine IC_{50} (μM)
$[^{125}I]$RTI-55-labeled DA transporters (DAT)	11.83 ± 0.39[b]	4.17 ± 0.19
$[^{125}I]$RTI-55-labeled 5-HT transporters (SERT)	3.85 ± 0.21	0.18 ± 0.01
$[^3H]$DA uptake rat caudate	10.03 ± 0.72	13.05 ± 0.72
$[^3H]$5-HT uptake rat whole brain minus cerebellum	3.15 ± 0.10	0.33 ± 0.02

[a]Binding assays used $[^{125}I]$RTI-55 to label DAT and SERT sites in rat caudate membranes. DAT binding was conducted in the presence of 50 nM paroxetine whereas SERT binding was conducted in the presence of 100 nM GBR12935. Uptake assays were performed in synaptosomes prepared from rat brain.

[b]Values are mean ± SD expressed as IC_{50} values determined from three independent experiments each performed in triplicate.

DAT sites in human brain tissue (see Table I). In order to further explore the interactions of *iboga* alkaloids with DATs, the effects of ibogaine and noribogaine in assays measuring DAT binding and [^{3}H]DA uptake in rat brain tissue preparations were tested. As depicted in Table III, ibogaine and noribogaine are low potency inhibitors of DAT binding when DAT sites are labeled with the cocaine analog [^{125}I]RTI-55. More importantly, both *iboga* alkaloids block the uptake of [^{3}H]DA into rat caudate synaptosomes with IC_{50} values in the range of 10 μM. These findings agree with the recent data of Wells *et al.* (*60*), who reported that ibogaine inhibits [^{3}H]DA uptake in rat striatal synaptosomes with an IC_{50} of 20 μM. These same investigators showed that ibogaine does not evoke appreciable release of preloaded [^{3}H]DA from nervous tissue. It is noteworthy that ibogaine and noribogaine possess similar IC_{50} values in assays measuring inhibition of DAT binding and inhibition of [^{3}H]DA uptake; this indicates that the binding-to-uptake ratios for these alkaloids are close to one. We have previously reported that drugs exhibiting binding-to-uptake ratios close to unity are pure uptake blockers (*61*). According to this classification scheme, ibogaine and noribogaine are low-potency DA uptake inhibitors *in vitro*.

There are a number of research methods available for studying DA neurochemistry *in vitro* and *in vivo*. Sershen *et al.* have published an excellent review (*62*, see Sershen *et al.* in this volume) summarizing the use of *in vitro* perfusion techniques to assess the effects of ibogaine on [^{3}H]DA release, so this topic will not be discussed further here. For whole animal studies, two basic neurochemical methods have been used to examine the effects of *iboga* alkaloids on DA function in rodents: (1) measurement of DA and its metabolites, dihydroxyphenylacetic acid (DOPAC) and homovanillic acid (HVA), in tissue samples from postmortem brain, and (2) measurement of DA and its metabolites in extracellular fluid from living brain using *in vivo* microdialysis sampling. It is important to realize that these two methods assess different aspects of DA neuronal function. For example, the amount of DA in brain extracellular fluid is presumably an index of synaptic DA that is, in turn, dependent on the combined processes of DA cell firing, DA release, and DA reuptake. Additionally, the amount of DA in extracellular fluid is a very small fraction of the total content of DA in brain tissue. In the present discussion, the term "tissue DA" will be used to designate DA levels measured in postmortem brain tissue, whereas the term "dialysate DA" will be used to designate DA levels measured in microdialysis samples.

It is well established that ibogaine administration causes dramatic changes in the tissue levels of DA and its metabolites in rodent brain (*18,30,31*), and these changes are dose and time dependent. In our laboratory, the time-course effects of ibogaine on DA metabolism in rats were examined (*31*). Groups of male rats received 50 mg/kg i.p. ibogaine and were sacrificed at various times thereafter for up to 24 hours. Brain regions were dissected, and tissue levels of DA, DOPAC,

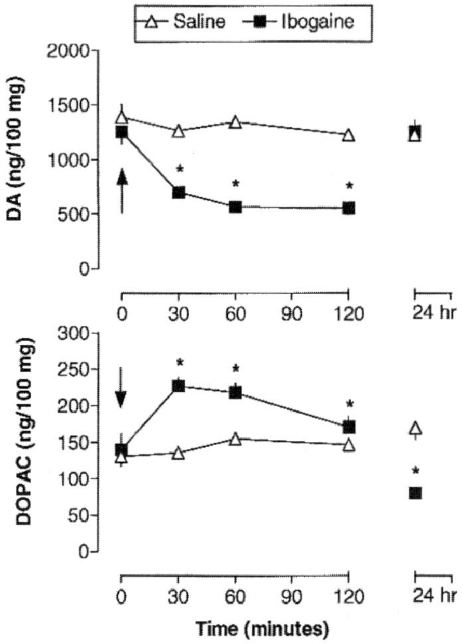

FIGURE 3. Time-course effects of ibogaine administration on postmortem tissue levels of DA and DOPAC in rat caudate. Rats received saline or ibogaine (50 mg/kg, i.p.) at time zero, and groups of rats (N = 6 per group at each time point) were sacrificed at various times thereafter. Tissue levels of DA and DOPAC were assayed using HPLC-EC.
* = $P < 0.05$ compared to saline control at specific time points (*31*).

and HVA were assayed by high-pressure liquid chromatography with electrochemical detection (HPLC-EC). Figure 3 illustrates the time-course effects of ibogaine on tissue DA and DOPAC in rat caudate. Ibogaine causes marked decreases (~50% reductions) in tissue DA levels that last for at least 2 hours. The acute reduction in tissue DA is accompanied by concomitant increases in tissue levels of DOPAC and HVA (*30,31*). By 24 hours postinjection, DA levels return to normal whereas metabolite levels are reduced significantly. The acute stimulation of DA utilization evoked by ibogaine, as indicated by increased DOPAC/DA ratios, has been observed in every rat brain region examined including, PFC, hypothalamus, olfactory tubercle, and NAC (*30,31,63-66*). Thus, ibogaine exerts a biphasic effect on DA utilization that is characterized by an initial transient increase in metabolism followed by a more persistent decrease.

Given the dramatic effects of ibogaine on DA metabolism, it is surprising that no studies have reported the effects of noribogaine on tissue levels of DA and its metabolites. To address this issue, the effects of ibogaine and noribogaine on DA metabolism were compared in mice (*65*). Groups of male mice received ibogaine or noribogaine (30 or 100 mg/kg, i.p.) and were sacrificed at 60 minutes postinjection. Brain regions were dissected, and tissue levels of DA, DOPAC, and HVA were determined by HPLC-EC. Figure 4 depicts the effects of ibogaine and noribogaine on tissue DA and DOPAC in mouse caudate. Similar to the effects of ibogaine in rats, ibogaine and noribogaine produce dose-dependent reductions in tissue DA in mouse caudate. Additionally, both drugs cause a parallel increase in DOPAC over the same time course. Ibogaine and noribogaine are essentially equipotent in their ability to stimulate DA metabolism in mice, and this property of the *iboga* alkaloids appears similar between mice and rats.

The effects of ibogaine and noribogaine on extracellular DA levels have been investigated extensively. Maisonneuve *et al.* (*67*) published the first study examining the influence of ibogaine (40 mg/kg, i.p.) on dialysate DA in rat brain, and their findings showed that ibogaine alters DA levels in a region-specific manner. When examined 1 hour after ibogaine injection, dialysate DA levels are

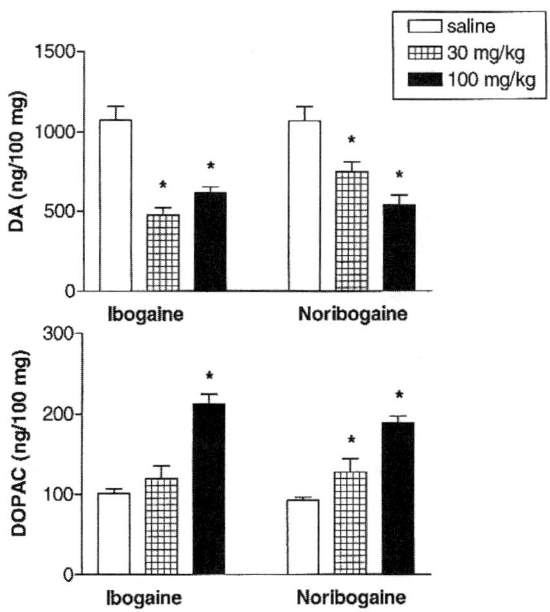

FIGURE 4. Dose-response effects of ibogaine and noribogaine on postmortem tissue levels of DA and DOPAC in mouse caudate. Groups of male mice (N = 6 per group at each dose) received saline, ibogaine, or noribogaine and were sacrificed 60 minutes later. Tissue levels of DA and DOPAC were determined using HPLC-EC.
* = P <0.05 with respect to saline controls (*65*).

increased in the frontal cortex, decreased in the caudate, and unchanged in the NAC. Broderick *et al.* (*68*) used *in vivo* microvoltammetry to confirm that ibogaine (40 mg/kg, i.p.) does not significantly alter extracellular DA levels in the NAC of male rats. Similarly, Benwell *et al.* (*66*) observed that dialysate DA in rat NAC is unchanged after ibogaine treatment. We have examined the acute effects of i.v. ibogaine and noribogaine (1 and 10 mg/kg) on extracellular DA in the NAC of male rats, and our findings show that neither drug significantly affects dialysate DA (*19,46*). In apparent contrast to the above-mentioned results, Glick *et al.* (*15,35*) reported that systemic administration of ibogaine, noribogaine, and other *iboga* alkaloids (40 mg/kg, i.p.) produces a significant decrease in dialysate DA in the NAC of *female* rats. Moreover, local infusion of high doses of ibogaine (100-400 μM) through the dialysis probe reduces extracellular DA concentrations in the NAC of both male and female rats (*69,70*). Taken together, the *in vivo* neurochemical data indicate that systemic doses of ibogaine and noribogaine have minimal effects on dialysate DA in the NAC of male rats, but the same systemic doses produce significant decreases in dialysate DA in females. The pharmacokinetic data discussed previously can explain the heightened sensitivity of females to ibogaine. Because females have higher brain levels of ibogaine after systemic injection, the effects of the drug on DA neurons are enhanced and mimic the effects of local high-dose drug infusion.

The precise mechanisms responsible for the effects of ibogaine and noribogaine on DA neurons are not clear. Any hypothesis attempting to explain these effects must account for a number of contradictory pieces of information. For example, *iboga* alkaloids bind to DAT sites with low μM affinity and block [^{3}H]DA reuptake *in vitro* (see Table III), yet these alkaloids do not uniformly elevate extracellular DA *in vivo* as measured by microdialysis. Such findings demonstrate a clear discrepancy between the *in vitro* and *in vivo* results. The available data from rats show that ibogaine-induced changes in central DA metabolism are not accompanied by elevations in extracellular DA levels *in vivo*. To complicate matters even more, there appear to be species differences in some dopaminergic actions of *iboga* alkaloids. Harsing *et al.* (*71*) reported that ibogaine stimulates release of preloaded [^{3}H]DA from mouse striata, while Wells *et al.* (*60*) showed that even high doses of ibogaine cause minimal [^{3}H]DA release from rat brain synaptosomes.

Staley *et al.* (*21*) have proposed that *iboga* alkaloids might promote a "reserpine-like" redistribution of intraneuronal DA from vesicular to cytoplasmic pools. While this hypothesis is purely speculative, there is evidence supporting the concept. Reserpine is known for its ability to disrupt the vesicular storage of monoamines, and the acute effects of ibogaine and noribogaine mimic the acute effects of reserpine on DA metabolism: all three drugs cause a depletion of tissue DA along with an increase in DOPAC and HVA (*30,31,72*). The effects of reserpine, however, are irreversible and long lasting, whereas the effects of *iboga* alkaloids are reversible and transient. Ibogaine and noribogaine both display low μM potency at vesicular monoamine transporters (VMAT) labeled with [^{125}I]-tetrabenazine (*21*). These intracellular transporter sites are crucial for the accumulation of DA into synaptic vesicles. If *iboga* alkaloids interact with VMAT to disrupt compartmentalization of DA within vesicles, then stored DA would be redistributed into the cytoplasm. Under such circumstances, rapid metabolism of transmitter by monoamine oxidase would account for the dramatic decrease in tissue DA content and the parallel increase in acid metabolites.

Behavioral findings are consistent with the notion that *iboga* alkaloids might impair vesicular storage of DA, at least transiently. Sershen *et al.* (*18*) showed that acute ibogaine pretreatment (40 mg/kg, i.p., -2 hours) blocks the locomotor activity produced by cocaine, but not amphetamine, in mice. Similarly, Broderick *et al.* (*68*) reported that ibogaine reduces cocaine-stimulated locomotion in rats. More recently, Maisonneuve *et al.* (*73*) reported that pretreatment with either ibogaine or noribogaine (40 mg/kg, i.p., -1 hour) significantly attenuates the locomotor activity caused by cocaine administration in rats. It is well established that cocaine-induced psychomotor stimulant effects are dependent on a reserpine-sensitive vesicular pool of DA, whereas the effects of amphetamine are not (*74*). Thus *iboga* alkaloids appear reserpine-like in their ability to distinguish between two types of stimulants: DA reuptake blockers (i.e. cocaine) and DA releasers (i.e. amphetamine).

Glick and Maisonneuve (*15*) have proposed a neuronal circuit model that describes how ibogaine-induced changes in DA transmission might contribute to the antiaddictive properties of the drug. The neurochemical data reviewed above are consistent with the notion that dysregulation of normal DA function by *iboga* alkaloids renders DA neurons refractory to the effects of subsequently administered drugs of abuse. More studies are needed to unravel the complex mechanisms responsible for the dopaminergic actions of ibogaine and noribogaine.

B. Effects on Serotonin Systems

5-HT is an important neurotransmitter in mammals, and abnormalities in 5-HT function have been implicated in the etiology of psychiatric diseases including depression, obsessive-compulsive disorder, and schizophrenia (*75,76*). In rodent brain, neurons that synthesize and release 5-HT have their cell bodies located in the brain stem raphe nuclei. In particular, 5-HT cells of the dorsal and median raphe send axonal projections that ascend through the median forebrain bundle *en route* to terminal fields in all regions of the forebrain, including the PFC, NAC, and striatum. A number of studies have shown that pharmacological treatments causing increased synaptic 5-HT can suppress drug-seeking behavior in rodents trained to self-administer drugs of abuse (*77,78*). For instance, pretreatment with the 5-HT selective reuptake inhibitor (SSRI), fluoxetine, decreases the self-administration of stimulants like cocaine and amphetamine (*79,80*). Similarly, the 5-HT releasing agent, fenfluramine, reduces the self-administration of methamphetamine and heroin (*81,82*). Other studies have shown that chronic exposure to alcohol and cocaine causes a 5-HT deficit syndrome that may contribute to the maintenance of a drug-dependent state (*83-85*). Thus, 5-HT neurons appear to have an important modulatory role in drug-seeking behavior and the development of drug addiction.

It seems reasonable to suspect that ibogaine exerts at least some of its effects via interaction with 5-HT neurons. The *iboga* alkaloids are chemically related to 5-HT since these alkaloids contain an indole moiety as part of their chemical structure (see Figure 1). In rodents, ibogaine administration causes tremors and forepaw treading (*7-9,86*), behaviors that are hallmark signs of the so-called "5-HT syndrome" initially described by Jacobs (*87*). While early radioligand binding studies indicated that ibogaine does not interact with 5-HT receptors *in vitro* (*22,88*), more recent data demonstrate that ibogaine displays low μM affinity (~5-10 μM) for 5-HT_2 receptor sites labeled with ketanserin (*20,89*). In support of the binding data, Helsley *et al.* (*90,91*, see Helsley *et al.* in this volume) used a drug discrimination paradigm in rats to show that stimulus properties of ibogaine are at least partially mimicked by 5-HT agonists with preferential affinity for 5-HT_{2A} and 5-HT_{2C} receptor subtypes. Collectively, such data suggest that ibogaine can act as a low potency 5-HT_2 agonist in rats. The role of 5-HT_2 sites in mediating the psychoactive properties of ibogaine in humans has not been evaluated, and this topic deserves further study.

Mash *et al.* (19) used radioligand binding methods to show that *iboga* alkaloids interact with SERT sites in the human occipital cortex. In their study, ibogaine and noribogaine displaced [^{125}I]RTI-55-labeled SERT binding with Ki values of 550 nM and 40 nM, respectively (see Table I). These data demonstrate that ibogaine and noribogaine exhibit much higher affinity for SERTs when compared to 5-HT receptor subtypes. Interestingly, Staley *et al.* (*21*) found that ibogaine and noribogaine are significantly more potent at SERT sites labeled with cocaine analogs (i.e., RTI-55) when compared to SERT sites labeled with SSRIs such as paroxetine. Thus, similar to the results from DAT binding studies, *iboga* alkaloids display selective affinity for a cocaine-binding site on SERT proteins. The significance of this finding is unknown but warrants investigation.

In order to evaluate the interaction of *iboga* alkaloids with SERTs in greater detail, the activity of ibogaine and noribogaine in assays measuring SERT binding and inhibition of [^{3}H]5-HT uptake in rat brain tissue preparations was examined. As shown in Table III, ibogaine and noribogaine were observed to inhibit [^{125I}]RTI-55 binding to rat SERT sites with IC_{50} values of 3.85 μM and 180 nM, respectively. While these IC_{50} determinations are slightly higher than those observed in the human brain, all data agree that noribogaine has an approx-

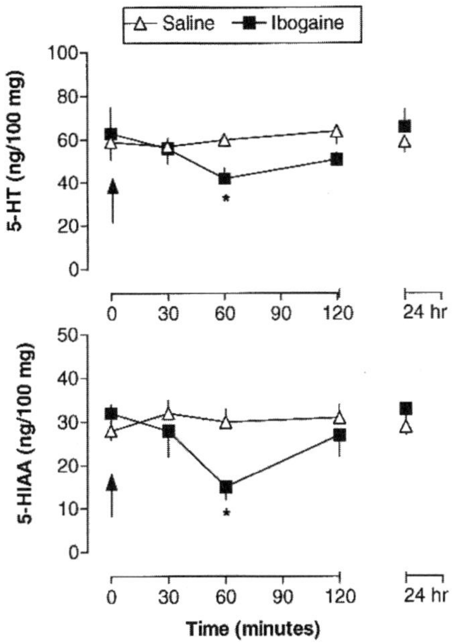

FIGURE 5. Time-course effects of ibogaine administration on postmortem tissue levels of 5-HT and 5-HIAA in rat caudate. Male rats received saline or ibogaine (50 mg/kg, i.p.) at time zero, and groups of rats (N = 6 per group at each time point) were sacrificed at various times thereafter. Tissue levels of 5-HT and 5-HIAA were assayed using HPLC-EC.
* = $P < 0.05$ compared to saline control at specific time points (*31*).

imately 10-fold greater affinity than ibogaine for binding to SERT. The data in Table III also show that ibogaine and noribogaine block [^{3}H]5-HT uptake in rat brain with IC_{50} values of 3.15 μM and 330 nM, respectively. These ibogaine results are consistent with the recent findings of Wells *et al.* (*60*), who reported that ibogaine inhibits [^{3}H]5-HT uptake in rat brain synaptosomes with an IC_{50} of 2.5 μM. These same investigators showed that ibogaine does not stimulate the release of preloaded [^{3}H]5-HT from nervous tissue; indeed, ibogaine antagonizes KCl-evoked release of [^{3}H]5-HT. Taken together, the findings indicate that ibogaine and noribogaine interact with SERT sites to inhibit 5-HT uptake and that noribogaine is 10 times more potent. Therefore, *iboga* alkaloids appear to affect 5-HT neurons *in vitro* in a manner similar to SSRIs.

A number of investigators have examined the effects of ibogaine on 5-HT neurotransmission in rodents (*18,31,66-68*). Most studies have used one of two methods to study 5-HT neurochemistry in whole animals: (1) measurement of 5-HT and its metabolite, 5-hydroxyindoleacetic acid (5-HIAA), in postmortem

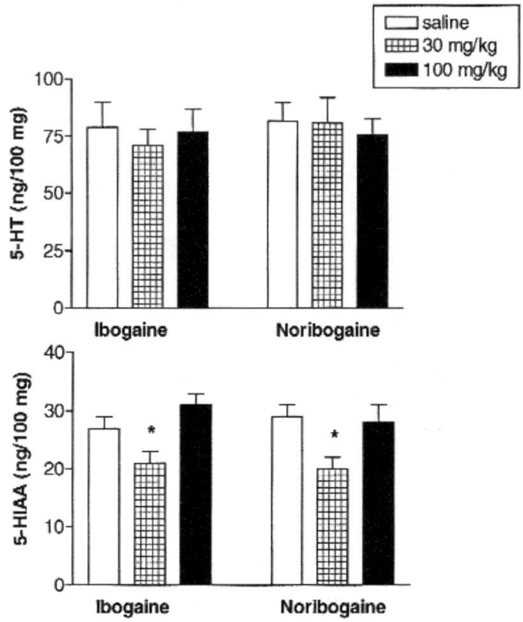

FIGURE 6. Dose-response effects of ibogaine and noribogaine on postmortem tissue levels of 5-HT and 5-HIAA in mouse caudate. Groups of male mice (N = 6 per group at each dose) received saline, ibogaine, or noribogaine and were sacrificed 60 minutes later. Tissue levels of 5-HT and 5-HIAA were determined using HPLC-EC.
* = P<0.05 with respect to saline controls.

brain tissue, or (2) measurement of extracellular 5-HT in living brain via *in vivo* microdialysis. We assessed the time-course effects of ibogaine administration on postmortem tissue levels of 5-HT and 5-HIAA in dissected rat brain regions (*31*). Figure 5 shows that i.p. administration of ibogaine (50 mg/kg) produces an acute and transient reduction in tissue levels of 5-HIAA and 5-HT in rat caudate, with the decline in 5-HIAA being significantly more robust. In agreement with our data, Benwell *et al.* (*66*) showed that ibogaine decreases tissue levels of 5-HIAA in rat caudate and NAC. Thus, ibogaine produces a short-lived and modest reduction in 5-HT metabolism, as indicated by a decrease in the ratio of 5-HIAA/5-HT.

No published studies have assessed the effects of noribogaine on postmortem tissue levels of 5-HT and 5-HIAA. With this in mind, the effects of ibogaine and noribogaine on 5-HT metabolism in mouse caudate were compared. Figure 6 shows that administration of ibogaine or noribogaine, at a dose of 30 mg/kg i.p., significantly decreases 5-HIAA without affecting 5-HT. Curiously, this effect is lost as the dose of either drug is increased to 100 mg/kg. Sershen *et al.* (*18*) demonstrated that i.p. administration of ibogaine to mice (40 mg/kg) produces a reduction in cortical tissue 5-HIAA that lasts for at least 24 hours after treatment. In general, the effects of *iboga* alkaloids on central 5-HT metabolism resemble the effects of SSRIs. Fluoxetine, for example, produces a consistent decrease in brain tissue 5-HIAA without affecting 5-HT (*92*). One notable exception to the fluoxetine-like effect of ibogaine is found in the medial PFC of rats. Benwell *et al.* (*66*) reported that ibogaine treatment (40 mg/kg, i.p.) causes a sustained increase in tissue 5-HIAA in the medial PFC that lasts for 7 days. More studies are needed to clarify the effects of acute ibogaine administration on 5-HT metabolism in discrete rat brain regions, especially in subdivisions of the cerebral cortex.

Broderick *et al.* (*68*) published the first study evaluating the effects of ibogaine on extracellular 5-HT *in vivo*. These investigators used *in vivo* microvoltammetry to show that ibogaine (40 mg/kg, i.p.) increases the concentration of extracellular 5-HT in rat NAC. We have used *in vivo* microdialysis methods to compare the effects of i.v. ibogaine and noribogaine on dialysate 5-HT levels in rat NAC (*19,46*). Our pharmacokinetic data show that metabolism of ibogaine to noribogaine is dramatically reduced when ibogaine is administered by the i.v. route (*46*). Thus, we used i.v. administration of ibogaine and noribogaine to assess the effects of the drugs without the complication of first-pass metabolism. In our experiments, indwelling jugular catheters and intracerebral guide cannulae aimed at the NAC were surgically implanted into anesthetized male rats. One week later, rats were subjected to microdialysis testing. Dialysate samples were collected at 20-minute intervals, and the dialysate concentrations of 5-HT were assayed using HPLC-EC. As depicted in Figure 7, i.v. injection of either ibogaine or noribogaine (1 and 10 mg/kg) causes a significant elevation in extracellular 5-HT in the NAC, and noribogaine is more potent in this regard. Ibogaine and noribogaine appear to display similar efficacy in their ability to increase dialysate 5-HT, since the maximal effect of both drugs is comparable (i.e., two- to threefold). The modest elevations in extracellular 5-HT produced by ibogaine and noribogaine are fully consistent with the ability of these drugs to bind to SERT sites and inhibit 5-HT reuptake.

In a recent publication, Wei *et al.* (*93*) reported that ibogaine and noribogaine produce large increases in dialysate 5-HT in the caudate and NAC of females rats. In their study, i.p. ibogaine (40 mg/kg) elicited a 20-fold increase in extracellular 5-HT in the NAC, whereas an equivalent dose of noribogaine caused an 8-fold increase. The authors concluded that ibogaine is a 5-HT releaser and noribogaine is a 5-HT uptake inhibitor. There are several caveats related to the findings of Wei *et al.* that deserve comment. First, the authors used female rats in their studies whereas other investigators have used males. The pharmacokinetic data described above show that females are more sensitive to the effects of ibogaine because of

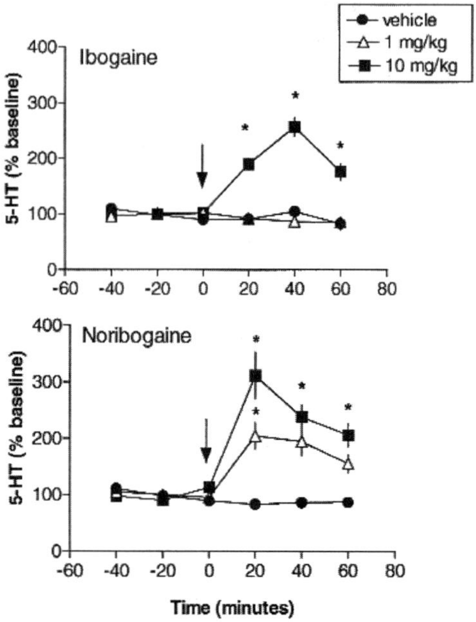

FIGURE 7. Dose-response effects of ibogaine and noribogaine on extracellular 5-HT in rat nucleus accumbens. After three baseline samples were collected, rats received i.v. injection of ibogaine, noribogaine, or ethanol:saline (10%) vehicle. Dialysate samples were collected at 20-minute intervals. Data are mean ± SEM for N = 7 rats/group expressed as a percentage of preinjection baseline. Baseline concentration of dialysate 5-HT was 0.44 ± 0.08 nM.
* $P < 0.05$ relative to vehicle controls (*46*).

higher blood and brain levels of the drug, so data generated from female subjects must be interpreted cautiously. Second, only one dose of drug was tested in the Wei *et al.* study, precluding determination of dose-response effects. We (*61*), and others (*94*), have assessed the dose-response profiles of 5-HT-releasing agents and rarely observe such massive (i.e., 20-fold) elevations in extracellular 5-HT. Finally, the results of Wei *et al.* are difficult to reconcile with the *in vitro* findings of Wells *et al.* (*60*), who demonstrated that ibogaine does not release [^{3}H]5-HT from rat brain synaptosomes. One possible explanation for these discrepancies is that an unidentified metabolite of ibogaine is formed after i.p. injection, and this metabolite is a powerful 5-HT-releasing agent.

Based on the SERT binding data, [^{3}H]5-HT uptake data, and the bulk of the *in vivo* neurochemical data, we hypothesize that ibogaine and noribogaine are 5-HT reuptake inhibitors with a mechanism of action similar to fluoxetine (*61,92,94*). In addition, most evidence agrees that noribogaine is much more potent than ibogaine as an indirect 5-HT agonist. A number of questions remain to be answered, however, with respect to the effects of *iboga* alkaloids on 5-HT neurons. For example, it seems peculiar that ibogaine and noribogaine display higher potency at SERT sites relative to DAT sites, yet the effects of the drugs on DA metabolism are more robust. Stated more simply, why do *iboga* alkaloids cause dramatic depletion of tissue DA, but not tissue 5-HT? Another important question relates to why *iboga* alkaloids do not uniformly decrease 5-HIAA/5-HT ratios in all regions of the brain. Further studies are needed to address these issues.

C. Effects on Opioid Systems

Opioid drugs of abuse, such as heroin and morphine, elicit their effects by binding to opioid receptors located in nerve cell membranes. Historically, three subtypes of opioid receptors were identified on the basis of pharmacological data: mu-, delta-, and kappa-opioid receptor subtypes (*95*). Recent advances in molecular biology have demonstrated that mu, delta, and kappa receptors are derived from three separate genes (*96*). Moreover, all three opioid receptors and their gene transcripts are found in abundance within the mesolimbic neuronal circuitry implicated in drug-seeking behavior and drug dependence (*97,98*). It is well accepted that heroin and morphine are preferential mu-opioid receptor agonists, and activation of mu receptors is associated with the positive reinforcing effects (i.e., euphoria) of habit-forming opioids (*99,100*). Activation of kappa-opioid receptors, on the other hand, is associated with aversive effects (i.e., dysphoria). Interestingly, mu agonist administration increases extracellular levels of DA in the NAC whereas kappa agonists decrease extracellular DA in this region (*101,102*). Thus, mood and behavioral effects of opioids may be mediated, at least in part, via changes in DA transmission in the mesolimbic system.

Substantial evidence indicates that ibogaine modulates opioid transmission. For example, preclinical studies have shown that ibogaine can decrease morphine self-administration (*7,9*) and reduce opioid withdrawal symptoms (*10,11*). The opioid antagonist, naloxone, evokes a dramatic withdrawal syndrome when administered to morphine-dependent rats—the constellation of naloxone-precipitated symptoms includes grooming, burying, teeth-chattering, diarrhea, and wet-dog shakes. Ibogaine pretreatment reduces specific symptoms of naloxone-precipitated withdrawal in morphine-dependent rats, whether ibogaine is administered centrally (i.c.v.)(*10*) or peripherally (i.p.)(*11*). Experiments in mice have shown that ibogaine can decrease naloxone-precipitated withdrawal signs in this species as well (*26,27*). Clinical experience demonstrates that ibogaine administration alleviates the opioid withdrawal syndrome in human opioid addicts (*12-14*). Indeed, Lotsof (*103*) initially filed patents for ibogaine as a pharmacotherapy for narcotics addiction (*Endabuse*) based on the ability of the drug to suppress opioid withdrawal symptoms in human heroin users. Alper *et al.* (*13*) recently collected data from a group of heroin addicts who received ibogaine treatment in nonmedical settings for the purpose of rapid opioid detoxification. In 25 of 33 patients, ibogaine eliminated opioid withdrawal symptoms, and this effect was sustained for 72 hours post-treatment. Finally, Mash *et al.* (*104*) have provided pharmacokinetic evidence suggesting noribogaine is responsible for the alleviation of opioid withdrawal in human patients treated with systemic ibogaine.

The fact that ibogaine suppresses opioid withdrawal symptoms in diverse animal species suggests that ibogaine and noribogaine might interact directly with opioid receptors. Early radioligand binding studies reported that ibogaine binds with low μM affinity (~ 2μM) to kappa-opioid receptors, but is inactive at mu and delta sites (*22*). More recent work has demonstrated that ibogaine displays low μM potency at mu-opioid receptors labeled with [^{3}H]DAMGO and kappa-opioid receptors labeled with [^{3}H]U69,593 (*20,23,24*). Codd (*105*) examined the effects of ibogaine on [^{3}H]naloxone binding in mouse forebrain and resolved two ibogaine binding sites with K_i values of 130 nM and 4.0 μM. On the basis of sodium shift experiments, this investigator postulated that ibogaine is an opioid agonist that exhibits submicromolar affinity for mu-opioid receptors. Pearl *et al.* (*23*) published the first study directly comparing the potencies of ibogaine and noribogaine in opioid binding assays. Their findings showed that noribogaine displays higher affinity than ibogaine for mu- and kappa-opioid receptors. Additionally, noribogaine binds to delta-opioid sites with appreciable affinity (~20 μM), whereas ibogaine does not. We recently compared the ability of ibogaine and noribogaine to displace [^{3}H]DAMGO binding to mu-opioid receptors in rat thalamic membranes (*24*). The data demonstrate that ibogaine and noribogaine display affinities for the mu receptor of 3.76 μM and 160 nM, respectively. Other investigations have confirmed that noribogaine possesses much higher affinity for kappa-opioid sites when compared to ibogaine (*21,27*). In one study, Layer *et al.* (*27*) found that noribogaine is 10 to 100 times more potent than ibogaine at binding to kappa receptors, and this difference in potency is species dependent. Collectively, the radioligand binding results show that ibogaine and noribogaine bind to mu- and kappa-opioid receptors. Noribogaine is significantly more potent in this respect, with the drug exhibiting submicromolar affinities for mu and kappa subtypes under some binding conditions (see Table I).

It seems clear that *iboga* alkaloids can interact directly with opioid receptors, but traditional binding methods cannot provide information on drug efficacy (i.e., the degree to which a drug can elicit functional responses). Opioid receptors are linked to their intracellular transduction enzyme, adenylyl cyclase, via guanine nucleotide-binding proteins (G-proteins) (*96*). Agonist binding to an opioid receptor stimulates binding of GTP to G-proteins, thereby activating the G-proteins and leading to a cellular response. Antagonist binding to an opioid receptor, in contrast, does not activate G-proteins and therefore does not lead to a cellular response. The findings of Codd (*105*) support the idea that *iboga* alkaloids might function as mu-opioid agonists. To assess the efficacy of ibogaine and noribogaine at mu-opioid receptors, we examined the effects of these drugs in the [^{35}S]GTPγS binding assay in rat thalamic membranes (*24*). In this assay, opioid agonists stimulate [^{35}S]GTPγS binding to G-proteins, whereas opioid antagonists do not. We observed that noribogaine stimulates [^{35}S]GTPγS binding in a naloxone-reversible manner with an EC_{50} of 320 nM. Ibogaine, in contrast, is inactive in this assay at concentrations up to 100 μM. The intrinsic activity of noribogaine in the [^{35}S]GTPγS binding assay is comparable to the prototypical mu-opioid agonist DAMGO, suggesting noribogaine displays properties of a full-efficacy agonist at mu-opioid receptors.

The mu-opioid effects of noribogaine might explain the ability of systemically administered ibogaine to block opioid withdrawal. Theoretically, noribogaine could suppress opioid withdrawal by "substituting" for the addictive mu-opioid drug during acute abstinence. Indeed, mu-opioid agonist substitution-type medications, such as methadone and buprenorphine, are the most effective means of opioid detoxification (*106*). In addition, preliminary findings indicate that noribogaine binds to mu-opioid sites in a persistent manner, with dissociation rates in the range of days (*107*). Thus, noribogaine appears to display a unique profile of activity at mu-opioid receptors that includes low-affinity, pseudo-irreversible binding, and full-agonist efficacy. One caveat to the aforementioned hypothesis is that neither ibogaine nor noribogaine has significant analgesic potency *in vivo* (*36,37*). This finding is not consistent with the mu-opioid agonist actions of *iboga* alkaloids determined *in vitro*. On the other hand, noribogaine pretreatment is able to enhance the analgesic potency of morphine in rats and mice (*36,37*). The underlying basis for this peculiar finding is unknown, but it may be related to the purported ability of noribogaine to enhance mu-opioid receptor-mediated inhibition of adenylyl cyclase (*108*).

There is some data suggesting that the effects of ibogaine on DA neurons might be mediated by kappa-opioid receptors *in vivo*. As mentioned previously, local infusion of high-dose ibogaine (100 to 400 μM) into the NAC decreases extracellular levels of DA (*69,70*). Reid *et al.* (*70*) reported that reductions in dialysate DA levels produced by local infusion of ibogaine (100 μM) are reversed by coinfusion of the nonselective opioid antagonist naloxone (1 μM) or the kappa-opioid selective antagonist norbinaltorphine (1-10 μM). Glick *et al.* (*109*) reported comparable findings in female rats. These investigators showed that i.p. administration of norbinaltorphimine (10 mg/kg, i.p.) attenuates ibogaine-induced decreases in extracellular DA in the NAC. Few studies have examined the effects of receptor-selective opioid antagonists on the pharmacological actions of ibogaine and noribogaine *in vivo*, and more studies of this type need to be carried out.

IV. Neuroendocrine Effects

Stress is a major factor contributing to the development of drug dependence (*110,111*). Studies in rats have shown that various types of stressors can facilitate acquisition of drug self-administration behavior (*112-114*) and trigger relapse during drug withdrawal (*115,116*). The effects of stress on drug-seeking behavior appear to be mediated by hormones of the hypothalamic-pituitary-adrenal (HPA) axis, particularly corticosterone (*117,118*). It is well known that corticosterone is secreted from the adrenal cortex in response to natural stressors or drugs of abuse. Corticosterone, in turn, can facilitate acquisition of drug self-administration

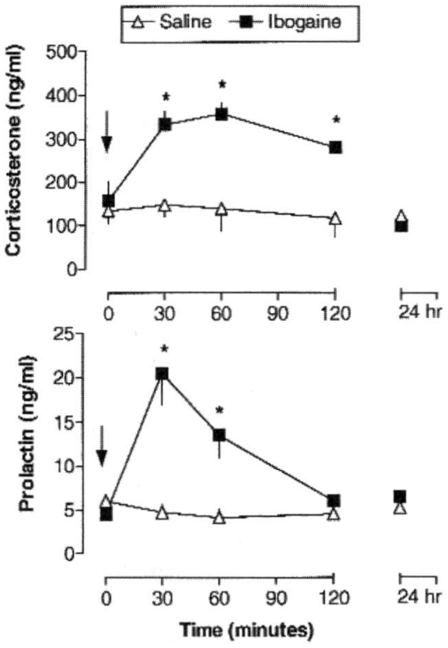

FIGURE 8. Time-course effects of ibogaine administration on circulating levels of corticosterone and prolactin in rats. Male rats received saline or ibogaine (50 mg/kg, i.p.) at time zero, and groups of rats (N = 6 per group at each time point) were sacrificed at various times thereafter. Plasma corticosterone and prolactin were assayed using radioimmunoassay (RIA) methods.
* = $P < 0.05$ compared to saline control at specific time points (*31*).

behavior, similar to the effects of stress (*117,118*). Reductions in circulating corticosterone, produced by surgical adrenalectomy or inhibition of corticosterone biosynthesis, decrease drug self-administration behavior (*119,120*). Interestingly, corticosterone itself appears to have intrinsic reinforcing properties because this hormone is readily self-administered under certain experimental conditions (*121,122*). Taken together, the preclinical data show that adrenal corticosteroids are important biological substrates mediating the ability of stress to influence the effects of abused drugs. Such findings may have clinical relevance, since drug addicts and clinicians alike would agree that stressful life events often contribute to relapse in human drug-dependent patients.

With reference to the preceding discussion, it seems pertinent to examine the neuroendocrine effects of *iboga* alkaloids. We have evaluated the time-course effects of ibogaine on the circulating levels of corticosterone and prolactin in rats (*31,63,64*). As discussed above, corticosterone is a major stress hormone of the

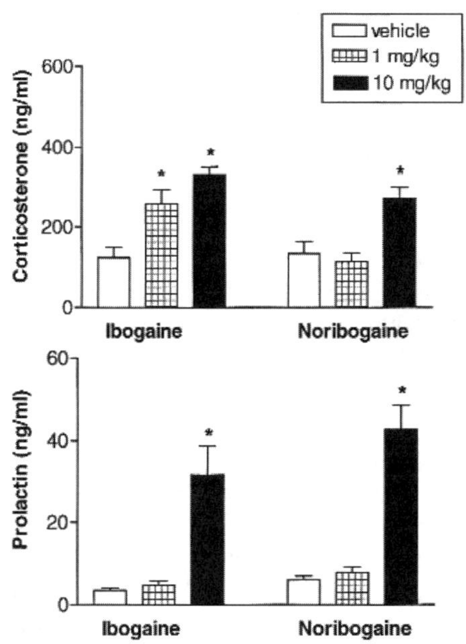

FIGURE 9. Dose-response effects of ibogaine and noribogaine on circulating corticosterone and prolactin in rats. Male rats bearing indwelling jugular catheters received i.v. injection of ibogaine, noribogaine, or ethanol:saline vehicle (10%), and blood samples were withdrawn via the catheters. Data are mean ± SEM expressed as peak plasma hormone levels for N = 7 rats/group. Peak hormone secretion occurred at 30 minutes for corticosterone and at 15 minutes for prolactin.
* = $P < 0.05$ with respect to vehicle controls (*46*).

HPA axis. Prolactin is a protein hormone synthesized in the anterior pituitary that is also secreted in response to stress, but regulation of prolactin is independent of the HPA axis (*123,124*). The data in Figure 8 show that i.p. ibogaine administration (50 mg/kg) causes a sustained increase in corticosterone secretion that lasts for at least 2 hours. This effect is normalized within 24 hours. Ibogaine also elevates plasma prolactin, but this effect is short-lived, with hormone levels returning to baseline by 2 hours postinjection.

The effects of i.v. ibogaine and noribogaine on neuroendocrine secretion in rats were also compared. In these experiments, indwelling jugular catheters were surgically implanted into anesthetized male rats (*46*). After one week of recovery from surgery, rats received i.v. injection of ibogaine, noribogaine, or vehicle. Repeated blood samples were withdrawn via the catheters, and plasma was assayed for hormone levels by double-antibody RIA. Previous studies from our laboratory have shown that the chronic-catheterized rat model allows for repeated blood sampling with a minimum of stress to the animal (*125*).

As shown in Figure 9, plasma corticosterone levels are significantly increased after i.v. administration of ibogaine or noribogaine, but ibogaine is significantly more potent as a stimulator of the corticosterone secretion. After a 1 mg/kg dose, for instance, ibogaine elevates corticosterone whereas noribogaine does not. Ibogaine and noribogaine produce comparable increases in circulating prolactin that are seen only after the 10 mg/kg dose. The drug-induced hormonal effects reported here are likely to be mediated via central pathways because i.c.v. administration of ibogaine to rats causes elevations in circulating corticosterone and prolactin similar to those observed after i.p. and i.v. administration (Baumann, unpublished).

In our initial studies, we proposed that neuroendocrine effects of ibogaine might be mediated by central 5-HT neurons based on the similar *in vivo* effects of ibogaine and the 5-HT releaser, fenfluramine (*31,63,125*). However, the neurochemical data reviewed above suggest that 5-HT neurons are not involved in ibogaine-induced corticosterone secretion. The microdialysis data, for example, show that i.v. ibogaine is less potent than i.v. noribogaine in its ability to elevate extracellular 5-HT in the brain. Likewise, ibogaine has lower affinity for SERT and lower potency at blocking 5-HT uptake when compared to noribogaine. Thus, ibogaine is less potent as an indirect 5-HT agonist, but much more potent as a stimulator of corticosterone secretion. The mechanisms underlying the effects of *iboga* alkaloids on plasma corticosterone are unclear, and they could be mediated by any number of targets, including NMDA-coupled ion channels or sigma-2 receptors (see Table I). Similar to ibogaine, the noncompetitive NMDA antagonist, MK-801, is an activator of the HPA axis in rats (*126*). Ibogaine inhibits [^{3}H]MK-801 binding in rat brain tissue and ibogaine mimics the electrophysiological, neuroprotective, and discriminative stimulus properties of MK-801 (*25-27*). Likewise, the prototypical sigma receptor drug phencyclidine

(PCP) elevates plasma corticosterone (*127*), and ibogaine displays submicromolar affinity for sigma-2 binding sites (16,17). It also seems possible that ibogaine-induced activation of the HPA axis may represent a nonspecific stress response secondary to the adverse behavioral consequences of ibogaine administration. Doses of ibogaine used in our studies cause tremors, forepaw treading, and ataxia (see below). Interestingly, noribogaine administration does not cause the same adverse behaviors as ibogaine, and noribogaine is less potent as an activator of the HPA axis in rats.

The mechanisms responsible for prolactin secretion elicited by ibogaine and noribogaine are not known, but may involve hypothalamic DA neurons (*123*). Under normal circumstances, secretion of pituitary prolactin is inhibited by tonic release of DA from tuberoinfundibular DA (TIDA) neurons in the mediobasal hypothalamus (*128*). TIDA nerve terminals in the median eminence release DA into the hypothalamic-hypohysial portal circulation in an endocrine fashion. DA, in turn, binds to DA D2 receptors on pituitary lactotrophs to directly inhibit prolactin secretion. The elevation of prolactin evoked by ibogaine and noribogaine might be mediated by a reduction in extracellular DA levels in the hypothalamus. Consistent with this proposal, ibogaine administration produces a significant decrease in tissue levels of DA in rat hypothalamus (*63*). However, no studies have examined the effects of *iboga* alkaloids on TIDA neuronal activity. It is noteworthy that mu- and kappa-opioid receptor agonists produce significant elevations in plasma prolactin, and this effect is mediated by suppression of DA release from TIDA neurons (*129,130*). Thus, the prolactin-releasing effect of ibogaine and noribogaine may involve activation of opioid receptors. No investigators have examined the ability of receptor antagonists to reverse the endocrine effects of ibogaine or noribogaine. Further studies are needed to determine the specific receptor sites involved in mediating the neuroendocrine actions of *iboga* alkaloids.

V. Adverse Consequences

A. Behavioral Effects

Ibogaine is known to produce adverse behavioral effects in both humans and animals. In humans, oral administration of ibogaine (5 to 25 mg/kg) produces dizziness, nausea, vomiting, and motor incoordination that last for hours. Naranjo *et al.* (*131,132*) reported that ibogaine evokes a hallucinogenic-like visual experience that resembles a dream, but without loss of consciousness. The neurobiological underpinnings of the so-called waking dream state are not known. In addition, the possible role of noribogaine in mediating specific aspects of the ibogaine-induced visual experience have not been studied. Some investigators have speculated that the psychoactive properties of ibogaine are important for the antiaddictive effects of the drug, and this hypothesis deserves to be investigated under controlled experimental conditions.

Administration of ibogaine to rats causes a spectrum of behaviors including tremors, forepaw treading, and ataxia (*7-9,86*). These motor behaviors are transient and resolve within the first few hours postinjection. Interestingly, the receptor mechanisms responsible for these ibogaine-induced behaviors have not been clarified. We have compared the effects of ibogaine and noribogaine on various motor behaviors in rats (*46*). As discussed previously, i.v. drug administration was used in these experiments because the i.v. route enables an assessment of drug-induced effects without the complication of significant first-pass metabolism. Rats received i.v. injection of ibogaine, noribogaine, or vehicle. Animals were observed for 90 second intervals at various times thereafter, and specific behaviors were scored using a graded scale: 0=absent, 1=equivocal, 2=present, 3=intense (*46,125*). Rats were given a single numerical score for each behavior that consisted of the summed total for that behavior across all time points.

The data in Figure 10 illustrate the effects of ibogaine and noribogaine on tremors and forepaw treading. Ibogaine produces a dose-related increase in the occurrence of tremors and forepaw treading; the ibogaine-induced tremorigenic effect consists of fine tremors of the face, head, and neck, as well as prominent shivering movements of the trunk. After the highest dose of ibogaine (10 mg/kg, i.v.), most rats display abnormal postures, body sway, and a staggering-type locomotion. In contrast to ibogaine, noribogaine does not elicit tremors or ataxia. Noribogaine does cause a slight stimulation of forepaw treading, but it is much less effective when compared to ibogaine. In addition, noribogaine increases the incidence of penile erections, a behavior that is rarely seen after ibogaine administration. It should be mentioned that behavioral effects elicited by both drugs are transient, with rats appearing completely normal by 30 minutes postinjection. Our findings with i.v. ibogaine are fully consistent with prior reports showing i.p. ibogaine elicits tremors and ataxia when administered to rats at typical antiaddictive doses (i.e., 40 mg/kg). In agreement with the i.v. noribogaine results, Glick *et al.* (*35*) demonstrated that i.p. noribogaine (40 mg/kg) is not tremorigenic when administered to female rats. Thus, ibogaine and noribogaine evoke quite different behavioral effects despite having similar chemical structures.

It might be assumed that ibogaine-induced actions are mediated by central 5-HT mechanisms because tremors and forepaw tapping are classic signs of the 5-HT behavioral syndrome (*86,87,125*). Irrespective of such similarities between the behavioral effects of ibogaine and certain 5-HT drugs (see *86*), the data

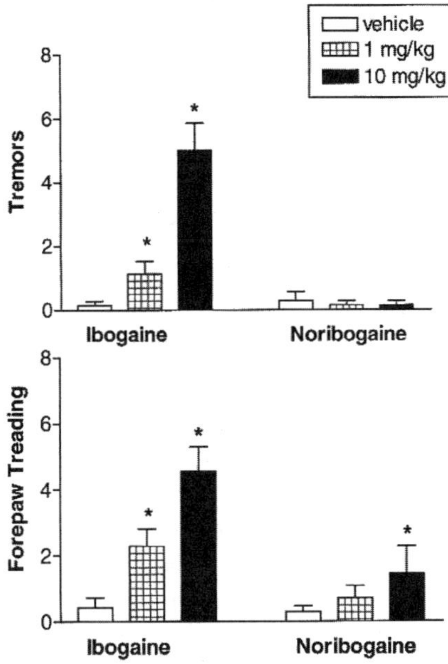

FIGURE 10. Dose-response effects of ibogaine and noribogaine on tremors and forepaw treading. Male rats bearing indwelling jugular catheters received i.v. injection of ibogaine, noribogaine, or ethanol:saline vehicle (10%). Behaviors were scored at 2, 10, 20, and 30 minutes postinjection. Rats were given a single numerical score that was the summed total for that behavior across all time points. Data are mean ± SEM for N = 7 rats/group.
* = $P < 0.05$ with respect to vehicle controls (*46*).

discussed herein indicate that 5-HT mechanisms are probably not involved in the motor effects of ibogaine. Ibogaine is less potent than noribogaine as an indirect 5-HT agonist, yet ibogaine is more potent as a stimulator of tremors and forepaw treading. It is tempting to speculate that ibogaine-induced motor effects might involve NMDA-coupled ion channels or sigma-2 receptors. Ibogaine is significantly more potent than noribogaine at these sites (*16,21,27,* see Table I), possibly explaining why ibogaine evokes more potent behavioral actions. Similar to ibogaine, drugs that interact with sigma receptors (i.e., PCP) and NMDA sites (i.e., MK-801) are known to cause forepaw tapping and ataxia in rats (*133,134*). The mechanisms underlying ibogaine-induced tremor remain to be determined.

B. Neurotoxicity

Probably the most serious impediment to the development of ibogaine as an antiaddictive medication is the reported neurotoxicity of the drug (*135-137*, see Molliver *et al.* in this volume). O'Hearn *et al.* (*135,136*) were the first to show that single or multiple injections of ibogaine (100 mg/kg, i.p.) cause glial cell proliferation and Purkinje cell degeneration in the rat cerebellar vermis. The effects of ibogaine in rat cerebellum are consistent with a trans-synaptic excitoxic lesion that involves sustained activation of the olivocerebellar pathway (*137*). O'Callaghan *et al.* (*138*) examined the effects of ibogaine (100 mg/kg, i.p.) on tissue levels of glial fibrillary acidic protein (GFAP) in dissected rat brain regions. Increased expression of GFAP, a hallmark sign of astrogliosis, is a generic indicator of neuronal injury in the brain. These investigators demonstrated that ibogaine increases expression of GFAP in the cerebellum and in other brain areas such as the striatum and hippocampus. Somewhat surprisingly, Scallet *et al.* (*139*) reported that ibogaine-induced cerebellar injury is observed in rats, but not in mice. Sanchez-Ramos and Mash (*140*) found no evidence for neurotoxic damage in African green monkeys treated repeatedly with orally administered ibogaine (5 to 25 mg/kg). Thus, the neurotoxic effects of ibogaine appear to be species dependent, suggesting extrapolation of toxicity data across species lines is ill advised. No studies have systematically investigated the neurotoxic potential of noribogaine, and such studies need to be carried out.

The neurotoxic effects of ibogaine are clearly dose related. For example, Molinari *et al.* (*141*) demonstrated that a single injection of the typical antiaddictive dose of ibogaine (40 mg/kg, i.p.) does not cause cerebellar damage in female rats. Furthermore, repeated administration of lower doses of ibogaine (10 mg/kg, i.p.), given every other day for 60 days, does not affect Purkinje cell number in male rats (142). Recently, Xu *et al.* (*143*) evaluated the dose-response effects of single i.p. injections of ibogaine on markers of cerebellar neurotoxicity in rats. In their study, ibogaine doses of 75 and 100 mg/kg produced evidence for cerebellar damage in all rats tested. On the other hand, a dose of 25 mg/kg had no effect. An ibogaine dose of 50 mg/kg produced no obvious Purkinje cell degeneration, but it did increase cerebellar GFAP staining in one-third of the rats studied. Collectively, the neurotoxicity data show that the margin of safety for ibogaine administration in rats is very narrow, since therapeutic doses of the drug (i.e., 40 mg/kg, i.p.) are very close to the minimum doses required for eliciting cerebellar damage (i.e., 50 mg/kg, i.p.).

The mechanism underlying ibogaine-evoked neurotoxicity is not known. Vilner *et al.* (*144, 145*) have provided evidence that sigma-2 receptors might be involved in the neurotoxic effects of ibogaine (see Bowen *et al.* in this volume). These investigators examined sigma-2 receptor-mediated cytotoxicity in human SK-N-SH neuroblastoma cells cultured *in vitro*. Incubation of neuroblastoma cells with sigma-2 drugs, including ibogaine, causes dose- and time-dependent morphological changes that are characterized by loss of processes, detachment, and ultimately cell death. The same sigma-2 ligands produce elevations in intracellular Ca^{++} concentrations in neuroblastoma cells that may be causally linked to cytotoxicity. Interestingly, noribogaine is not toxic to neuroblastoma cells in culture, consistent with the lack of affinity of noribogaine for sigma-2 receptors (*16*, Table I). The exact relationship between sigma-2 cytotoxicity *in vitro* and ibogaine-induced cerebellar degeneration *in vivo* remains to be established. One caveat to the sigma-2 receptor hypothesis of ibogaine neurotoxicity relates to the effects of the beta-carboline compound, harmaline. Harmaline causes cerebellar neurotoxicity in rats analogous to the effects of ibogaine (*135-137*), but harmaline has little affinity for sigma-2 binding sites (*16*). More studies are needed to determine the precise mechanisms responsible for the neurotoxic effects of *iboga* alkaloids. In particular, direct comparisons of the neurotoxic potential of ibogaine, noribogaine, and other *iboga* alkaloids need to be carried out.

VII. Conclusions

The data reviewed in this chapter show that ibogaine interacts with multiple neurotransmitter systems known to modulate drug addiction (see Table I). The *in vivo* pharmacological activity of ibogaine is further complicated by the metabolic conversion of ibogaine to its active *O*-desmethyl metabolite, noribogaine, in rats, monkeys, and humans (*19,21,29,34,41,46*). After ibogaine administration to rats (40 mg/kg, i.p.), concentrations of noribogaine in blood and brain tissue are equal to, or exceed, the levels of ibogaine itself. Moreover, noribogaine has a much longer biological half-life than ibogaine, with pharmacologically relevant concentrations of noribogaine persisting in the bloodstream for at least one day after ibogaine treatment. These findings suggest the possibility that noribogaine contributes to the antiaddictive properties of systemically administered ibogaine. Moreover, noribogaine might be the active compound mediating the long-term actions of ibogaine. Indeed, Glick *et al.* (*35*) have shown that ibogaine and noribogaine are equipotent in their ability to produce long-lasting reductions in the self-administration of cocaine and morphine in rats.

Gender and gonadal steroid hormones can influence ibogaine pharmacokinetics and metabolism in rats and possibly other species (*29*). When ibogaine is administered to female rats during the preovulatory phase of the reproductive cycle (i.e., high estrogen levels), concentrations of ibogaine in blood and brain tissue are much higher than those of similarly treated males (see Table II). We speculate that elevated levels of circulating estrogen enhance the absorption of ibogaine from peripheral compartments into the bloodstream. Such gender-dependent alterations in ibogaine pharmacokinetics may serve to explain the heightened sensitivity of female rats to the neurochemical, behavioral, and neurotoxic effects of ibogaine (*29,138*).

Perhaps the most dramatic neurochemical effects produced by *iboga* alkaloids are on DA metabolism in the brain. Ibogaine and noribogaine are equipotent in their ability to evoke a transient stimulation of DA metabolism that is characterized by profound depletion of tissue DA (~50% reduction) in mesolimbic, mesocortical, and mesostriatal terminal projection areas (*18,30,31,63-65*). The reserpine-like depletion of DA is short-lived, but this effect could have long-term consequences. For example, dysregulation of DA function produced by ibogaine and noribogaine may alter the responsiveness of DA neurons to the effects of subsequently administered drugs of abuse. Consistent with this notion, ibogaine pretreatment can block the drug-induced elevations in extracellular DA in rat NAC normally produced by cocaine, morphine, and nicotine (*66-68*). Thus, alterations in DA transmission may represent a common mechanism underlying the ability of *iboga* alkaloids to suppress the self-administration of diverse types of addictive drugs.

Ibogaine and noribogaine bind to SERT sites and inhibit 5-HT uptake *in vitro*; noribogaine is at least 10 times more potent than ibogaine in this regard (*19,21*, see Table III). The blockade of 5-HT uptake afforded by ibogaine and noribogaine leads to elevations in extracellular 5-HT in the NAC *in vivo*. The serotonergic actions of *iboga* alkaloids resemble the actions of SSRIs such as fluoxetine. It seems plausible that fluoxetine-like effects of ibogaine, and especially noribogaine, might contribute to the antiaddictive properties of systemically administered ibogaine. Elevations in synaptic 5-HT could help to prevent relapse by alleviating withdrawal-associated depression, reducing drug craving, and decreasing the impulse to use drugs (*146-148*). In addition, elevations in synaptic 5-HT may enhance the mu-opioid activity of noribogaine, similar to the dual mechanism of action of the novel opioid, tramadol (*149,150*). Such 5-HT-opioid synergism could contribute to the ibogaine-induced suppression of opioid withdrawal symptoms that has been observed in heroin-dependent subjects.

Emerging evidence shows that ibogaine and noribogaine interact with opioid receptors in the brain, and noribogaine has much higher affinity at mu-, delta-, and kappa-opioid receptor subtypes (*21-24,27*). Recent findings suggest that noribogaine displays a unique profile of activity at mu-opioid receptors including low affinity, persistent binding, and full agonist efficacy. Thus, noribogaine may function as a methadone-like substitution medication that attenuates opioid withdrawal symptoms via direct agonist actions at mu-opioid receptors. The possible role of delta- and kappa-opioid receptors in mediating the antiaddictive properties of ibogaine and noribogaine is largely unexplored. It is noteworthy that kappa opioid agonists have been considered as potential pharmacotherapies for stimulant addiction based on a growing body of preclinical literature (*151,152*).

The collective findings suggest that noribogaine might be superior to ibogaine as an antiaddictive medication, due to the higher affinity of noribogaine at SERT sites and multiple opioid receptor subtypes. In addition, noribogaine appears to exhibit a superior side-effects profile when compared to ibogaine. Ibogaine causes tremors and ataxia in rats, whereas noribogaine does not. Possibly because noribogaine does not produce adverse behavioral effects, noribogaine is less potent as a stimulator of the HPA axis. A lack of sigma-2 receptor activity may render noribogaine free from the cerebellar toxicity associated with ibogaine. Thus, based on the data reviewed in this chapter, we propose that noribogaine may be a more effective and safer alternative to ibogaine as a candidate for medication development. Future studies should examine the antiaddictive potential of ibogaine and noribogaine in drug-dependent human patients under well-controlled experimental conditions.

Acknowledgments

The authors are indebted to Mario Ayetas, Artensie Carter, Chris Dersch, Joy Jackson, and Jennifer Phillips for their expert technical assistance during these studies. The authors wish to thank the National Institute on Drug Abuse, Intramural Research Program, for financial support of this work.

References

1. "Epidemiological Trends in Drug Abuse," Vol. II: Proceedings of the IEWG on Drug Abuse, NIDA, NIH Publication No. 00-4530, 1999.
2. K. Morris, *Lancet* **354,** 1883 (1999).
3. B.I. Koerner, *US News World Report* **127,** Dec. 6, 82 (1999).
4. R. Goutarel, O. Gollnhofer, and R. Silans, *Psychedel. Monogr. Essays* **6,** 71 (1993).
5. P. Popik, R.T. Layer, and P. Skolnick, *Pharmacol. Rev.* **47,** 235 (1995).
6. P. Popik and P. Skolnick, *in* "The Alkaloids," (G. A. Cordell, ed.), Vol. 52, p. 197. Academic Press, New York, NY, 1999.
7. S.D. Glick, K. Rossman, S. Steindorf, I.M. Maisonneuve, and J.N. Carlson, *Eur. J. Pharmacol.* **195,** 341 (1991).
8. S.L. Cappendijk and M.R. Dzoljic, *Eur. J. Pharmacol.* **241,** 261 (1993).
9. S.D. Glick, M.E. Kuehne, J. Raucci, T.E. Wilson, D. Larson, R.W. Keller, and J.N. Carlson, *Brain Res.* **657,** 14 (1994).
10. E.D. Dzoljic, C.D. Kaplan, and M.R. Dzoljic, *Arch. Int. Pharmacodyn. Ther.* **294,** 64 (1988).
11. S.D. Glick, K. Rossman, N.C. Rao, I.M. Maisonneuve, and J. Carlson, *Neuropharmacology* **31,** 497 (1992).
12. S.G. Sheppard, *J. Subst. Abuse Treat.* **11,** 379 (1994).
13. K.R. Alper, H.S. Lotsof, G.M.N. Frenken, D.J. Luciano, and J. Bastiaans, *Am. J. Addictions* **8,**

234 (1999).
14. D.C. Mash, C.A. Kovera, B.E. Buck, M.D. Norenberg, P. Shapshak, W.L. Hearn, and J. Sanchez-Ramos, *Ann. N.Y. Acad. Sci.* **844,** 274 (1998).
15. S.D. Glick and I.M. Maisonneuve, *Ann. N.Y. Acad. Sci.* **844,** 214 (1998).
16. W.D. Bowen, B.J. Vilner, W. Williams, C.M. Bertha, M.E. Kuehne, and A.E. Jacobson, *Eur. J. Pharmacol.* **279,** R1 (1995).
17. R.H. Mach, C.R. Smith, and S.R. Childers, *Life Sci.* **57,** PL57 (1995).
18. H. Sershen, A. Hashim, L. Harsing, and A. Lajtha, *Life Sci.* **50,** 1079 (1992).
19. D.C. Mash, J.K. Staley, M.H. Baumann, R.B. Rothman, and W.L. Hearn, *Life Sci.* **57,** PL45 (1995).
20. P.M. Sweetnam, J. Lancaster, A. Snowman, J.L. Collins, S. Perschke, C. Bauer, and J. Ferkany, *Psychopharmacology* **118,** 369 (1995).
21. J.K. Staley, Q. Ouyang, J. Pablo, W.L. Hearn, D.D. Flynn, R.B. Rothman, K.C. Rice, and D.C. Mash, *Psychopharmacology* **127,** 10 (1996).
22. D.C. Deecher, M. Teitler, D.M. Sonderlund, W.G. Bornmann, M.E. Kuehne, and S.D. Glick, *Brain Res.* **571,** 242 (1992).
23. S.M. Pearl, K. Herrick-Davis, M Teitler, and S.D. Glick, *Brain Res.* **675,** 342 (1995).
24. J.P. Pablo and D.C. Mash, *NeuroReport* **9,** 109 (1998).
25. P. Popik, R.T. Layer, and P. Skolnick, *Psychopharmacology* **114,** 672 (1994).
26. P. Popik, R.T. Layer, L.H. Fossom, M. Benveniste, B. Geter-Douglss, J.M. Witkin, and P. Skolnick, *J. Pharmacol. Exp. Ther.* **275,** 753 (1995).
27. R.T. Layer, P. Skolnick, C.M. Bertha, U.K. Bandarage, M.E. Kuehne, and P. Popik, *Eur. J. Pharmacol.* **309,** 159 (1996).
28. L.B. Hough, S.M. Pearl, and S.D. Glick, *Life Sci.* **58,** PL119 (1996).
29. S.M. Pearl, L.B. Hough, D.L. Boyd, and S.D. Glick, *Pharmacol. Biochem. Behav.* **57,** 809 (1997).
30. I.M. Maisonneuve, K.L. Rossman, R.W. Keller, and S.D. Glick, *Brain Res.* **575,** 69 (1992).
31. S.F. Ali, G.D. Newport, W. Slikker, R.B. Rothman, and M.H. Baumann, *Brain Res.* **737,** 215 (1996).
32. H.I. Dhahir, "A Comparative Study of the Toxicity of Ibogaine and Serotonin." Ph.D. thesis, Indiana University, 1971.
33. G. Zetler, G. Singbarth, and L. Schlosser, *Pharmacology* **7,** 237 (1972).
34. W.L. Hearn, J. Pablo, G.W. Hime, and D.C. Mash, *J. Anal. Toxicol.* **19,** 427 (1995).
35. S.D. Glick, S.M. Pearl, J. Cai, and I.M. Maisonneuve, *Brain Res.* **713,** 294 (1996).
36. A.A. Bagal, L.B. Hough, J.W. Nalwalk, and S.D. Glick, *Brain Res.* **741,** 258 (1996).
37. H.N. Bhargava, Y.-J. Cao, and G.-M. Zhao, *Brain Res.* **752,** 234 (1997).
38. M.H. Baumann, J. Jackson, A. Carter, D.C. Mash, and R.B. Rothman, *NIDA Res. Monogr.* **174,** 213 (1997).
39. S. Helsley, R.A. Rabin, and J.C. Winter, *Life Sci.* **60,** PL147 (1997).
40. S. Sunder Sharma and H.N. Bhargava, *Pharmacology* **57,** 229 (1998).
41. C. Zubaran, M. Shoaib, I.P. Stolerman, J. Pablo, and D.C. Mash, *Neuropsychopharmacology* **21,** 119, (1999).
42. C.A. Gallagher, L.B. Hough, S.M. Keefner, A. Seyed-Mozaffari, S. Archer, and S.D. Glick, *Biochem. Pharmacol.* **49,** 73 (1995).
43. M.E. Alburges, R.L. Foltz, and D.E. Moody, *J. Anal. Toxicol.* **19,** 381 (1995).
44. F.R. Ley, A.R. Jeffcoat, and B. Thomas, *J. Chromatogr. A* **723,** 101 (1996).
45. R.S. Obach, J.P. Pablo, and D.C. Mash, *Drug Metabol. Disp.* **26,** 764 (1998).
46. M.H. Baumann, J.P. Pablo, S.F. Ali, R.B. Rothman, and D.C. Mash, *Ann. N.Y. Acad. Sci.* **914,** 354, (2000).
47. M.H. Baumann, J. Pablo, J.M. Phillips, R.B. Rothman, and D.C. Mash, *NIDA Res. Monogr.* **180,** 287 (2000).
48. R.A. Wise and P.P. Rompre, *Ann. Rev. Psychol.* **40,** 191 (1989).
49. R. Spanagel and F. Weiss, *Trends Neurosci.* **22,** 521 (1999).

50. G. Di Chiara and A. Imperato, *Proc. Natl. Acad. Sci. U.S.A.* **85,** 5274 (1988).
51. F.E. Pontieri, G. Tanda, and G. Di Chiara, *Proc. Natl. Acad. Sci. U.S.A.* **92,** 12304 (1995).
52. A. Imperato, A. Mele, M.G. Scrocco, and S. Puglisi-Allegra, *Eur. J. Pharmacol.* **212,** 299 (1992).
53. Z.L. Rossetti, Y. Hmaidan, and G.L. Gessa, *Eur. J. Pharmacol.* **221,** 227 (1992).
54. D.W. Self, *Ann. Med.* **30,** 379 (1998).
55. G.F. Koob, *Ann. N.Y. Acad. Sci.* **877,** 445 (1999).
56. P.A. Broderick, F.T. Phelan, and S.P. Berger, *NIDA Res. Monogr.* **119,** 285 (1992).
57. R.A.Vaughan, *Mol. Pharmacol.* **47,** 956 (1995).
58. G. Rudnick and J. Clark, *Biochem. Biophys. Acta* **1144,** 249 (1993).
59. S.G. Amara and M.J. Kuhar, *Ann. Rev. Neurosci.* **16,** 73 (1993).
60. G.B.Wells, M.C. Lopez and J.C. Tanaka, *Brain Res. Bull.* **48,** 641 (1999).
61. R.B. Rothman, M.A. Ayestas, C.M. Dersch, and M.H. Baumann, *Circulation* **100,** 869 (1999).
62. H. Sershen, A. Hashim, and A. Lajtha, *Brain Res. Bull.* **42,** 161 (1997).
63. M.H. Baumann, R.B. Rothman, and S.F. Ali, *Ann. N.Y. Acad. Sci.* **844,** 252 (1998).
64. M.H. Baumann, R.B. Rothman, and S.F. Ali, *Drug. Alcohol Depend.* **59,** 143 (2000).
65. S.F. Ali, R.B. Rothman, and M.H. Baumann, *NIDA Res. Monogr.* **180,** 286 (2000).
66. M.E. Benwell, P.E. Holton, R.J. Moran, and D.J. Balfour, *Br. J. Pharmacol.* **117,** 743 (1996).
67. I.M. Maisonneuve, R.W. Keller, and S.D. Glick, *Eur. J. Pharmacol.* 199, 35 (1991).
68. P.A. Broderick, F.T. Phelan, F. Eng, and R.T. Wechsler, *Pharmacol. Biochem. Behav.* **49,** 711 (1994).
69. S.D. Glick, K. Rossman, S. Wang, N. Dong, R.W. Keller, *Brain Res.* **628,** 201 (1993).
70. M.S. Reid, K. Hsu, K.H. Souza, P.A. Broderick, and S.P. Berger, *J. Neural Transm.* **103,** 967 (1996).
71. L.G. Harsing, H. Sershen, and A. Lajtha, *J. Neural Transm.* **96,** 215 (1994).
72. K. Umezu and K.E. Moore, *J. Pharmacol. Exp. Ther.* **208,** 49 (1978).
73. I.M. Maisonneuve, K.E. Visker, G.L. Mann, U.K. Bandarage, M.E. Kuehne, and S.D. Glick, *Eur. J. Pharmacol.* **336,**123 (1997).
74. R. Kuczenski and D.S. Segal, *in* "Amphetamine and Its Analogs: Pharmacology, Toxicology, and Abuse" (A.K. Cho and D.S. Segal, eds.), p. 115. Academic Press, San Diego, CA, 1994.
75. "Serotonin in Major Psychiatric Disorders" (E.F. Coccaro and D.L. Murphy, eds.), American Psychiatric Press, Washington, DC, 1990.
76. I. Lucki, *Biol. Psychiatry* **44,** 151 (1998).
77. E.M. Sellers, G.A. Higgins, D.M. Tomkins, M.K. Romach, and T. Toneatto, *J. Clin. Psychiatry* **52(suppl.),** 49 (1991).
78. S.L. Walsh and K.A. Cunningham, *Psychopharmacology* **130,** 41 (1997).
79. D.S. Yu, F.L. Smith, D.G. Smith, and W.H. Lyness, *Life Sci.* **13,** 1383 (1986).
80. M.E. Carroll, S.T. Lac, M. Asencio, and R. Kragh, *Pharmacol. Biochem. Behav.* **35,** 237 (1990).
81. Y. Wang, N. Joharchi, P.J. Fletcher, E.M. Sellers, and G.A. Higgins, *Psychopharmacology* **120,** 134 (1995).
82. P. Munzar, M.H. Baumann, M. Shoaib, and S.R. Goldberg, *Psychopharmacology* **141,** 287 (1999).
83. L.H. Parsons, G.F. Koob, and F. Weiss, *J. Pharmacol. Exp. Ther.* **274,** 1182 (1995).
84. F. Weiss, L.H. Parsons, G. Schulteis, P. Hyytia, M.T. Lorans, F.E. Bloom, and G.F. Koob, *J. Neurosci.* **16,** 3474 (1996).
85. M.H. Baumann and R.B. Rothman, *Biol Psychiatry* **44,** 578 (1998).
86. R.S. Sloviter, E.G. Drust, B.P. Damiano, and J.D. Connor, *J. Pharmacol. Exp. Ther.* **214,** 231 (1980).
87. B.L. Jacobs, *Life Sci.* **19,** 777 (1976).
88. P.M. Whitaker and P. Seeman, *Proc. Natl. Acad. Sci. USA* **75,** 5783 (1978).
89. D.B. Repke, D.R. Artis, J.T. Nelson, and E.H. Wong, *J. Org. Chem.* **59,** 2164 (1994).
90. S. Helsley, D. Fiorella, R.A. Rabin, and J.C. Winter, *Pharmacol. Biochem. Behav.* **59,** 419

(1998).
91. S. Helsley, R.A. Rabin, and J. C. Winter, *Prog. Neuro-Psychopharmacol. Biol. Psychiat.* **23,** 317 (1999).
92. R.W. Fuller, H.D. Snoddy, and D.W. Robertson, *Pharmacol. Biochem. Behav.* **30,** 715 (1988).
93. D. Wei, I.M. Maisonneuve, M.E. Kuehne, and S.D. Glick, *Brain Res.* **800,** 260 (1998).
94. C. Gundlah, K.F. Martin. D.J. Heal, and S.B. Auerbach, *J. Pharmacol. Exp. Ther.* **283,** 581 (1997).
95. G.W. Pasternak, *Clin. Pharmacol.* **16,** 1 (1993).
96. H. Akil, C. Owens, H. Gutstein, L. Taylor, E. Curren, and S. Watson, *Drug Alcohol Depend.* **51,**127 (1998).
97. A. Mansour, H. Khatchaturian, M.E. Lewis, H. Akil, and S.J. Watson, *Trends Neurosci.* **11,** 308 (1988)
98. A. Mansour, C.A. Fox, H. Akil, and S.J. Watson, *Trends Neurosci.* **18,** 22 (1995).
99. T.S. Shippenberg, A. Herz, R. Spanagel, R. Bals-Kubik, and C. Stein, *Ann. N.Y. Acad. Sci.* **654,** 347 (1992).
100. T.S. Shippenberg and G.I. Elmer, *Crit. Rev. Neurobiol.* **12,** 267 (1998).
101. G. Di Chiara and A. Imperato, *J. Pharmacol. Exp. Ther.* **244,** 1067 (1988).
102. D.P. Devine, P. Leone, D. Pocock, and R.A. Wise, *J. Pharmacol. Exp. Ther.* **266,** 1236 (1993).
103. H.S. Lotsof, *U.S. Patent* **4,499,096**; *Chem. Abstr.* **102,** 160426w (1985).
104. D.C. Mash, C.A. Covera, J.P. Pablo, R.F. Tyndale, F.D. Ervin, I.C. Williams, E.G. Singleton, and M. Mayor, *Ann. N.Y. Acad. Sci.* **914,** 394 (2000).
105. E.E. Codd, *Life Sci.* **57,** PL315 (1995).
106. M.J. Kreek, *Ann. N.Y. Acad. Sci.* **909,** 186 (2000).
107. J.P. Pablo and D.C. Mash, *Soc. Neurosci. Abstr.* **26,** 1593 (2000).
108. R.A. Rabin and J.C. Winter, *Eur. J. Pharmacol.* **316,** 343 (1996).
109. S.D. Glick, I.M. Maisonneuve and S.M. Pearl, *Brain Res.* **749,** 340 (1997).
110. P.V. Piazza and M. Le Moal, *Ann. Rev. Pharmacol. Toxicol.* **36,** 359 (1996).
111. P.V. Piazza and M. Le Moal, *Trends Pharmacol. Sci.* **19,** 67 (1998).
112. M.A. Bozarth, A. Murray, and R.A. Wise, *Pharmacol. Biochem. Behav.* **33,** 903 (1989).
113. N.E. Goeders and G.F. Guerin, *Psychopharmacology* **114,** 63 (1994).
114. M. Haney, R. Maccari, M. Le Moal, H. Simon, and P.V. Piazza, *Brain Res.* **698,** 46 (1995).
115. Y. Shaham and J. Stewart, *Psychopharmacology* **119,** 334 (1995).
116. S. Erb, Y. Shaham, and J. Stewart, *Psychopharmacology* **128,** 408 (1996).
117. P.V. Piazza, M. Maccari, J.M. Deminiere, M. Le Moal, P. Mormede and H. Simon, *Proc. Natl. Acad. Sci. USA* **88,** 2088 (1991).
118. P.V. Piazza and M. Le Moal, *Brain Res. Rev.* **25,** 359 (1997).
119. N.E. Goeders and G.F. Guerin, *Neuroendocrinology* **64,** 337 (1996).
120. V. Deroche, M. Marinelli, M. Le Moal, and P.V. Piazza, *J. Pharmacol. Exp. Ther.* **281,** 1401 (1997).
121. P.V. Piazza, V. Deroche, J.M. Deminiere, S. Maccari, M. Le Moal and H. Simon, *Proc. Natl. Acad. Sci. USA* **90,** 11738 (1993).
122. V. Deroche, P.V. Pizza, J.M. Deminiere, M. Le Moal, and H. Simon, *Brain Res.* **622,** 315 (1993).
123. M.E. Freeman, B. Kanyicska, A. Lerant, G. Nagy, *Physiol. Rev.* **80,** 1523 (2000).
124. L.D. Van de Kar and M.L. Blair, *Front. Neuroendocrinol.* **20,** 1 (1999).
125. M.H. Baumann, M.A. Ayestas, and R.B. Rothman, *J. Neurosci.* **18,** 9069 (1998).
126. R.N. Pechnick, R. George, and R.E. Poland, *Eur. J. Pharmacol.* **164,** 257 (1989).
127. R.N. Pechnick, R. George, R.E. Poland, M. Hiramatsu, and A.K. Cho, *LJ. Pharmacol. Exp. Ther.* **250,** 534 (1989).
128. N. Ben-Jonathan, *Endocrine Rev.* **6,** 564 (1985).
129. A. Pfeiffer and A. Herz, *Horm. Metabol. Res.* **16,** 386 (1984).
130. P.L. Wood and S. Iyengar, *in* "The Opiate Receptors" (G. Pasternak, ed.), p. 307. Humana Press, Clifton, NJ, 1988.

131. C. Naranjo, *Clin. Toxicol.* **2,** 209 (1969).
132. C. Naranjo, *in* "The Healing Journey. New Approaches to Consciousness," Pantheon, New York, NY, 1973.
133. W. Koek, J.H. Woods, and G.D. Winger, *J. Pharmacol. Exp. Ther.* **245,** 969 (1988).
134. M. Hiramatsu, A.K. Cho, and T. Nabeshima, *Eur. J. Pharmacol.* **166,** 359 (1989).
135. E. O'Hearn, D.B. Long, and M.E. Molliver, *NeuroReport* **4,** 299 (1993).
136. E. O'Hearn and M.E. Molliver, *Neurosci.* **55,** 303 (1993).
137. E. O' Hearn and M.E. Molliver, *J. Neurosci.* **17,** 8828 (1997).
138. J.P. O'Callaghan, T.S. Rogers, L.E. Rodman, and J.G. Page, *Ann. N.Y. Acad. Sci.* **801,** 205 (1996).
139. A.C. Scallet, X. Ye, R. Roundtree, P. Nony, and S.F. Ali, *Ann N.Y. Acad. Sci.* **801,** 217 (1996).
140. J. Sanchez-Ramos and D.C. Mash, *Bull. M.A.P.S.* **4,** 11 (1994).
141. M. Molinari, I.M. Maisonneuve, and S.D. Glick, *Brain Res.* **737,** 255 (1996).
142. S.Helsley, C.A. Dlugos, R.J. Pentney, R.A. Rabin, and J.C. Winter, *Brain Res.* **759,** 306 (1997).
143. Z. Xu, L.W. Chang, W. Slikker Jr., S.F. Ali, R.L. Roundtree, and A.C. Scallet. *Toxicol. Sci.* **57,** 95 (2000).
144. B.J. Vilner, U.K. Bandarage, M.E. Kuehne, C.M. Bertha, and W.D. Bowen, *NIDA Res. Monogr.* **178,** 235 (1998).
145. B.J. Vilner and W.D. Bowen, *J. Pharmacol. Exp. Ther.* **292,** 900 (2000).
146. N.S. Miller and J.C. Guttman, *J. Psychoactive Drugs* **29,** 249 (1997).
147. T.R. Kosten, A. Markou, and G.F. Koob, *J. Nerv. Ment. Dis.* **186,** 737 (1998).
148. R. Ciccocioppo, *Alcohol Alcohol.* **34,** 244 (1999).
149. R.B. Raffa, *Am. J. Med.* **101,** 40S (1996).
150. L. Radbruch, S. Grond, and K.A. Lehmann, *Drug Saf.* **15,** 8 (1996).
151. S.D. Glick, I.M. Maisonneuve, J. Raucci, and S. Archer, *Brain Res.* **681,** 147 (1995).
152. M.J. Kreek, *J. Addict. Dis.* **15,** 73 (1996).

—CHAPTER 6—

CHARACTERIZATION OF MULTIPLE SITES OF ACTION OF IBOGAINE

HENRY SERSHEN, AUDREY HASHIM,
AND ABEL LAJTHA

Nathan Kline Institute
Orangeburg, New York 10962

I. Introduction

Ibogaine, the principal alkaloid of *Tabernathe iboga*, has been studied for the past 100 years. Early in the past century (1900) it was isolated in crystalline form and was later marketed in Europe as the mild stimulant "Lambarène" (8 mg tablet). There was renewed interest in indole alkaloids with the discovery of reserpine, and the structural similarity of ibogaine to serotonin was the basis for Dhahir's 1970s thesis studies (*1*). However, the hallucinogenic properties of ibogaine had moved the FDA to ban its use in the United States. The renewed interest in ibogaine in the past 10 years is related to its putative antiaddictive properties. Several review articles have been published recently describing the historical and pharmacological perspective of ibogaine (*2-4*).

THE ALKALOIDS, Vol.56
0099-9598/01 $35.00

Unfortunately, the reports of ibogaine's antiaddictive effects have been termed "anecdotal" for the past 10 to 15 years, and although there have been over 150 publications related to its purported effects and action, clinical trials have not been forthcoming. There have been concerns related to its hallucinogenic effects and possible cerebellar toxicity (*5-9*). The First International Conference on Ibogaine brought together the addict, the researcher, and grant-funding source, in the hope of reviewing the current findings and status of ibogaine in the treatment of substance abuse.

Our own studies focused on the behavioral and biochemical effects of ibogaine related to cocaine administration and pharmacological responses. Our results suggest that ibogaine can act at multiple sites and that attempts to focus on one site as the primary site of action can be misleading. Interaction at several sites is more than likely to be important for its antiaddictive properties. In addition to being an overview of these studies, this chapter attempts to demonstrate that to understand the action of ibogaine one must also consider the multifaceted pharmacology of the drugs of abuse themselves. Most recent conceptual views accept that drugs of abuse involve multiple neural mechanisms. Any given behavior is likely to be influenced by a number of neurotransmitter systems, and transmitter systems do not work independently, but rather interact with one another by stimulating, inhibiting, or modulating each other. Various brain structures and components, receptors, and neurotransmitters are involved. Their participation in the reward mechanism is not the same for all drugs of abuse. Genetic risk factors in drug abuse have also been identified. The action of ibogaine could be an important paradigm for further characterizing the action of drugs of abuse.

It is also important to recognize that there are multiple and complex behavioral responses associated with acute and chronic drug administration, and that there are different behaviors associated with drug initiation, maintenance, withdrawal, and extinction. Each of these responses is probably mediated by a different neural mechanism and varies with different drugs, and therefore it is not surprising that a number of varied receptor type agonists and antagonists appear to have some remediation of a particular drug response. A therapeutic approach that targets more than one system is possibly more efficacious, if addiction is a multifactorial disease. This chapter will describe findings that indicate support for the use of ibogaine, its metabolite, and/or ibogaine-related compounds in the treatment of addiction, based on their ability to target relevant multiple neurotransmitter sites appropriate for the drug of abuse examined. Because of the multiple components of reward systems, a "dirty" drug like ibogaine that affects multiple neurotransmitter systems should not be excluded from consideration. Indeed, it is a likely positive attribute.

II. Issues Related to Ibogaine in the Treatment of Drug Dependence

Although the results have been discrepant at times, in the majority of studies, ibogaine has been proposed to have antiaddictive properties, modifying behavioral effects of various drugs and their self-administration in rodent models. Based on radioligand binding and other *in vivo/in vitro* studies, and several behavioral assays, to characterize its effects, ibogaine has been reported to have affinities to at least the dopamine and serotonin transporters, and to the glutamatergic (NMDA), sigma, kappa- and mu-opioid, and nicotinic acetylcholine receptors (see the references listed later in Table I). This raises the question of whether the action of ibogaine at a single site relates to its antiaddictive properties, or whether multiple sites are implicated in its action. Alternatively, ibogaine's affinity to ligand binding sites may not necessarily indicate the functionally relevant site.

A. Dopamine as a Primary Site of Drug-Mediated Responses

Despite the pronounced involvement of dopamine in stimulant drug-mediated behavioral effects, it is important to recognize that many of the addictive drugs have affinity to multiple neurotransmitter sites; for example, cocaine is not a selective dopamine reuptake inhibitor. Cocaine also binds and inhibits the uptake of serotonin and norepinephrine, with equal potency. "Knockout" models of rodents missing dopamine reuptake transport still self-administer cocaine (*10,11*). We should also recognize that the neurobiology associated with addictive behaviors (cognition, reward, withdrawal, craving, sensitization) involve multiple neurotransmitter systems and their interactions. For example, serotonin transmission and the subsequent activation of serotonin receptor(s) (numbering 14 serotonin receptor subtypes) have a strong modulatory role, either stimulatory or inhibitory, in dopaminergic neurotransmission. Although nicotine and cocaine both increase dopamine, their actions are not similar, and we recently reported that selective neurotransmitter antagonists can block response to one and not the other (*12*).

It needs to be considered that although the prevailing theory is that elevated extracellular dopamine is the primary mediator of cocaine's reinforcing effects, this has been challenged by the finding that in mice lacking the dopamine transporter who still self-administer cocaine (*10,11*), cocaine has no effect on dopamine levels, further supporting the involvement of other neurotransmitter systems in drug behavior. Serotonin, acting through many receptors can modulate the activity of neural reward pathways and thus the effects of various drugs of abuse. Mice lacking one of the serotonin-receptor subtypes, the 5-HT_{1B} receptor, display increased locomotor responses to cocaine, and they are more motivated to self-administer cocaine (*13*). In mice in which the beta-2 subunit of the nicotinic receptor is lacking, the normal increase in dopamine after nicotine injection is not seen, and nicotine fails to be self-administered, but cocaine is self-administered (*14*), showing differences between nicotine and cocaine reward mechanisms.

B. Ibogaine and Its Metabolite and Acute Versus Long-Term Effect

There are other issues to consider; for example, what is the importance of ibogaine's acute versus long-lasting effects on transmitter functioning? Why and how does ibogaine produce its long-lasting effect? Is it just slow release of a metabolite from lipid stores or long-term block/conformational change in some receptor? Understanding apparent gender and genetic differences in behavioral responses to and metabolism of drugs and ibogaine is also of importance. The issue of increased sensitivity of females to ibogaine has been raised. Female rodents have a higher brain level of ibogaine after administration (*15*), and female mice show increased locomotor responses to cocaine (*16*). Gender differences were also observed in kappa-opioid and NMDA-mediated dopamine release (*16*) and in human reactions to nicotine and cocaine. In humans, genetic differences in nicotine metabolism have been observed (*17*).

The data and discussions presented emphasize the importance of investigating the interaction of multiple neurotransmitter systems and multiple neuronal pathways in the mediation of drug-induced behaviors, with differences among the various drugs of abuse justifying the use of drugs that target multiple sites in protocols for drug-dependence treatment. The difficulties in devising appropriate therapy are compounded by genetic and sex variations in drug responsiveness.

C. Single or multiple Sites of Action of Ibogaine

Ibogaine has been suggested to inhibit the physiological and psychological effects of a number of drugs of abuse: heroin, morphine, amphetamine, cocaine, alcohol, and nicotine. This suggests a common site(s) of action of the drugs of abuse and that of ibogaine, or that ibogaine acts at some common pathway(s), secondary to the initial site of drug action, that affects some common behavior associated with addictive drugs. Alternatively, ibogaine may act at multiple sites one or more of which may "coincidentally" involve a common site of action of several addictive drugs. As studies move away from the simplistic approach based on the notion that a drug acts at only one specific site and that drug behaviors involve individual neural systems, to one that explores more complex multiple interactive neural systems, we will be able to better understand the action of ibogaine and that of drugs of abuse.

III. Effect of Ibogaine on Drug-Induced Behavior

Initial studies of the effects of ibogaine on drug self-administration behavior in animals were received with some skepticism, as were the varied case reports on human experiences. Early NIDA-funded projects did not find any effects of ibogaine in rodent models, or the effects consisted of nonspecific inhibition of overall activity, for example, inhibition of food consumption at high doses.

Dworkin *et al.* (*18*) found suppression of responding to cocaine or heroin at 60 minutes after treatment with high doses of ibogaine, but responding to food was also suppressed, suggesting nonspecific effects. No long-term effects were seen, except at the 80 mg/kg dose with 60-minute pretreatment, where cocaine self-administration was suppressed at 48 hours. The literature is also mixed on ibogaine reduction of naloxone-precipitated morphine withdrawal; in some cases it blocked expression of withdrawal, or it had no effect (*19-22*). Locomotor activity is reportedly either inhibited or enhanced after stimulant drugs such as cocaine and amphetamine (*23-25*).

Clearly, the initial behavioral responses to ibogaine (high dose) were disruptive to overall behavior and could not be clearly interpreted, but some long-term effects have been suggestive of antiaddictive properties (*24-29*). It is not known why there was such variability in results. Species and sex differences, and treatment protocols have been suggested. Possibly the batches (pure or semisynthetic extract) of ibogaine were somewhat different. However, the potency of samples of ibogaine obtained from Sigma Chemical Company or NIDA appears to have been similar (*30*), which would suggest that there are no significant differences between batches of ibogaine.

IV. Binding Site Activity

There have been a number of studies reporting on the "affinity" of ibogaine and some analogs to known receptor systems utilizing a radiolabeled ligand that has specificity for a binding site of a particular receptor site. These affinities have been reviewed elsewhere (*2,4,31*). Additionally, *in vitro* assays to measure functional changes, for example, transmitter release or channel blockade, have been used to assess the site of action of ibogaine (*3,25*). The most recent addition is the report that ibogaine has affinity to the nicotinic-acetylcholine receptor (*32,33*). These results are summarized in Table I.

Clearly, the diversity of potential interactions of ibogaine can be inferred from these binding site affinities. However, a question to be asked is how does the binding site affinity of ibogaine relate to its pharmacological action? Although ibogaine has affinity to the kappa-opioid receptor, it was concluded that it does not produce such an action by interacting directly with multiple opioid receptors. Ibogaine injected 10 minutes before the opioid drugs did not modify the antinociceptive actions of morphine, kappa-opioid agonist U-50,488H, or delta-opioid agonist DPDPE. However, the metabolite of ibogaine enhances the antinociception of morphine, but not of U-50,488H or DPDPE. Thus, it was concluded that that there is an interaction of ibogaine with the mu-opioid receptor following its metabolism to noribogaine (*34*).

Brain levels of ibogaine or its metabolites have been estimated to be in the micromolar range, sufficiently high to affect those systems showing affinities in the low micromolar range. However, ibogaine is metabolized very rapidly, raising the question of a long-lasting metabolite (that would also have to be at a sufficiently high level to affect some receptors) (*35*). Since the half-life of ibogaine is relatively short, how this would relate to its long-term effects is not

TABLE I.
REPORTED SITES OF ACTIVITY OF IBOGAINE FROM BINDING AND FUNCTIONAL ASSAYS: MULTIPLE NEUROTRANSMITTER SITES

Receptor Systems	Binding Aassays	Functional Assays
DA (receptors/transporter)	(*36,39,40*)	(*25,41-45*)
5-HT (receptors subtypes/transporter)	(*36,39,40,46*)	(*42,45-52*)
NMDA	(*39,40,53-58*)	(*16,53,58-61*)
Kappa-opioid	(*39,40*)	(*16,51,60,62*)
Mu-opioid	(*39,63*)	(*34,64,65*)
Sigma (1 and 2)	(*54,66,67*)	(*68,69*)
Na^+ Channel	(*40*)	
Muscarinic	(*40*)	
Nic-ACh		(*32,33,70,71*)
Adrenergic	(*40*)	(*72*)
Purinergic		(*73*)
Neuropeptides		(*74-76*)
Early genes		(*77*)

Receptor/neurotransmitter system sites showing binding affinities (in the range of levels reached by ibogaine) for ibogaine or metabolite(s) and suggested sites of action based on functional assays, for example, *in vitro/in vivo* transmitter release, isolated tissue contractions, discriminative stimulus, and anticonvulsant efficacies (*indicated references*).

clear. Possibly a long-lasting metabolite, for example, noribogaine (10-hydroxy-ibogamine), is present, or its slow release from lipid depots may play a role (*4,36,37*).

Are there long-lasting changes in any of these receptor systems or second messenger systems to account for its long-lasting effects? Such studies have not been conducted. Again, ibogaine itself has several pharmacological effects, for example, its stimulatory or hallucinogenic effects, in addition to, or part of, its antiaddictive properties, that each may involve single or multiple interactions at several neurotransmitter sites. Alternatively, ibogaine or its metabolite may act to alter the receptor, similarly to metaphit, a proposed phencyclidine receptor acylator (*38*). It is still unclear how one or two doses of ibogaine can produce such long-lasting effects.

Even the depot release of a metabolite(s) is difficult to accept as having profound and long-lasting effects (effects reported to last for months in humans). Most rodent studies have not been conducted beyond a duration of one week. One could also speculate that the long-lasting effect(s) of ibogaine "restores" neurotransmitter interactions back to some pre-drug, pre-craving, or pre-withdrawal level, resulting from its diversified effects on multiple neurotransmitter systems, somewhat similar to the diversified effects of electro-convulsive therapy (ECT) in the treatment of depression unresponsive to standard antidepressant therapy. Although unknown, the mechanism of action is thought to result from distinct combinations of neuropeptide and neurotransmitter changes and changes in gene expression in selected neuronal populations (*78-81*). For example, a single electroconvulsive shock (ECS) pretreatment suppresses the inhibition of dopamine release mediated by kappa-opioid receptors, suggesting that a single ECS treatment modifies the sensitivity of the kappa-opioid receptors located on the presynaptic dopamine terminals in the rat striatum (*82*). The simultaneous action of ibogaine at multiple sites induces a major resetting of transmitter interactions, and there is no need for it to be present long term. Effects of ibogaine on changes in second messenger systems and gene expression need to be examined as mechanisms of its long-lasting effects.

A. Relevant Site of Action

There have been a number of studies attempting to determine which neurotransmitter system is *most* affected by ibogaine or a metabolite that relates to its antiaddictive property. The dopamine transporter is a target for cocaine; we reported affinity of ibogaine for the transporter in the low micromolar range. This affinity, however, is ten times higher (weaker) than that of cocaine (*29*). The studies of Popik *et al.* (*57,58*) indicated that the NMDA receptor plays a major role, whereas Glick's (*25*) laboratory suggest a strong involvement of both the kappa-opioid and NMDA receptor (*60*). Helsley *et al.* (*69*) reported some interaction with the sigma-2 and opiate receptors, while the NMDA antagonist activities do not play a major role in the ibogaine discriminative stimulus. Their later studies also suggest multiple interactions and a role of the 5-HT_{2c} receptor in ibogaine discriminative stimulus (*47*). The antagonist action of ibogaine at the nicotinic receptor may be involved in reducing nicotine preference and action at the serotonin transporter affecting alcohol consumption (*25*). Mah *et al.* (*71*) suggested that ibogaine at an initially high concentration acts at multiple sites and then, after metabolism to lower levels, has a selective action at the nicotinic acetylcholine receptor to inhibit catecholamine release. We also reported that ibogaine can block cocaine-mediated effects on serotonin transmission and block the kappa-opioid inhibitory effect on dopamine and serotonin release (*62*). Mash *et al.* (*36,83*) have suggested involvement of the serotonin transporter and NMDA receptor site in the action of ibogaine and its metabolite (noribogaine). Noribogaine has an affinity to the serotonin transporter 50-fold more potent than to the dopamine transporter (*36*). However, studies with the ibogaine congener, 18-methoxycoronaridine (18-MC), suggested that the serotonin system might not be essential for 18-MC antiaddictive action, although the serotonin system may be involved in the action of ibogaine and its metabolite (*52*). The NMDA receptor and D1 dopamine receptor are suggested to be involved in the release of neurotensin by ibogaine, and that neurotensin may contribute to the interaction of ibogaine and the dopamine system (*75*).

Clearly, complex interactions occur, each probably related to some different aspect of drug-induced behavior. Whether the dopamine system is the final common denominator—that is, can ibogaine act at some site(s), the final action of which is to reduce drug-induced changes in dopamine release without affecting overall dopaminergic responses?—is far from understood.

V. Functional Activity

Binding to a specific site suggests sites of action, but does not indicate functional activity. The functional effects of ibogaine were studied in our laboratory by utilizing an *in vitro* perfusion technique that enabled us to study mechanisms of regulation and modulation of dopamine transmitter release processes. The results are summarized in Table II.

At the nerve terminal level, ibogaine added *in vitro* released dopamine from the cytoplasmic pool (*43*). This release was not subject to presynaptic autoreceptor regulation (dopamine D2 antagonist sulpiride-stimulated dopamine release is not affected) (*50,43*). Cocaine as a reuptake blocker increases the level of dopamine; this response was not affected by ibogaine. However, the cocaine-induced increase in serotonin level (reuptake blockade?) was blocked by ibogaine. The NMDA-mediated dopamine release was partially inhibited by ibogaine (*61*). The kappa-opioid agonist-induced inhibitions of dopamine and serotonin release were both blocked by ibogaine pretreatment (*51*). The sigma agonist-stimulated dopamine release was inhibited 50% by ibogaine (*61*). A strong serotonergic component of ibogaine's effects was also reported, involving both the reuptake transporter and 5-HT_{1b} receptor, increasing the exchange of dopamine for serotonin via the dopamine transporter and inhibition of serotonin 5-HT_{1b} agonist-mediated inhibition of dopamine release (*3,51*). The studies of Mah *et al.* (*71*) showed that ibogaine also blocks the nicotinic receptor-mediated stimulation (acetylcholine) of catecholamine (norepinephrine) release in bovine chromaffin cells. This is also supported by microdialysis studies showing attenuation of nicotine-induced dopamine release (*84,85*). Glick's (*25*) *in vivo* studies also show stimulation with amphetamine or cocaine and block with nicotine or morphine of dopamine release by ibogaine (Table II, bottom). Utilizing other methods, Broderick *et al.* (*26*) and French *et al.* (*90*) suggested either a decrease or no effect of ibogaine on cocaine-mediated dopamine increase. Our results show a number of interactions of ibogaine with various neurotransmitter systems that can regulate dopamine release. It is interesting that although many of the studies were conducted with the addition of ibogaine to an *in vitro* preparation, most showed the same effect when animals were treated *in vivo* with ibogaine and tissue responses tested later *in vitro*. Since the tissue preparation is extensively washed in the latter experiments, it is unlikely that ibogaine or its metabolite is present during the release portion of the study. This could be suggestive of some receptor conformational change that is long lasting, beyond the period of exposure to ibogaine or the "resetting" ability.

From studies over the past 10 years, it is clear that ibogaine can act at different neural sites (via neurotransmitters and ion channels), which can modulate terminal dopamine release (Table II and Figure 1). Figure 1 is a model of a dopamine terminal, which is offered to diagrammatically represent these multiple interactions on presynaptic terminal dopamine responses. Receptor-induced stimulation (+) or inhibition (–) of dopamine release is shown. There are also interactions/modulation between different receptors; for example, the kappa-opioid receptor is inhibitory on the NMDA and acetylcholine receptors, inhibitory and excitatory on the serotonin system, and can inhibit calcium channels. Ibogaine effects on these receptor responses are indicated. In most cases, ibogaine inhibits (–) these receptor-mediated excitatory or inhibitory responses. The resultant effects of ibogaine on drugs that increase extracellular dopamine level are indicated on the right. The responses are either further stimulated by ibogaine (cocaine and amphetamine) or inhibited by ibogaine (nicotine and morphine). The literature is mixed on the effect of ibogaine on cocaine-mediated increase in dopamine.

VI. Stimulant Drug Actions/Behaviors

Psychostimulants act predominantly to elevate brain dopamine, either by their ability to release dopamine, as is the case for amphetamine, or by blockade of the

TABLE II.
SUMMARY OF EFFECTS OF IBOGAINE ON [^{3}H]DOPAMINE AND [^{3}H]SEROTONIN RELEASE: *In Vitro* PERFUSION AND BRAIN MICRODIALYSIS STUDIES ON REGULATION OF TRANSMITTER RELEASE

	[^{3}H]Dopamine		[^{3}H]Serotonin	
	Control	Ibogaine	Control	Ibogaine
In Vitro Studies				
DA autoreceptor (electrical-evoked) (Sulpiride)	↑ Increase	No effect		
DA transporter (electrical- evoked) (Cocaine)	↑ Increase	No effect	↑ Increase	Block
NMDA receptor (basal) (NMDA)	↑ Increase	Partial inhibition		
Kappa-opioid receptor (electrical-evoked) (U-62066)	↓ Inhibit	Block	↓ Inhibit	Block
Sigma receptor (electrical-evoked) (pentazocine)	↑ Increase	Inhibit 50%		
Serotonin transporter (basal) (serotonin)	↑ Increase	↑↑ Increase		
Serotonin 5-HT$_{1b}$ receptor (basal) (CGS-12066A)	↑ Increase	Block		
Nicotinic receptor (acetylcholine)	↑ Increase (Norepinephrine release in chromaffin cells)	Block		
In Vivo Studies				
Cocaine	↑ Increase	↑↑ Increase ↓ Inhibit, no effect		
Amphetamine	↑ Increase	↑↑ Increase		
Morphine	↑ Increase	Block		
Nicotine	↑ Increase	Block		

Summary of *in vitro* studies on the effect of ibogaine on electrical stimulation or drug-induced release of dopamine and serotonin release in the presence of selective neurotransmitter system agonists/antagonists (*3,16,43,50,51,61,62*) and *in vivo* brain microdialysis studies examining drug-induced changes in dopamine level (*25-28,37,85-90*).

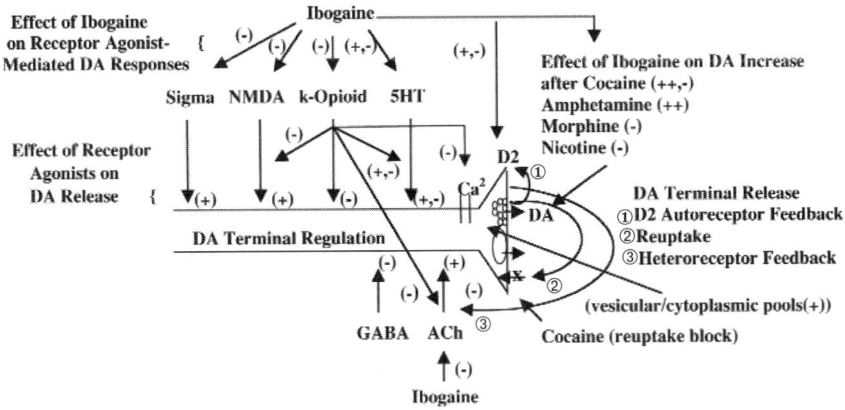

FIGURE 1. REPORTED MULTIPLE SITES OF ACTION OF IBOGAINE ON DOPAMINERGIC FUNCTION. The effect of ibogaine on stimulatory and inhibitory modulation of dopamine terminal release by sigma, NMDA, kappa-opioid, 5-HT, and ACh receptor agonists, and the effect of ibogaine on the increase in dopamine level after stimulant drug administration. The above figure represents a dopamine terminal showing inhibitory (-) and stimulatory (+) interactions of multiple neurotransmitter systems that modulate dopamine release (from Table II). Dopamine release is under excitatory (+) modulation by agonists to the sigma (*61*), NMDA (*61*), ACh (*32*), and 5-HT (*3,51*) receptors or inhibitory (-) modulation by kappa-opioid agonists (*51*), and also 5-HT (*51*). There are interactions/modulation between receptors; the kappa-opioid system interacts with the NMDA, 5-HT, and ACh receptors, and calcium channels (*51,16*). The effect of ibogaine on these receptor system interactions is shown (predominantly inhibition or blockade, except some stimulation of 5-HT function) (see Table II). Also indicated are the stimulatory and inhibitory effects of ibogaine on dopamine release (*43*) and extracellular level after stimulant drugs (right side) (*25*). Terminal DA release is subject to inhibitory ① auto- and ③ heteroreceptor feedback control and ② reuptake. Updated from Sershen *et al.* (*3*).

reuptake transporter, as in the case of cocaine. The elevation of dopamine resulting from release or reuptake inhibition is thought to be the basis of the rewarding effects of stimulant drugs. However, some direct reuptake blockers are not self-administered, for example, mazindol, suggesting that either other sites of action are also involved, or that there are different sites on the dopamine transporter which, depending on the conformational sites that are occupied, might determine the potential for self-administration. Stimulant drugs can also act at other neurotransmitter systems. As mentioned earlier, dopamine transporter knockout studies raise questions as to whether the dopamine transporter is solely responsible for self-administration. However, caution should be taken with interpretation of these studies because they do not take into consideration compensatory changes that occur during development in the knockout animal.

The final common pathway may be dopamine, but most likely other pathways are also involved in different drug-induced behaviors. The pathways indicated in

Figure 1 are all affected in some way by ibogaine and are briefly discussed in this chapter. Another important area for future studies will be changes in gene expression as involved in the long-term effects of drugs and ibogaine action.

In addition to targeting the dopamine transporter directly, a number of studies have attempted to target neurotransmitter sites that can modulate the dopaminergic response, in an effort to attenuate the stimulant drug-induced increase in dopamine. Glutamate antagonists (MK-801) can antagonize cocaine stimulant responses (*91*). The inhibitory neurotransmitter GABA, elevated by administering gamma-vinyl-GABA, can also attenuate effects of cocaine in increasing extracellular dopamine (*92,93*).

In addition to stimulating the dopamine reward system, stimulant drugs produce other behaviors. Sensitization-tolerance are behavioral responses generally observed with repeated stimulant administration, either an enhanced response to subsequent exposure as in the case of sensitization, or less of a response, requiring more drug to produce a similar behavioral response as in the case of tolerance. The dopamine receptor exists as several subtypes; some of them, the D1 and D4 dopamine receptors, have been implicated in sensitization, either its initiation or maintenance. Other neurotransmitter systems can alter this process, for example, the serotonin, NMDA, and kappa-opioid receptors (*94*). With craving/reinforcing effects of drugs, the dopamine, serotonin, glutamate, opioid, GABA, and cAMP systems have all been implicated. Drug withdrawal symptoms have been associated with a transmitter depletion response after removal of a drug. Implicated in this behavior are the dopamine and serotonin systems, excitatory amino acids (NMDA), and interactions with nitric oxide (NO), and cGMP.

Behavioral studies involving diverse drugs of abuse suggest that ibogaine may affect multiple neurotransmitter systems which are involved in the modulation of dopaminergic responses to stimulants:

Opioid Withdrawal: Noribogaine has been shown to have a lower affinity than, but an increased intrinsic activity over, buprenorphine as a mu-agonist. In addition, it was reported that noribogaine has weak intrinsic activity (partial agonist) or antagonist actions at kappa-opioid receptors; together suggesting that the ability of ibogaine to inhibit opiate withdrawal symptoms may be explained by a mixed mu- and kappa-opioid receptor profile and an affinity for the serotonin transporter of the active metabolite noribogaine (*65*). Pablo and Mash (*65*) also suggested that the capacity for noribogaine to reset multiple opioid receptors and the serotonin transporter mechanism may explain the reportedly easy transition after only a single dose of ibogaine following the abrupt discontinuation of opiates.

Drug Discrimination: Drug discrimination studies with ibogaine did not show substitution with mu- or kappa-opioid receptor agonists, although sigma-2 receptors may be involved (*69*). At low doses of ibogaine, NMDA receptor antagonists did not show any substitution. For the metabolite noribogaine, NMDA antagonists did not show substitution in discriminative effects (*95*).

Cocaine and Morphine Self-Administration: Ibogaine effects on both the kappa-opioid and NMDA receptor have been shown to be involved in its effects on cocaine self-administration (*25*).

Alcohol Consumption: Rezvani *et al.* (*96,97*) reported that ibogaine reduces alcohol consumption, although mechanisms involved were not determined. It was found that the novel, nontoxic ibogaine analog 18-methoxycoronaridine also reduces alcohol consumption (*96*). Although, Glick and Maisonneuve suggested that the serotonergic effects of ibogaine might mediate some of the shorter-lasting effects of ibogaine, for example, effects on alcohol intake (*25*), they also report that 18-methoxycoronaridine had no effect on the serotonin transporter (*52*). The opioid antagonist naltrexone and serotonin uptake inhibitor fluoxetine have been used for treatment of alcohol abuse. Rezvani (*98*) has shown that combination therapy (naltrexone, fluoxetine, and a TRH analogue (TA091)) reduces ethanol intake in rats. Opioid antagonists in combination with isradipine (Ca^{2+} channel blocker) showed sustained effects in reducing cocaine and alcohol intake (*99*). The kappa-opioid receptor appears to mediate inhibition of dopamine release via a decrease in calcium conductance (*100*). The action of ibogaine at the kappa-opioid receptor may be mediated by this effect. Acamprosate for the treatment of alcohol abuse is thought to act at the NMDA receptor and to reduce calcium fluxes through voltage-dependent channels (*101*). It is also thought to inhibit GABA B receptors (*102*). Interestingly, ibogaine has been reported to act at all these sites.

These results suggest that stimulant drugs have multiple actions and behaviorial effects, and that targeting sites that can modulate dopamine responses is one approach to treatment development. Such sites may be involved directly in modulating the dopaminergic response or act via other neurotransmitters.

VII. Current Non-Ibogaine Drug Treatment Protocols

Further support for a multiple-site-target approach to drug treatment development can be inferred from recent treatment protocols tested against different behaviors associated with drug use. With cocaine abuse, a variety of approaches have been proposed, depending on the behavior being studied. The dopaminergic, serotonergic, GABAergic, opioid, and excitatory amino acid receptors have received the most attention. For example, treatment for cocaine addiction has focused on the dopamine transporter, developing drugs that can bind to the receptor without elevating synaptic dopamine. Dopamine knockout-mouse studies have suggested the importance of the serotonin system (*10,13*). Cocaine is also a serotonin and norepinephrine uptake blocker.

The development of effective pharmacotherapy of substance abuse and dependence considers specific drug-related behaviors, for example, medication for the withdrawal syndromes. Treatment must also consider craving, especially early during the withdrawal period. Effective anticraving medication has been limited. The opioid antagonists have been tested, since the opioid receptors are associated with the reward pathways. Methadone and other long-lasting opiates, such as buprenorphine or levo-alpha-acetylmethadol (LAAM), induce tolerance to the effects of opiates (*103*). Naltrexone is used to block the euphoria that occurs when opiates are administered (*104*). The euphoria component for drug behavior has also been targeted by the use of calcium channel blockers; verapamil reduces the subjective effects of morphine in humans (*105*). Attempts at maintenance therapy have used such drugs as amantadine, bromocriptine, and methylphenidate that act to release dopamine (*106*). The use of dopamine antagonists is based on the premise that stimulant drug euphoria appears to be mediated by a rapid increase in dopamine; blockade of specific dopamine receptors may change stimulant effects. Studies have suggested that dopamine receptor subtypes play a role in the reinforcing effect of cocaine. In general, the D1 and D2 antagonists can maintain cocaine responding, whereas D1 and not D2 agonists have been reported to block cocaine self-administration. However, chronic dopamine antagonist treatment may lead to receptor supersensitivity and enhanced responses to stimulants (*107*).

Dopamine hypofunction and depletion occurring during stimulant withdrawal have been the basis for dopamine agonist (or drugs that release, block reuptake, or inhibit dopamine metabolism) treatment. A recent review of preclinical trials by McCance (*108*) suggested that agonist-type treatments have low efficacy against stimulant dependence. Cocaine-type antagonists such as mazindol to block dopamine reuptake; carbamazepine, an anticonvulsant, to block kindling; and naltrexone, an opioid antagonist to block some of the opiate pathways involved in reinforcing effects of cocaine had no effect. However, fluoxetine to block serotonin reuptake had some effectiveness. A D1 antagonist (SCH22390) and an NMDA antagonist (dextrophan) have some effect in animal models.

Studies of antagonism of the different serotonin receptor subtypes have yielded mixed results. Many of the serotonin drugs are also thought to treat depression, anxiety, and obsessive-compulsive behaviors that may underlie cocaine abuse. A number of studies have examined the effects of altering serotonin levels, for example, with L-tryptophan (serotonin precursor) or specific serotonin reuptake inhibitors (SSRI, a class of antidepressants) such as sertraline. Serotonin reuptake inhibitors have been reported to decrease cocaine self-administration, but may also decrease food-maintained behavior. Continuous cocaine administration induces tolerance to its behavioral effects (*109,110*) and a functional down-regulation of accumbens 5-HT_3 receptors. Agonists at the 5-HT_{1b} receptor partly generalize to cocaine in drug-discrimination experiments (*111*) and enhance the reinforcing effects of cocaine (*112*). Mice lacking the 5-HT_{1b} receptor consume more ethanol than controls (*13*). Undoubtedly, one or more of the 5-HT receptor subtypes could appear as a key component in drug dependence.

Since there has been association of anxiety with cocaine use, GABAergic agents have been tested. Anticonvulsants have also shown some clinical or anecdotal effectiveness. The blockade of the NMDA glutamate receptors in the nucleus accumbens appears to reduce the reinforcing effects of cocaine. As reviewed recently (*113*), to date none of the medications have singly been accepted as efficacious for treating cocaine abuse. This may be because there are several different aspects to the problem of cocaine abuse, each potentially treatable by different medications (*113*).

Since it has been shown that the neural systems involved are complex in drug behaviors, it is surprising that strategies for drug treatments have not, until recently, targeted multiple sites.

VIII. Conclusions

Ibogaine has a history of at least 100 years from its discovery and isolation in the early 1900s. Its use in Africa for ritual ceremonies may well extend before this. Its use as a mild stimulant was not much noticed, but its reported psychedelic properties in the 1960s gave it renewed interest. Although banned by the FDA, ibogaine has had a curious attraction over the past 20 to 30 years, suggesting it may have antiaddictive properties. While concerns have been raised regarding potential neurotoxicity and hallucinogenic properties, such concerns must be weighed against the devastating morbidity associated with drug dependence. Case reports in humans and animal data indicating significant potential would appear to argue in favor of the further development of ibogaine, especially in view of the high cost of the disorder that it is intended to treat. The possibility of a novel treatment of drug addiction deserves attention, and studies have to go beyond the anecdotal.

The primary aims of our studies were to examine ibogaine in rodent models to see whether there is any validity to its use, and how it works, and also to enhance our understanding of mechanisms that are involved in drug dependence. That ibogaine works can be further suggested from the reported summary of results of a subset of patients treated in nonmedical settings for acute opioid withdrawal with ibogaine between 1962 and 1993; these case studies appeared to provide some evidence for the efficacy of ibogaine in acute opioid withdrawal (*114*). Maybe further studies with ibogaine would give suggestions for the development of other drug-treatment protocols.

Our current understanding of dopaminergic function and response suggests that there are many complex modulatory influences on dopamine release, and that many neural systems are involved in the different behaviors associated with drug dependence. These modulatory regulations can be both stimulatory and inhibitory. Certain drugs, for example, stimulants like amphetamine and cocaine, unlike opioids like morphine, may act at some of the same sites, but also at different sites. Clearly, drug abuse is a complex behavioral and neurobiological process that lends itself to complex treatment protocols. Maybe what we learn from the action of ibogaine will lead us in the direction of new treatment approaches.

References

1. H.I. Dhahir, *Diss. Abstr. Int.* **32/04-B**, 2311 (1971).
2. P. Popik, R.T. Layer, and P. Skolnick, *Pharmacol. Rev.* **47**, 235 (1995).
3. H. Sershen, A. Hashim, and A. Lajtha, *Brain Res. Bull.* **42**, 161 (1997).
4. P. Popik and P. Skolnick, *in* "The Alkaloids" (G.A. Cordell, ed.), Vol. 52, 197. Academic Press, San Diego, CA, 1999.
5. J.P. O'Callaghan, T.S. Rogers, L.E. Rodman, and J.G. Page, *Ann. N.Y. Acad. Sci.* **801**, 205 (1996).
6. E. O'Hearn, D.B. Long, and M.E. Molliver, *Neuroreport* **4**, 299 (1993).
7. E. O'Hearn and M.E. Molliver, *J. Neurosci.* **17**, 8828 (1997).
8. E. O'Hearn and M.E. Molliver, *Neuroscience* **55**, 303 (1993).
9. E. O'Hearn, P. Zhang, and M.E. Molliver, *Neuroreport* **6**, 1611 (1995).
10. B.A. Rocha, F. Fumagalli, R.R. Gainetdinov, S.R. Jones, R. Ator, B. Giros, G.W. Miller, and M.G. Caron, *Nat. Neurosci.* **1**, 132 (1998).
11. Y. Itzhak, J.L. Martin, S.F. Ali, and M.D. Norenberg, *Neuroreport* **8**, 3245 (1997).
12. I. Sziraki, H. Sershen, M. Benuck, A. Hashim, and A. Lajtha, *Ann. N.Y. Acad. Sci.* **877**, 800 (1999).
13. B.A. Rocha, K. Scearce-Levie, J.J. Lucas, N. Hiroi, N. Castanon, J.C. Crabbe, E.J. Nestler, and R. Hen, *Nature* **393**, 175 (1998).
14. M.R. Picciotto, M. Zoli, R. Rimondini, C. Lena, L.M. Marubio, E.M. Pich, K. Fuxe, and J.P. Changeux, *Nature* **391**, 173 (1998).
15. S.M. Pearl, L.B. Hough, D.L. Boyd, and S.D. Glick, *Pharmacol. Biochem. Behav.* **57**, 809 (1997).
16. H. Sershen, A. Hashim, and A. Lajtha, *Brain Res.* **801**, 67 (1998).
17. M A. Rossing, *Environ. Health Perspect.* **106**, 231 (1998).
18. S.I. Dworkin, S. Gleeson, D. Meloni, T.R. Koves, and T.J. Martin, *Psychopharmacology*

(Berl) **117**, 257 (1995).
19. S.L. Cappendijk, D. Fekkes, and M.R. Dzoljic, *Behav. Brain Res.* **65**, 117 (1994).
20. E.D. Dzoljic, C.D. Kaplan, and M.R. Dzoljic, *Arch. Int. Pharmacodyn. Ther.* **294**, 64 (1988).
21. B. Frances, R. Gout, J. Cros, and J.M. Zajac, *Fundam. Clin. Pharmacol.* **6**, 327 (1992).
22. L.G. Sharpe and J.H. Jaffe, *Neuroreport* **1**, 17 (1990).
23. J.R. Blackburn and K.K. Szumlinski, *Behav. Brain Res.* **89**, 99 (1997).
24. H. Sershen, L.G. Harsing, Jr., A. Hashim, and A. Lajtha, *Life Sci.* **51**, 1003 (1992).
25. S.D. Glick and I.S. Maisonneuve, *Ann. N.Y. Acad. Sci.* **844**, 214 (1998).
26. P.A. Broderick, F.T. Phelan, F. Eng, and R.T. Wechsler, *Pharmacol. Biochem. Behav.* **49**, 711 (1994).
27. I.M. Maisonneuve and S.D. Glick, *Eur. J. Pharmacol.* **212**, 263 (1992).
28. I.M. Maisonneuve, R.W. Keller, Jr., and S.D. Glick, *Brain Res.* **579**, 87 (1992).
29. H. Sershen, A. Hashim, L. Harsing, and A. Lajtha, *Life Sci.* **50**, 1079 (1992).
30. M.D. Schechter and T.L. Gordon, *Eur. J. Pharmacol.* **249**, 79 (1993).
31. D.C. Deecher, M. Teitler, D.M. Soderlund, W.G. Bornmann, M.E. Kuehne, and S.D. Glick, *Brain Res.* **571**, 242 (1992).
32. A.S. Schneider, J.E. Nagel, and S.J. Mah, *Eur. J. Pharmacol.* **317**, R1 (1996).
33. B. Badio, W.L. Padgett, and J.W. Daly, *Mol. Pharmacol.* **51**, 1 (1997).
34. H.N. Bhargava, Y.J. Cao, and G.M. Zhao, *Brain Res.* **752**, 234 (1997).
35. L.B. Hough, S.M. Pearl, and S.D. Glick, *Life Sci.* **58**, L119 (1996).
36. D.C. Mash, J.K. Staley, M.H. Baumann, R.B. Rothman, and W.L. Hearn, *Life Sci.* **57**, L45 (1995).
37. S.D. Glick, S.M. Pearl, J. Cai, and I.M. Maisonneuve, *Brain Res.* **713**, 294 (1996).
38. H. Sershen, P. Berger, A.E. Jacobson, K.C. Rice, and M.E. Reith, *Neuropharmacology* **27**, 23 (1988).
39. J.K. Staley, Q. Ouyang, J. Pablo, W.L. Hearn, D.D. Flynn, R.B. Rothman, K.C. Rice, and D.C. Mash, *Psychopharmacology (Berlin)* **127**, 10 (1996).
40. P.M. Sweetnam, J. Lancaster, A. Snowman, J.L. Collins, S. Perschke, C. Bauer, and J. Ferkany, *Psychopharmacology (Berl)* **118**, 369 (1995).
41. M.H. Baumann, R.B. Rothman, and S.F. Ali, *Ann. N.Y. Acad. Sci.* **844**, 252 (1998).
42. Z. Binienda, M.A. Beaudoin, B.T. Thorn, D.R. Prapurna, J.R. Johnson, C.M. Fogle, W. Slikker, Jr., and S.F. Ali, *Ann. N.Y. Acad. Sci.* **844**, 265 (1998).
43. L.G. Harsing, Jr., H. Sershen, and A. Lajtha, *J. Neural Transm. Gen. Sect.* **96**, 215 (1994).
44. M.S. Reid, K. Hsu, Jr., K.H. Souza, P.A. Broderick, and S.P. Berger, *J. Neural Transm.* **103**, 967 (1996).
45. G.B. Wells, M.C. Lopez, and J.C. Tanaka, *Brain Res. Bull.* **48**, 641 (1999).
46. S. Helsley, D. Fiorella, R.A. Rabin, and J.C. Winter, *Pharmacol. Biochem. Behav.* **59**, 419 (1998).
47. S. Helsley, R.A. Rabin, and J.C. Winter, *Prog. Neuropsychopharmacol. Biol. Psych.* **23**, 317 (1999).
48. E. Mocaer-Cretet, M. Prioux-Guyonneau, F. Redjimi, Y. Cohen, and C. Jacquot, *Naunyn Schmiedebergs Arch. Pharmacol.* **326**, 287 (1984).
49. P.A. Palumbo and J.C. Winter, *Pharmacol. Biochem. Behav.* **43**, 1221 (1992).
50. H. Sershen, A. Hashim, and A. Lajtha, *Neurochem. Res.* **19**, 1463 (1994).
51. H. Sershen, A. Hashim, and A. Lajtha, *Brain Res. Bull.* **36**, 587 (1995).
52. D. Wei, I.M. Maisonneuve, M.E. Kuehne, and S.D. Glick, *Brain Res.* 800, 260 (1998).
53. K. Chen, T.G. Kokate, S.D. Donevan, F.I. Carroll, and M.A. Rogawski, *Neuropharmacology* **35**, 423 (1996).
54. Y. Itzhak and S.F. Ali, *Ann. N.Y. Acad. Sci.* **844**, 245 (1998).
55. R.T. Layer, P. Skolnick, C.M. Bertha, U.K. Bandarage, M.E. Kuehne, and P. Popik, *Eur. J. Pharmacol.* **309**, 159 (1996).
56. D.C. Mash, J.K. Staley, J.P. Pablo, A.M. Holohean, J.C. Hackman, and R.A. Davidoff, *Neurosci. Lett.* **192**, 53 (1995).

57. P. Popik, R.T. Layer, and P. Skolnick, *Psychopharmacology (Berl)* **114**, 672 (1994).
58. P. Popik, R.T. Layer, L.H. Fossom, M. Benveniste, B. Geter-Douglass, J.M. Witkin, and P. Skolnick, *J. Pharmacol. Exp. Ther.* **275**, 753 (1995).
59. B. Geter-Douglass and J.M. Witkin, *Psychopharmacology (Berlin)* **146**, 280 (1999).
60. S.D. Glick, I.M. Maisonneuve, and S.M. Pearl, *Brain Res.* **749**, 340 (1997).
61. H. Sershen, A. Hashim, and A. Lajtha, *Brain Res. Bull.* **40**, 63 (1996).
62. H. Sershen, A. Hashim, and A. Lajtha, *Pharmacol. Biochem. Behav.* **53**, 863 (1996).
63. E.E. Codd, *Life Sci.* **57**, L315 (1995).
64. Y.J. Cao and H.N. Bhargava, *Brain Res.* **752**, 250 (1997).
65. J.P. Pablo and D.C. Mash, *Neuroreport* **9**, 109 (1998).
66. W.D. Bowen, B.J. Vilner, W. Williams, C.M. Bertha, M.E. Kuehne, and A.E. Jacobson, *Eur. J. Pharmacol.* **279**, R1 (1995).
67. R.H. Mach, C.R. Smith, and S.R. Childers, *Life Sci.* **57**, L57 (1995).
68. S. Couture and G. Debonnel, *Synapse* **29**, 62 (1998).
69. S. Helsley, R.A. Filipink, W.D. Bowen, R.A. Rabin, and J.C. Winter, *Pharmacol. Biochem. Behav.* **59** , 495 (1998).
70. J.D. Fryer and R.J. Lukas, *J. Pharmacol. Exp. Ther.* **288**, 88 (1999).
71. S.J. Mah, Y. Tang, P.E. Liauw, J.E. Nagel, and A.S. Schneider, *Brain Res.* **797**, 173 (1998).
72. R.C. Miller and T. Godfraind, *Eur. J. Pharmacol.* **96**, 251 (1983).
73. M.K. Mundey, N.A. Blaylock, R. Mason, S.D. Glick, I.M. Maisonneuve, and V.G. Wilson, *Br. J. Pharmacol.* **129**, 1561 (2000).
74. M.E. Alburges and G.R. Hanson, *Brain Res.* **847**, 139 (1999).
75. M.E. Alburges and G.R. Hanson, *Brain Res.* **818**, 96 (1999).
76. M.E. Alburges, B.P. Ramos, L. Bush, and G.R. Hanson, *Eur. J. Pharmacol.* **390**, 119 (2000).
77. S.F. Ali, N. Thiriet, and J. Zwiller, *Brain Res. Mol. Brain Res.* **74**, 237 (1999).
78. S.M. Pearl, K. Herrick-Davis, M. Teitler, and S.D. Glick, *Brain Res.* **675**, 342 (1995).
79. K. Ishihara and M. Sasa, *Jpn. J. Pharmacol.* **80**, 185 (1999).
80. A.A. Mathe, *J. ECT.* **15**, 60 (1999).
81. J.J. Mann, *J. ECT.* **14**, 172 (1998).
82. F. Passarelli, E. Carmenini, L. Calo, F. Orzi, and F.E. Pontieri, *J. Neural Transm.* **104**, 805 (1997).
83. D.C. Mash, C.A. Kovera, B.E. Buck, M.D. Norenberg, P. Shapshak, W.L. Hearn, and J. Sanchez-Ramos, *Ann. N.Y. Acad. Sci.* **844**, 274 (1998).
84. M.E. Benwell, P.E. Holtom, R.J. Moran, and D.J. Balfour, *Br. J. Pharmacol.* **117**, 743 (1996).
85. I.M. Maisonneuve, G.L. Mann, C.R. Deibel, and S.D. Glick, *Psychopharmacology (Berlin)* **129**, 249 (1997).
86. S.D. Glick, K. Rossman, S. Wang, N. Dong, and R.W. Keller, Jr., *Brain Res.* **628**, 201 (1993).
87. S.D. Glick, M.E. Kuehne, J. Raucci, T.E. Wilson, D. Larson, R.W. Keller, Jr., and J.N. Carlson, *Brain Res.* **657**, 14 (1994).
88. S.D. Glick, M.E. Kuehne, I.M. Maisonneuve, U.K. Bandarage, and H.H. Molinari, *Brain Res.* **719**, 29 (1996).
89. S.D. Glick, I.M. Maisonneuve, K.E. Visker, K.A. Fritz, U.K. Bandarage, and M.E. Kuehne, *Psychopharmacology (Berl)* **139**, 274 (1998).
90. E.D. French, K. Dillon, and S.F. Ali, *Life Sci.* 59, L199 (1996).
91. I.T. Uzbay, C.J. Wallis, H. Lal, and M.J. Forster, Behav. *Brain Res.* **108**, 57 (2000).
92. C.R. Ashby, Jr., R. Rohatgi, J. Ngosuwan, T. Borda, M.R. Gerasimov, A.E. Morgan, S. Kushner, J.D. Brodie, and S.L. Dewey, *Synapse* **31**, 151 (1999).
93. M.R. Gerasimov, W.K. Schiffer, J.D. Brodie, I.C. Lennon, S.J. Taylor, and S.L. Dewey, *Eur. J. Pharmacol.* **395**, 129 (2000).
94. G.F. Koob and F. Weiss, *Recent Dev. Alcohol* **10**, 201 (1992).
95. C. Zubaran, M. Shoaib, I.P. Stolerman, J. Pablo, and D.C. Mash, *Neuropsychopharmacology* **21**, 119 (1999).
96. A.H. Rezvani, D.H. Overstreet, and Y.W. Lee, *Pharmacol. Biochem. Behav.* **52**, 615 (1995).

97. A.H. Rezvani, D.H. Overstreet, Y. Yang, I.M. Maisonneuve, U.K. Bandarage, M.E. Kuehne, and S.D. Glick, *Pharmacol. Biochem. Behav.* **58**, 615 (1997).
98. A.H. Rezvani, "Combination Pharmacotherapy: Implications for Drug Addiction," *33rd Annual Winter Conference on Brain Research,* Snowmass, Colorado, January 28, 2000.
99. C.M. Cramer, L.R. Gardell, K.L. Boedeker, J.R. Harris, C.L. Hubbell, and L.D. Reid, *Pharmacol. Biochem. Behav.* **60**, 345 (1998).
100. M.H. Heijna, F. Hogenboom, A.H. Mulder, and A.N. Schoffelmeer, *Naunyn Schmiedebergs Arch. Pharmacol.* **345**, 627 (1992).
101. R. Spanagel and W. Zieglgansberger, *Trends Pharmacol. Sci.* **18**, 54 (1997).
102. F. Berton, W.G. Francesconi, S.G. Madamba, W. Zieglgansberger, and G.R. Siggins, *Alcohol Clin. Exp. Res.* **22**, 183 (1998).
103. W. Ling, R.A. Rawson, and M.A. Compton, *J. Psychoactive Drugs* **26**, 119 (1994).
104. C.P. O'Brien, A.R. Childress, A.T. McLellan, J. Ternes, and R.N. Ehrman, *J. Clin. Psychiatry* **45**, 53 (1984).
105. D.B. Vaupel, W.R. Lange, and E.D. London, *J. Pharmacol. Exp. Ther.* **267**, 1386 (1993).
106. N.W. Withers, L. Pulvirenti, G.F. Koob, and J.C. Gillin, *J. Clin. Psychopharmacol.* **15**, 63 (1995).
107. P.A. LeDuc and G. Mittleman, *Psychopharmacology (Berl)* **110**, 427 (1993).
108. E.F. McCance, *NIDA Res. Monogr* **175**, 36 (1997).
109. M.E. Reith, M. Benuck, and A. Lajtha, *J. Pharmacol. Exp. Ther.* **243**, 281 (1987).
110. G.R. King, C. Joyner, T. Lee, C. Kuhn, and E.H. Ellinwood, Jr., *Pharmacol. Biochem. Behav.* **43**, 243 (1992).
111. P.M. Callahan and K.A. Cunningham, *Neuropharmacology* **36**, 373 (1997).
112. L.H. Parsons, F. Weiss, and G.F. Koob, *J. Neurosci.* **18**, 10078 (1998).
113. J.R. Glowa, M. LeSage, D. Stafford, and F. Vocci, *Exp. Opin. Ther. Patents* **7**, 1459 (1997).
114. K.R. Alper, H.S. Lotsof, G.M. Frenken, D.J. Luciano, and J. Bastiaans, *Am. J. Addict.* **8**, 234 (1999).

—CHAPTER 7—

CHANGES IN GENE EXPRESSION AND SIGNAL TRANSDUCTION FOLLOWING IBOGAINE TREATMENT

EMMANUEL S. ONAIVI

Departments of Psychiatry and Pharmacology
Vanderbilt University School of Medicine
Nashville, TN 37232

Molecular Neurobiology Branch
NIDA Intramural Research Program
Baltimore, MD 21224

B. EMMANUEL AKINSHOLA

Department of Pharmacology
Howard University College of Medicine
Washington D.C. 20059

SYED F. ALI

Neurochemistry Laboratory
National Center for Toxicological Research
Food and Drug Administration
Jefferson, AR 72079

0099-9598/01 $35.00

I. Introduction

A. Ibogaine: Changes in Programs of Gene Expression and Signal Transduction

Drug addiction may not be mediated by one neurochemical pathway and brain structure, but by a complex interaction of programs of gene expression with specific signal transduction pathways and environmental factors. The putative antiaddictive effect of ibogaine may result from the restoration of altered or disrupted programs of central and peripheral neuroadaptative processes involving programmed genes and their associated signaling mechanisms. As discussed in this chapter, because the pharmacotherapy of drug addiction has been largely disappointing, it is now more important than ever to consider new hypotheses. One new hypothesis being explored is that both the peripheral and central actions of abused substances contribute to drug addiction. It is suggested that an effective therapeutic agent for addiction may be obtained only when both peripheral and central actions of the processes contributing to addictions are considered. There is evidence from animal studies, and from anecdotal human studies, that the alkaloid ibogaine, and perhaps its metabolites, alter or regulate gene expression and signal transduction pathways and restore altered neuroadaptive processes arising from the loss of control due to drug addiction. We and others have observed that treatment with cocaine influences the regulation of certain genes in the brain, as indicated by the activation and inhibition of the expression sequence tags (ESTs) that have been isolated. The behavioral data presented here supports the notion that ibogaine restores the behavioral and neurogenetic alterations resulting from exposure to cocaine.

The recent advances in neurobiology have enabled the complex biological mechanisms underlying drug and alcohol addiction to be investigated at the cellular and molecular levels (*1*). Abused substances exert biological effects by interacting with cell membranes and receptors, and modify the function of proteins, which regulate signal transduction, intracellular pathways, and gene expression. Adaptation to the effects of abused substances is known to constitute a major determinant of the development of increased tolerance, withdrawal syndrome, and dependence. Thus, important targets for alcohol and other abused substance include second messengers, gene transcription factors, transmitter and voltage-regulated ion channels, GTP-binding proteins, and metabolizing enzymes (*1*). It is hoped that this research focus will identify important molecules for the development of drug and alcohol addiction, as this will certainly lead to identifying genes that are most critical in mediating addiction. A consideration of the contribution of environmental factors to addictions should not be underestimated. However, to develop novel pharmacologic therapies for treating or preventing drug abuse, addiction, craving, or withdrawal symptoms when an individual is attempting to quit, we must first identify neural substrates and understand the mechanism by which abused drugs act at these target sites.

The involvement of coordinated programs of gene expression appears to be critical for many brain functions, including long-term memory and drug addiction (*2*). As shown in Figure 1, the cascade of intracellular signals mediated by receptors interacting with G proteins initiates the communication between extracellular signals and the nucleus to trigger specific patterns of gene expression (*3*). We have assumed that the initiation of compulsive drug use may

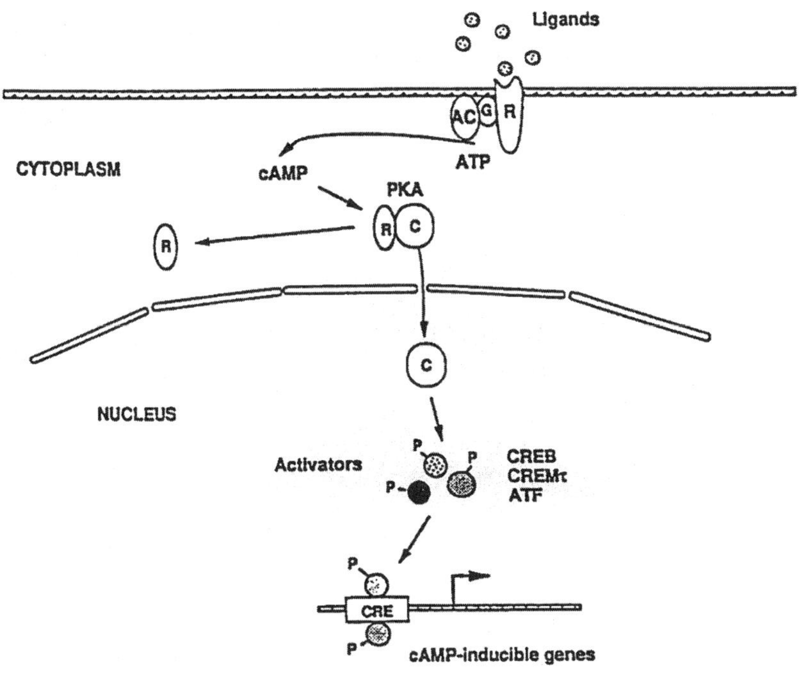

FIGURE 1. THE cAMP SIGNAL TRANSDUCTION PATHWAY. Schematic representation of the route whereby ligands at the cell surface interact with, and thereby activate, membrane receptors (R) and result in altered gene expression. Ligand binding activates coupled G-proteins (G), which, in turn, stimulate the activity of the membrane-associated adenylyl cyclase (AC). This converts ATP to cAMP, which causes the dissociation of the interactive tetrameric protein kinase A (PKA) complex into the active catalytic subunits (C) and the regulatory subunits (R). Catalytic subunits migrate into the nucleus where they phosphorylate (P) and thereby activate transcriptional activators such as Ca^{2+}/cAMP response element binding protein (CREB), cAMP response element modulator (CREM), and activating transcription factor (ATF). These then interact with the cAMP response enhancer element (CRE) found in the promoters of cAMP-responsive genes to activated transcription.

involve the binding of the drug to a receptor resulting in activation or inhibition of the cAMP-dependent pathway. This will ultimately influence the transcriptional regulation of various genes through distinct promoter responsive sites (*4*). The cAMP-responsive element binding protein (CREB), the first CRE-binding factor to be characterized (*5*), is a transcription factor of general importance in both neuronal and other cells (*3*). CREB phosphorylation on Ser-133 promotes the activation of genes with an upstream CRE element (*6*). CREB phosphorylation and downstream gene expression can, in principle, be regulated by protein kinases under the control of cAMP (*7*), Ca^{2+} (*8-11*), or both (*12*). Alteration of CREB function specifically affects long-term synaptic changes and long-term memory, while sparing short-term changes (*8, 13-16*). Thus, the final transcriptional response to cAMP is the outcome of a complex interplay of nuclear targets activated by signal transduction events. In a stress model of anhedonia, our preliminary data (not shown) indicate a decrease in CREB phosphorylation. This is of significance and tremendous importance as anxiety and stress factors play a major role in the precipitation and maintenance of drug-seeking behavior (*17*). Our preliminary data (*18*) are supported by the growing consensus that genes influence behavior in both humans and animals, along with complex interactions with the environment. However, because any behavioral trait is likely to be affected by many genes acting in concert, the attempts to pin down which genes influence which behavior have proven difficult. Recent advances in genetic studies of human disease have linked some genes to some aspects of human biology, behavior, or disease. It is therefore timely that the tools are now available to discover the programs of gene expression that make the individual vulnerable to drug dependency.

Experience in life, and encounters with pain or pleasure, may leave indelible impressions. The pleasurable experience from abused drugs may cause a change in the programs of gene expression. The nature and consequences of addictive disorders and other CNS disturbances may appear to be aberrant programs of gene expression, suggesting that the CNS of the drug-dependent individual may be "locked" into programs that no longer respond to appropriate external circumstances. The intensity of the pleasure or experience may be irreversible, and one cannot forget the pleasure because the rate of firing of neurons during the experience has been altered from the normal pattern of neuronal firing. The extent of dependency and the loss of control and the ultimate loss of plasticity—that is, incomplete loss or partial loss—may allow for reconditioning and regaining of control. Willpower to quit addiction may be difficult to mimic in the laboratory so as to study its mechanism and rate limiting facets. It appears that interference with the dopamine system may not restore the altered function and plasticity in drug dependence. The initial trigger or switch that leads to changes in programs of gene expression may be different for the different drugs of abuse.

B. Genetic Markers and Signaling Proteins in Addiction

Over the years, significant effort has been made to uncover valid genetic markers for the risk of drug and alcohol addiction. It is now recognized that drug and alcohol dependence is a chronic brain disease and a long-lasting form of neuronal plasticity. At the cellular and molecular level, there is a growing body of evidence that substances that cause addiction affect hormones and neurotransmitter-activated signal transduction leading to short-term changes in regulation of cellular functions and long-term changes in gene expression.

Complex, but defined, processes are emerging for the mechanisms leading to the development of drug tolerance and dependence arising from adaptations in post-receptor signaling pathways with the accompanying transcriptional regulators. This may initiate a cascade of altered programs of gene expression that underlie the long-term consequences of withdrawal and relapse that leads to drug seeking behavior. The number of post-receptor events shown in Tables I and II underscores the complexity of the processes leading to drug and alcohol dependence. Numerous studies have therefore demonstrated that chronic drug administration drives the production of adaptations in post-receptor signaling by

TABLE I.
Signal Transduction Pathways Possibly Associated with Substance Abuse and Putative Ibogaine Effects

Post-Receptor Mediated Signal Transduction
Ca2+ signaling ions and other receptor and voltage operated channels • *K^+, Na^+, Ca^2+ (P, T, N and L-types) ions* • *(Na^+K)-ATPase alpha-subunit*
G protein-mediated signal transduction • *cAMP signal transduction* • *Phosphorylation of adenylyl cyclase* • *cAMP-PKA-PKC systems* • *CREB-dependent gene transcription* • *AP1 transcription factor gene expression (IEGs, c-fos, fos B, jun-B and* • *c-jun, zif 268, krox-20.*
Poly-phosphoinositide (PI) signaling pathway
Post-transcriptional palmitoylation of Gsα
Peripheral phospholipase D activit
Lipid peroxidation
Microtubule associated protein
Nuclear transcription factors

TABLE II.
CANDIDATE GENES POSSIBLY REGULATED BY ABUSED SUBSTANCES AND PUTATIVE IBOGAINE EFFECTS

Neurotransmitters and hormones, receptor gene expression
AMPA, NMDA, mGluR5, GABAA, dopamine, serotonin, dynorphin, mu- and kappa-opioid, beta endorphin, enkephalin and substance P, NGFI-β, cannabinoid receptor 1, CART peptide, Orphan receptor 1, corticotrophin-releasing factor gene, nitric oxide, cholecystokinin, α-2a-adrenoceptor, HSP-72, RGS2, and RGS3, mitochondrial 12 SrNA
Monoamine transporter genes
Metabolizing enzyme genes
Angiotensin converting enzyme, Arginine vasopressin, tyrosine hydroxylase, tryptophan hydroxylase, monoamine oxidase, glutamyl transpeptidase, liver β-galactoside α2, 6-sialyl transferase, trytophan pyrrolase, hepatic cytochrome P450-dependent microsomal monoxygenase system, mitochondrial NADH dehydrogenase, alcohol dehydrogenase, aldehyde dehydrogenase

a tangled, but precise, web of signal transducers. The G protein-mediated signal transduction pathway may be crucial in the pathophysiology of drug dependency, so that actions involving G proteins may be candidate markers in the addiction process. The activity of the adenylate cyclase enzyme in the signal transduction pathway of many of the G protein coupled receptors appears to be involved in drug-seeking behavior. Furthermore, brain region-specific changes in multiple signaling pathways of activator protein (AP-1 transcription factor changes in gene expression in c-fos, fosB, jun-B, and c-jun) have also been linked to a long-lasting form of neuronal plasticity associated with drug and alcohol dependence. Receptor and voltage operated channels and G-protein-mediated phosphoinositide and protein kinases are among the other signal transduction pathways that may underlie the clinical manifestations of drug/alcohol dependence, tolerance, withdrawal, and addiction.

C. IBOGAINE: BEYOND DOPAMINE IN THE NUCLEUS ACCUMBENS

The problem of drug addiction continues to affect modern society with severe consequences. Despite extensive research efforts, the neurobiological mechanisms responsible for compulsive and uncontrollable drug use remain poorly understood (*21*). Therefore the long-term use of ingestion of psychostimulants, like cocaine and amphetamines, narcotics like heroin, benzodiazepines like diazepam, and the recreational use of alcohol, nicotine in smoking, and cannabinoids in marijuana may cause addiction with craving and withdrawal syndrome acting as a deterrent from cessation from drug and/or alcohol use. For many years, the central dopamine hypothesis of reward and reinforcement in drug addiction has been associated with elevated dopamine levels in the nucleus accumbens (Acb), which has therefore been suggested as the central and final common neuroanatomical target for abused drugs in the brain (*22-26*). As discussed recently (*27*), if the dopamine hypothesis of reward and the Acb brain structure associated with reinforcement were all but proven, then manipulation of the dopamine system should provide medications for drug and alcohol addicts in the clinic. But since pharmacological treatment of drug and alcohol addiction has largely been disappointing, new therapeutic approaches and hypothesis are needed. Although, it has been previously recognized that the reward centers in the brain consist of multiple systems and neuroanatomical sites, emerging data have started to challenge the dopamine hypothesis of reward involving the Acb circuitry (*28*). The studies in normal and cocaine addicts using positron emission tomography (PET) are associated with metabolic abnormalities in the orbitofrontal cortex and the striato-thalamic-orbitofrontal circuit, which has now been postulated as a common mechanism underlying drug and alcohol addiction (*28,29*). Furthermore, there are other dopamine independent mechanisms involving other neurotransmitters like glutamate (*30,31*), γ-aminobutyric acid (GABA), dynorphin, serotonin (5-HT), and cholecystokinin (CCK), in the Acb and in other brain regions like the frontal cortex, hippocampus, locus coerulus, lateral hypothalamus, or the periaqueductal gray, that are potential neural substrates for the rewarding properties of psychostimulants, benzodiazepines, barbiturates, opiates, and phencyclidine hydrochloride (PCP) (*32*). Since the usefulness in treating any addiction with dopaminergic agents has been limited (*33,34*), one emerging potential, yet controversial, therapeutic agent is ibogaine, an indole alkaloid isolated from the bark of the African shrub, *Tabernanthe iboga*.

Ibogaine is used by some African societies in high doses during initiation ceremonies and rituals, and in low doses to combat hunger and fatigue. In Western cultures, new claims indicate that a single dose of ibogaine eliminates withdrawal symptoms and reduces drug cravings for extended periods of time (*35*). The mechanism of action associated with the ability of ibogaine to block drug-seeking behavior is currently incompletely understood, and a number of studies suggest that ibogaine has a broad spectrum of action on multiple systems (*36,37*). It is speculated that this broad spectrum of activity on opiate, serotonin, dopamine, choline, glutamate, *N*-methyl-D-aspartate (NMDA), sigma, noradrenegic, monoamine transporters, neurotensin, kappa-opioid, and other hormonal systems may, in part, contribute to the putative antiaddictive properties of ibogaine. As it is now doubtful that the mesolimbic dopamine acts by directly producing feelings of pleasure or euphoria (*38*), we have to move beyond the nucleus accumbens and dopamine hypothesis of reward. In place of the dopamine hypothesis, our working hypothesis is that the molecular events that underlie the development of compulsive drug-seeking behavior involve multiple brain sites and systems in drug reinforcement. These molecular switches lead to neuroplastic alterations in specific signal transduction systems that turn on/off subsets of genes that precipitate the behavioral manifestations of loss of control and compulsive drug or alcohol use. Alcohol and abused drugs turn on the switch, and withdrawal, craving, or relapse turns off the switch that creates the neuroadaptive addiction cycle. It is therefore possible that ibogaine, through its multiple actions, can restore the hedonic homeostatic dysregulation caused by drug and alcohol abuse (*39*). In these continuing studies we are using *in vivo* and *in vitro* systems to study the effects of cocaine and other abused substances and to determine whether these effects can be blocked by ibogaine.

II. In Vitro Action of Cocaine on Ca^{2+} and Protein Kinase C Signaling

As discussed below, ibogaine was shown to block some of the actions of cocaine *in vivo*. While an intact organism is desirable for studying *in vivo* effects, the *in vitro* system is also valuable in exploring the mechanism of action in isolated preparations. Thus, a number of investigators have used PC-12 cells to study the effects of alcohol, but not that of cocaine on cell function (*40-42*). It was demonstrated that chronic alcohol exposure increases protein kinase C (PKC) activity and regulation of Ca^{2+} channels in PC12 cell lines. The PC 12 clonal cell line of neural crest origin possesses the ability to secrete dopamine and other neurotransmitters that are known to be affected by cocaine. These cells are coupled to the second messenger systems necessary for signal transduction in response to a variety of stimuli. PKC consists of a family of closely related isoforms, which differ in their localization and pharmacological properties. It is a major mediator of transducing signals to the interior of the cell, and it is activated *in vivo* by Ca^{2+} and diacylglycerol. The activity and translocation of PKC has been implicated in a number of cellular and neuronal functions. Previous studies have therefore suggested a role of PKC in the modulation of ethanol effects on receptor function in cells of central nervous origin. The aim of the present study was to determine the activity and expression of PKC isoforms, along with changes in Ca^{2+} levels following incubation of PC 12 cells with cocaine. While alcohol-induced increases in PKC levels have been associated with the up-regulation of Ca^{2+}, we demonstrate the ability of cocaine to disrupt signal transduction of PC 12 cells.

The treatment of PC 12 cells with cocaine (0.01-3.0 mM) modified the activity and expression of the PKC isoforms and increased the intracellular levels of Ca^{2+} in the cells. SDS-PAGE and Western blotting analysis of the PC 12 cell

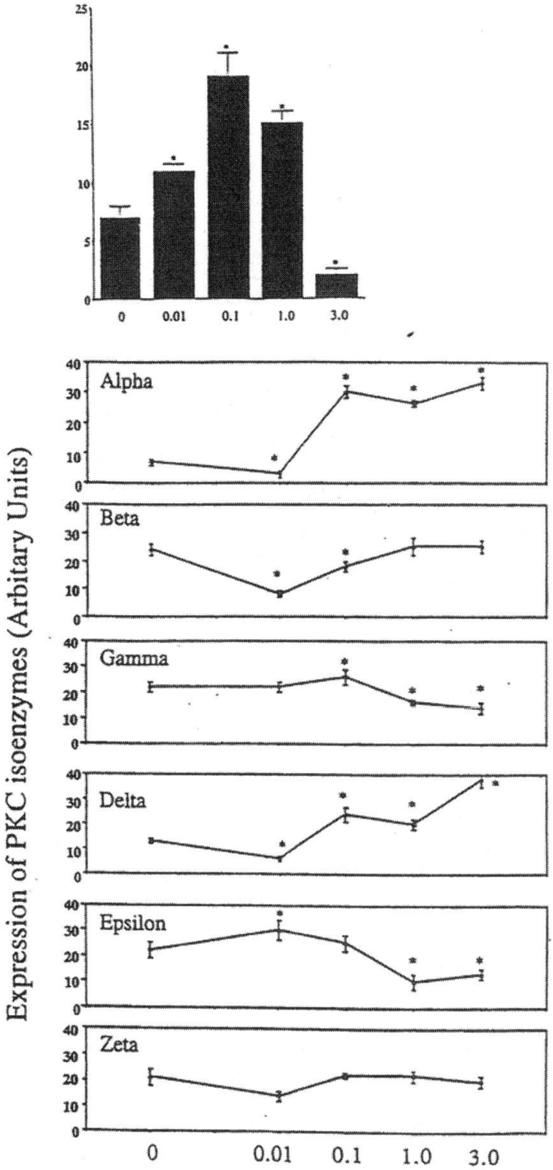

FIGURE 2. TOTAL ACTIVITY OF PROTEIN KINASE C (PKC) (TOP PANEL) AND THE DIFFERENTIAL EXPRESSION OF PKC ISOFORMS. For the PKC activity, the PC 12 cells were treated with or without cocaine for 6 days, and the total activity in the homogenates was analyzed as described. The immunoblotts derived from Western analysis were scanned and quantified, and the expression of the PKC isoforms presented in arbitrary units in shown in the lower panel.

homogenates with antibodies against PKC α, β, γ, δ, ε, and ζ after incubation with doses of cocaine are shown in Figure 2. The spectrum of the effects of increasing doses of cocaine varied according to the isoforms. At doses up to 7.0 mM, cocaine was lethal to the PC 12 cells and inhibited the expression of all the isoforms examined (data not shown). There was inhibition of the expression of PKCα at low dose of 0.01 mM and increased expression at higher doses (0.1-3.0 mM) of cocaine. Immunoblotting with the anti-PKCβ antibody detected an 80-kDa protein, whose expression increased as the dose of cocaine increased. While the expression of PKCδ also increased with increasing doses of cocaine, the expression of PKCγ and PKCε decreased with increasing doses of cocaine, with the expression of PKCζ remaining significantly unchanged. The incubation in the presence or absence of the antigenic peptides allowed identification of the PKC isoforms by the occurrence of immunolabeled bands, which were not seen when antigenic control peptide was present. Interestingly, the total activity of PKC increased with increasing concentrations of cocaine and declined with concentrations above 3.0 mM when compared to PC 12 cells that were not exposed to cocaine, as shown in Figure 2. The levels of Ca^{2+} in the PC 12 cell homogenates with or without incubation with cocaine were measured using fura-2 and analyzed with the SPEX AR-CM fluorometer. The Ca^{2+} levels significantly increased with increasing concentrations of cocaine in the PC 12 cells compared to the controls, as shown in Table III. These data, therefore, confirm that the antibodies used can detect PKCs α, β, γ, δ, ε, and ζ in the PC 12 cells, and show that cocaine differentially affects the expression of the subtypes of protein kinase C.

These results demonstrate the ability of cocaine to affect the activity and expression of PKC isoenzymes in PC 12 cells. The effect of cocaine was dose dependent and specific for the different isoforms of PKC. The differential expression of PKC isoforms was accompanied by increased total PKC enzyme activity and Ca^{2+} levels. These effects of cocaine in the expression and activity of PKC in the PC 12 cells share some similarities and differences with the results previously obtained with ethanol in PC 12 cells (*42*). It was demonstrated that

TABLE III.
$[Ca^{2+}]_i$ Measurements in PC 12 Cells Using Fura-2

Treatment	$\Delta[Ca^{2+}]_i$ nM	% Control
Control	43.70 ± 5	100
0.01 mM cocaine	349.30 ± 25	799
0.10 mM cocaine	270.40 ± 19	619
1.00 mM cocaine	19273.00 ± 2001	44103
3.00 mM cocaine	21137.00 ± 1998	48368

chronic ethanol exposure increased the levels of PKCδ and ε and PKC-mediated phosphorylation in cultured neural cells (*40,42*). The results obtained indicated that, like alcohol, cocaine increased the levels of PKCδ, but, unlike alcohol, increasing doses of cocaine decreased the expression of PKCε. Furthermore, a number of studies have suggested a role of PKC in the modulation of alcohol effects on receptor function in cells of CNS origin, and the basis of some of these pharmacological effects may be related to the PKC-derived transduction mechanisms (*41,45,46*). PKC activity has also been linked with tolerance to the effects of alcohol (*42*) and morphine (*47*). In the CNS, alcohol and cocaine are known to disrupt a number of hormonal and neurotransmitter systems, including dopaminergic mechanisms that may be associated with compulsive alcohol and drug use and relapse. Since not all the physiological actions of the multiple dopamine systems can be explained by the modifications of the cAMP-dependent pathway, some studies have suggested an involvement of the inositol phosphate-generating system (*48*). There is increasing experimental evidence for the existence of cross-talk or interaction between multiple signal transduction systems (*40,48*) in the action of drugs that modulate the dopamine system, including psychostimulants like amphetamine. The data obtained in this study suggest some role of PKC in the effects of cocaine and lend further support to the probable existence of cross-talk between multiple signal transduction systems.

It is known that PKC is a soluble enzyme in its active state, and translocation to the plasma membrane is required for its activation by Ca^{2+} and phospholipids (*49*). However, not all PKC isoenzymes are calcium dependent. It was not surprising that the levels of Ca^{2+} were increased by the treatment with cocaine, because any one or a combination of the following mechanisms can be speculated to be involved: (1) by increasing Ca^{2+} channels as reported for ethanol (*42*); (2) by inhibition of the plasma membrane Ca^{2+}ATPase pump, endoplasmic reticulum Ca^{2+} pump, mitochondrial Ca^{2+} uptake; and/or (3) due to the stimulation of the release of Ca^{2+} from internal storage by opening intracellular Ca^{2+} channels. Cocaine may also activate phospholipase C, which hydrolyzes phosphoinositol biphosphate (PIP2) yielding inositol triphosphate (IP3) and diacylglycerol (DAG), which, in turn, promotes translocation of PKC to the membrane, enhancing activation. The IP3 can release Ca^{2+} from internal storage sites and thus increase intracellular $Ca^{2+}([Ca^{2+}]_i)$. The high levels of Ca^{2+} may also stimulate Ca^{2+} dependent proteases to degrade membranes and inhibit translation and transcription. Some or all of these proposed mechanism may be implicated in cocaine-induced apoptosis, observed at higher doses of cocaine. Although the mechanisms by which cocaine increases Ca^{2+} levels remain to be established, it is attractive to speculate that just like alcohol, cocaine may also increase the number and function of Ca^{2+} channels in the neural PC 12 cell line (*42*). Of course, cocaine may also be acting by mechanisms independent of the voltage-dependent Ca influx. In summary, this part of the study showed that cocaine differentially altered the expression of PKC isoforms accompanied by increased levels of Ca^{2+} and total PKC activity. It is suggested that the differential expression of PKC isozymes may demonstrate distinct roles of PKC isoforms in the actions of cocaine. Thus, the PC 12 cell model may be exploited to further understand the neurobiology of cocaine's action in neural systems. We are currently looking at the effects of ibogaine on a number of signaling pathways. Some experimental evidence appears to suggest that ibogaine's action on signal transduction is more robust when that signal has been altered by abused substances. For example, ibogaine was reported to potentiate the inhibition of adenylyl cyclase by serotonin (*50*). The mechanism by which ibogaine and noribogaine elicited a concentration-dependent increase in receptor-mediated inhibition of adenylyl cyclase activity is unclear. Since ibogaine and noribogaine did alter adenylyl cyclase activity, the enhanced inhibition of enzyme activity appears to represent functional antagonism (*50*).

III. Effects of Ibogaine on the Action of Cocaine *In Vivo*

For many years, the powerful reinforcing effects of psychostimulants, including cocaine and other abused drugs, have been linked to the mesocorticolimbic dopamine system and its connections (*22-26*). Although dopamine is still thought to play a critical role in motivation and reward, it is doubtful that the mesocorticolimbic dopamine acts directly as the brain reward center (*38*). However, it is now conceptualized that rather than signaling pleasure as previously thought, the neurotransmitter dopamine may be released by neurons to highlight significant stimuli (*38*). The neurobehavioral effects of cocaine may be linked to a number of factors, including the route of administration, the dose of cocaine, the environmental cues, and the co-administration of other substances, including alcohol. It is unlikely that the overall neurobehavioral effects of cocaine are due to a single neurotransmitter action in one pathway in the central and peripheral nervous system. It is more likely that they are the result of the effects on multiple systems. The broad spectrum of action of ibogaine, therefore, makes it attractive to investigate whether it will functionally block an *in vivo* action of cocaine that is linked to emotionality/stress, which may be a factor in drug dependency.

The acute and subacute effects during treatment and withdrawal from ibogaine on the performance of ICR mice in the elevated plus-maze test did not show any clear dose-response profile of action following the acute administration of ibogaine, as shown in Figure 3. The data obtained following treatment with ibogaine, or the combination with cocaine, were compared to the effect of

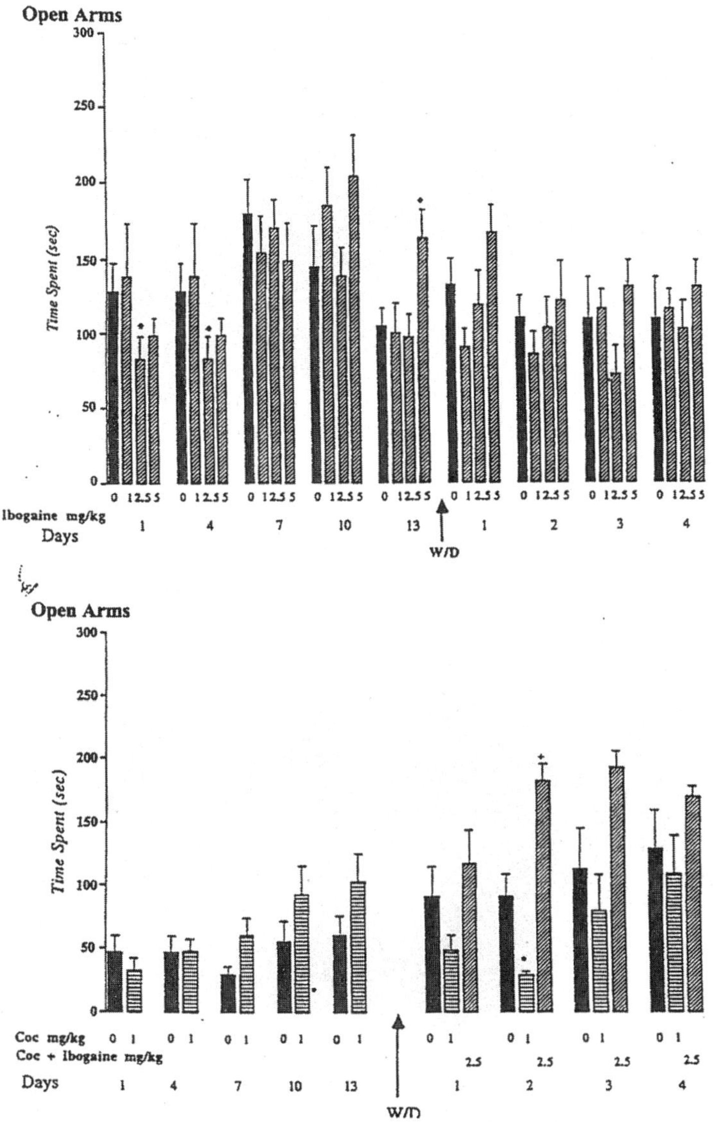

FIGURE 3 THE ACUTE AND SUBACUTE EFFECTS DURING TREATMENT AND WITHDRAWAL (W/D) FROM IBOGAINE (TOP PANEL), AND THE INFLUENCE OF IBOGAINE ON COCAINE WITHDRAWAL IN THE ELEVATED PLUS-MAZE TEST SYSTEM. The time spent in the open arms (sec) of the plus-maze following the 5-minute test session is shown. Following withdrawal from cocaine (1.0 mg/kg), ibogaine (2.5 mg/kg) was administered daily for 4 days accompanied by daily testing.
Significant differences from vehicle-treated animals are indicated as *$p<0.05$ or +$p<0.05$ (one-way ANOVA followed by Dunnet's test).

vehicle-treated control mice. While the decrease in time spent in the enclosed arms at the lowest dose of 1.0 mg/kg may indicate an antiaversive action, the 2.5 mg/kg dose induced an aversive action in the open arms characterized by a decrease in the time spent in these arms. Ibogaine at the doses used had no significant effect on the time spent in the central platform. The number of entries into the open and closed arms was reduced only at the 2.5 mg/kg dose acutely (data not shown). At the highest dose of ibogaine used, the time spent and number of entries into the open and enclosed arms and the central platform were not different from controls. The repeated treatment of mice with ibogaine induced aversive and antiaversive behavior to the open arms on day 4 and day 13, respectively. An antiaversive behavior of mice injected with a 5-mg/kg dose was also recorded on day 10, as shown by the reduced time spent in the enclosed arms. This subacute treatment with ibogaine did not affect the time the animals spent in the central platform. Following withdrawal from the 14-day treatment with ibogaine, there were no differences in the time spent and number of entries into the open arms, enclosed arms, and central platform in comparison to control animals, as shown in Figure 3.

The influence of ibogaine (2.5 mg/kg) on cocaine withdrawal in the plus-maze test showed that on withdrawal from 1.0 mg/kg cocaine, an intense aversion into the open arms was blocked by ibogaine (see Figure 3). The data obtained add to the growing evidence that ibogaine, its congeners, and perhaps its metabolites, may have value in the treatment of drug and alcohol dependency. This conclusion is supported by other animal and human anecdotal and clinical evidence (*37*). Although there are some negative data, a number of animal studies indicate that ibogaine reduces some of the behavioral manifestations associated with cocaine administration and withdrawal (*37*). For example, in a study with mice, ibogaine reduced cocaine consumption in a drinking preference model (*51*). In another study, ibogaine did not reduce the withdrawal manifestations following naloxone-precipitated withdrawal in morphine-dependent mice (*52*). In rats, ibogaine has been shown to decrease morphine self-administration (*53,54*), reduce the severity of withdrawal induced by naloxone (*53,55*), and decrease intravenous cocaine self-administration (*56*). The effect of ibogaine on alcohol consumption was also investigated in animals, and it was demonstrated that ibogaine and one of its metabolites, noribogaine, reduces alcohol consumption in a number of alcohol-preferring rat lines (*57-59*). There are suggestions that the use of ibogaine in the treatment of drug and alcohol abuse be viewed with some degree of caution (*60*) because of its hallucinogenic properties and perhaps toxicity, but it is difficult to ignore the balance of evidence now emerging from animal and human data. In a preliminary study of seven individuals addicted to opiates, three who had at least 1.0 gm ibogaine, had remained drug free for 14 weeks (*61*). Therefore, there is some merit in the further investigation of the value of ibogaine in drug and alcohol dependence, which may form a template for the development of novel compounds for substance abuse pharmacotherapy.

While the neuronal and molecular basis for the putative antiaddictiive properties of ibogaine, its congeners, and its metabolites are incompletely understood, it has been postulated that these ibogaine-like compounds may be countering the multiple actions of abused substances (*59*). Most abused substances are known to have effects and interactions with one or multiple systems, including activation of gene expression and signal transduction, serotonin, dopamine, GABA, glutamate, noradrenergic, opiates, and hormones, particularly stress hormones (*59*). Because of the promiscuity of action of ibogaine, it is not surprising that it has shown promise preclinically and in the clinic. Thus, several approaches including pharmacological, histochemical, biochemical, behavioral, radioligand binding, toxicological, spectrometry, synthesis, and more recently molecular biology and genetics have been used to probe the action of this alkaloid. Overall, consensus data support the multiple effects of ibogaine. There is increasing interest in the genetic and signaling molecules that are important in the multiple actions of ibogaine. It appears that ibogaine may be restoring the intricate interactions within and between signaling pathways that are disrupted by abused substances. Intriguingly, because of the tangled web of cellular signaling mechanisms, there is no doubt that the more we know about signal transduction, the more we realize that more has yet to be discovered (*62*). Therefore, as hypothesized above, ibogaine may be switching off a subset of genes that have been turned on by alcohol and abused drugs. One transduction cascade that has been associated with the chronic administration of opiates or psychostimulants is the cAMP signal transduction pathway, which leads to CREB phosphorylation and downstream gene expression that, in principle, can be regulated by protein kinases (*63,64*).

IV. Expression of Genes Regulated by Ibogaine Using cDNA Microarray Analysis

Microarray technology has been described as a minirevolution in science and medicine and holds tremendous potential in unraveling programs of expression in normal and disease states (*19,20*). Because of its relative simplicity and power, it has been referred to as the new frontier in gene discovery and expression analysis and can be used to study programs of gene expression and profiling gene expression patterns of many genes in a single experiment. Microarray analysis has already been used in a number of laboratories to answer different kinds of research questions relating to gene expression (*19,20*). This study utilizes commercially available Atlas mouse cDNA expression arrays on which 588 mouse cDNAs have been immobilized. Two Atlas arrays are used, along with the reagents needed to make the cDNA probes. These cDNA probes are prepared from RNA isolated from the brains of mice that have been treated with cocaine, or ibogaine, or vehicle control groups. The expression levels of these known genes following the treatment in the three groups can then be compared and analyzed. It must be recognized that this technology has its limitations, and this study did not involve the detailed preparation of the DNA arrays, but only uses the commercially available arrays. Examples of differential gene expression patterns in two groups using cDNA expression arrays are presented in Figure 4, while Tables I and II list the putative genes and signaling molecules regulated by abused drugs/alcohol and ibogaine. Further research is required to characterize the most important genes regulated by ibogaine. Two different technologies, both with tremendous potential application in ibogaine research, human disease, biology, and behavior, are the use of gene-targeting approaches and differential display polymerase chain reaction (DDPCR). We have used the DDPCR procedure and obtained some preliminary data. Our preliminary DDPCR data

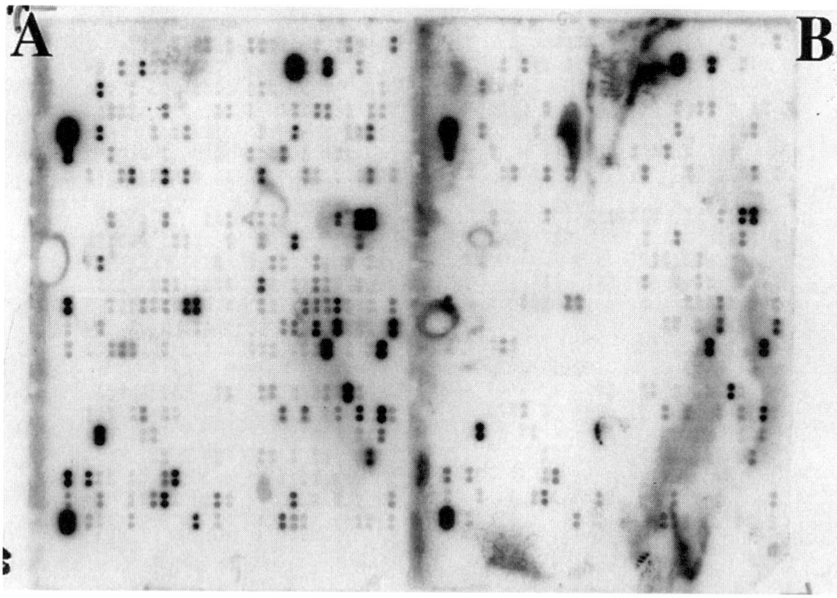

FIGURE 4. DIFFERENTIAL GENE EXPRESSION PATTERNS IN TWO GROUPS USING cDNA EXPRESSION ARRAYS. Example of differential gene expression patterns with subtle changes that can be detected during data analysis. Putative changes observed following administration of a number of abused substances may be normalized by treatment with ibogaine.

may suggest that abused substances are involved in the regulation of certain genes in the brain. If ibogaine reverses the action of cocaine on gene expression in the brain, as it does on cocaine-induced behaviors, characterization of the ESTs we have obtained may lead to the isolation and identification of genes induced or inactivated by ibogaine. Targeted gene disruption and gene manipulation technologies have been applied to neuroscience research. A number of novel genes have recently been identified and cloned, but the regulation of their expression is unknown. Homologous recombination enables the study of the physiological consequences of the absence of a specific gene. Recently, the function of a number of genes was studied by invalidating their genes. Once the complete genes activated or inactivated by ibogaine are known, then the functions of the genes can be analyzed by homologous recombination.

V. Conclusions

It is currently recognized that addiction is a chronic relapsing brain disease (*65*), for which behavioral and effective treatment is urgently needed. Unfortunately, effective drug-abuse treatment continues to be elusive, and the efficacies of new treatments for drug and alcohol addiction have been largely disappointing. The discovery that ibogaine, an emerging, controversial, potential treatment for alcohol and drug addiction, along with the recognition that the mesocorticolimbic dopamine may not, after all, underlie the reward pathway as previously hypothesized, may facilitate and aid rapid progress in substance-abuse research beyond the nucleus accumbens and dopamine hypothesis of reward. As reported recently, since ibogaine's excitatory effect on ventral tegmental area neurons was not long lasting, nor does it persistently alter cocaine- or morphine-induced changes in dopamine neuron impulse (*66*), it was concluded that other mechanisms must be explored to account for the proposed antiaddictive properties of ibogaine. For the *in vitro* studies, it was reported that cocaine disrupts signal transduction in PC 12 cells by altering the expression and activity of PKC isoforms and Ca^{2+} levels. Since cocaine differentially altered the expression of PKC isoforms accompanied by increased levels of Ca^{2+} and total PKC activity, it remains to be determined if ibogaine will block the effects of cocaine on the expression of PKC isozymes and activity. For the *in vivo* studies, it was demonstrated acutely that ibogaine induced an aversive behavior in the ICR mice in the plus-maze test. Ibogaine did not by itself precipitate withdrawal anxiogenesis in the mouse model, but it reversed the withdrawal aversions caused by cessation from cocaine administration. Therefore, it was concluded that if anxiety or stress is a factor in drug dependency, then the antiaddictive property of ibogaine *in vivo* may be associated with modifying the CNS neurotransmission that may be involved in anxiety. The ability of ibogaine to alter drug-seeking behavior may thus be due to the combined actions of the parent drug and metabolite at key pharmacological targets that modulate the activity of drug-reward circuits. Thus, further studies are required to establish the efficacy of ibogaine and the design of ibogaine-like compounds for substance treatment that lack the toxicity and hallucinogenic profile of ibogaine. Finally, the mapping of the human genome will enable us to identify all the potential gene products that could be involved in addictions and the action of ibogaine.

References

1. C. Alling, *Lakartidningen* **96,** 3248 (1999).
2. H. Bito, K. Deisseroth, and R.W. Tsien, *Cell* **87,** 1203 (1996).
3. T. Hunter, *Cell* **80,** 225 (1995).
4. J.P. Montmayeur and E. Borrelli, *Proc. Natl. Acad. Sci.* **88,** 3135 (1991).
5. J.P. Hoeffler, T.E. Meyer, Y. Yun, J.L. Jameson, and J.F. Habener, *Science* **242,** 1430 (1988).
6. P.K. Brindle and M.R. Montminy, *Curr. Opin. Genet. Dev.* **2,** 199 (1992).
7. G.A. Gonzalez and M.R. Montminy, *Cell* **59,** 675 (1989).
8. P.K. Dash, K.A. Karl, M.A. Colicos, R. Prywes, and E.R. Kandel, *Proc. Natl. Acad.Sci.* **88,** 5061 (1991).
9. M. Sheng, M.A. Thompson, and M.E. Greenberg, *Science* **252,** 1527 (1991).
10. H. Bading, D.D. Ginty, and M.E. Greenberg, *Science* **260,** 181 (1993).
11. K. Deisseroth, H. Bito, and R.W. Tsien, *Neuron* **16,** 89 (1996).
12. S. Impey, M. Mark, E.C. Villacres, S. Poser, C. Chavkin, and D.R. Storm, *Neuron* **16,** 973 (1996).
13. R. Bourtchuladze, B. Frenguelli, J. Blendy, D. Cioffi, G. Schutze, and A.J. Silva, *Cell* **79,** 59 (1994).
14. J.C.P. Yin, J.S. Wallach, M. Del Vecchio, E.L. Wilder, H. Zhou, W.G. Quinn, and T. Tully, *Cell* **79,** 49 (1994).
15. J.C.P Yin, M. Del Vecchio, H. Zhou, and T. Tully, *Cell* **81,** 107 (1995).
16. D. Bartsch, M. Ghirardi, P.A. Skehel, K.A. Karl, S.P. Herder, M. Chen, C.H. Bailey, and E.R. Kandel, *Cell* **83,** 979 (1995).
17. N.E. Goeders, *Ann. N. Y. Acad. Sci.,* **28,** 357 (1992).
18. Chakrabarti, S.P. Rossby, D.H. Manier, C. Perrin, E.S. Onaivi, and F. Sulser, *Soc. Neurosci. Abstr.* **22,** 809.15 (1996).
19. NHGRI sponsored review with different authors, The chipping forecast. Supp. to *Nature Genetics* **21,** 1, (1999).
20. K.M. Kurian, C.J. Watson, and A.H. Wyllie, *J. Pathol.* **187,** 267 (1999).
21. M.H. Baumann and R.B. Rothman, *Biol. Psychiat.* **44,** 578 (1998).
22. G.F. Koob and F.E. Bloom, *Science* **242,** 715 (1988).
23. G. Di Chiara and A. Imperato, *Proc. Natl. Acad. Sci.* **85,** 5274 (1988).
24. M.J. Kuhar, M.C. Ritz, and J.W. Boja, *Trends Neurol. Sci.* **14,** 299 (1991).
25. D.W. Self and E.J. Nestler, *Annu. Rev. Neurosci.* **18,** 463 (1995).
26. R.A. Wise, *Annu. Rev. Neurosci.* **19,** 319 (1996).
27. E.S. Onaivi, S.F. Ali, and A. Chakrabarti, *Ann. N.Y. Acad. Sci.* **844,** 227 (1998).

28. N.D. Volkow, G.J. Wang, J.S. Fowler, J. Logan, S.J. Gatley, R. Hitzemann, A.D. Chen, S.L. Dewey, and N. Pappas, *Nature* **386,** 830 (1997).
29. J.D. Modell, J.M. Mountz, and T.P. Beresford, *J. Neuropsych.* **2,** 123 (1990).
30. L. Pulvienti, R. Maldonaldo-Lopez, and G.F. Koob, *Brain Res.* **594,** 327 (1992).
31. W.A. Carlezon and R.A. Wise, *J. Neurosci.* **16,** 3112 (1996).
32. R.A. Wise, *Neurosci. Biobehav. Rev.* **13,** 129 (1989).
33. F.H. Gawn, *Science* **251,** 1580 (1991).
34. D.W. Self, W.J. Barnhart, D.A. Lehman, and E.J. Nestler, *Science* **271,** 1586 (1996).
35. D.C. Mash, C.A. Kovera, B.E. Buck, M.D. Norenberg, P. Shapshak, W.L. Hearn, and J. Sanchez-Ramos, *Ann. N.Y. Acad. Sci.* **844,** 274 (1998).
36. P.R. Popik, T. Layer, and P. Skolnick, *Pharmacol. Rev.* **47,** 235 (1995).
37. H. Sershen, A. Hashim, and A. Lajtha, *Brain Res. Bull.* **42,** 161 (1997).
38. Wickelgren, *Science* **278,** 35 (1997).
39. G.F. Koob and M. Le Moal, *Science* **278,** 52 (1997).
40. R.O. Messing, C.L. Carpenter, I. Diamond, and D.A. Greenberg, *Proc. Natl. Acad. Sci.* **83,** 6213 (1986).
41. S. Rabe and F.F. Weight, J. Pharmacol. Exp. Ther. 244, 417 (1988).
42. R.O. Messing, P.J. Petersen, and C.J. Henrich, *J. Biol. Chem.* **266,** 23428 (1991).
43. Y. Uehara-Kunugi, S. Shimohama, H. Kobayakawa, H. Tamura, and T. Taniguchi, *J. Neurochem.* **63,** 125 (1994).
44. R. Gopalakrishna, Z.H. Chen and U. Gundimeda, *Analyt. Biochem.* **206,** 24 (1992).
45. S. Skwish and W. Shain, *Life Sci.* **47,** 1037 (1990).
46. M. Virmani and B. Ahluwalia, *Alcohol* **27,** 393 (1992).
47. M. Narita, M. Makimura, Y. Feng, B. Hoskins, and I.K. Ho, *Brain Res.* **650,** 175 (1994).
48. T. Giambalvo and R.L. Wagner, *J. Neurochem.* **63,** 169 (1994).
49. R. Saxena, B.A. Saksa, K.S. Hawkins, and M.B. Ganz, *FASEB J.* **8,** 646 (1994).
50. R.A. Rabin and J.C. Winter, *Eur. J. Pharmacol.* **316,** 343 (1996).
51. H. Sershen, A. Hashim, and A. Lajtha, *Pharmacol. Biochem. Behav.* **47,** 13 (1994).
52. B. Frances, R. Gout, J. Cros, and J.M. Zajac, *Fundam. Clin. Pharmacol.* **6,** 327 (1992).
53. S.D. Glick, K. Rossman, N.C. Rao, I.M. Maisonneuve, and J.N. Carlson, *Neuropharmacol.* **31,** 497 (1992).
54. S.D. Glick, K. Rossman, S. Steindorf, I.M. Maisonneuve, and J.N. Carlson, *Eur. J. Pharmacol.* **195,** 341 (1991).
55. D.Dzoljic, C.D. Kaplan, and M.R. Dzoljic, *Arch. Int. Pharmacol. Ther.* **294,** 64 (1988).
56. S.L. Cappendijk and M.R. Dzoljic, *Eur. J. Pharmacol.* **241,** 261 (1991).
57. H. Rezvani, D.H. Overstreet, and D.S. Janowsky, *Alcohol Clin. Exp. Res.* **19,** 15A
58. H. Rezvani, D.H. Overstreet, and Y.W. Lee, *Pharmacol. Biochem. Behav.* **52,** 615 (1995).
59. H. Rezvani, D.H. Overstreet, Y. Yang, I.M. Maisonneuve, U.K. Bandarage, M.E. Kuehne, and S.D. Glick, *Pharmacol. Biochem. Behav.* **58,** 615 (1996).
60. S.I. Dworkin, S. Gleeson, D. Meloni, T.R. Koves, and T.J. Martin, *Psychopharmacol.* **117,** 257 (1995).
61. S.G. Sheppard, *J. Sub. Abuse Treat.* **11,** 379 (1994).
62. L.B. Ray, *Science* **284,** 756 (1999).
63. S.B. Lane-Ladd, J. Pineda, V.A. Boundy, T. Pfeuffer, J. Krupinski, G.K. Aghajanian, and E.J. Nestler, *J. Neurosci.* **17,** 7890 (1997).
64. J. Nestler and G.K. Aghajanian, *Science* **278,** 58 (1997).
65. I. Leshner, *Science* **278,** 45 (1997).
66. D. French, K. Dillion, and S.F. Ali, *Life Sci.* **59,** 199 (1996).

—CHAPTER 8—

IBOGAINE IN THE TREATMENT OF HEROIN WITHDRAWAL

DEBORAH C. MASH, CRAIG A. KOVERA, AND JOHN PABLO

Departments of Neurology and Pharmacology
University of Miami School of Medicine
Miami, FL 33124

RACHEL TYNDALE

Center for Addictions and Mental Health
University of Toronto
Toronto, Canada

FRANK R. ERVIN

Department of Psychiatry and Human Genetics
McGill University
Montreal, Canada

JEFFREY D. KAMLET

Mt. Sinai Medical Center
Addiction Treatment Program
Miami Beach, FL 33140

W. LEE HEARN

Miami-Dade County Medical Examiner Dept.
Miami, FL 33299

THE ALKALOIDS, Vol.56
0099-9598/01 $35.00

I. Introduction

Ibogaine, is a naturally occurring, psychoactive indole alkaloid derived from the roots of the rain forest shrub *Tabernanthe iboga*. Indigenous peoples of Western Africa use ibogaine in low doses to combat fatigue, hunger, and thirst, and in higher doses as a sacrament in religious rituals (*1*). The use of ibogaine for the treatment of drug dependence has been based on anecdotal reports from groups of self-treating addicts that the drug blocked opiate withdrawal and reduced craving for opiates and other illicit drugs for extended time periods (*2-4*). Preclinical studies have supported these claims and provided proof-of-concept in morphine-dependent rats (*5,6*). While ibogaine has diverse CNS effects, the pharmacological targets underlying the physiological and psychological actions of ibogaine in general, or its effects on opiate withdrawal in particular, are not fully understood. Pharmacological treatments for heroin addiction currently employ two treatment strategies: detoxification followed by drug-free abstinence or maintenance treatment with an opioid agonist. Because agonist maintenance with methadone usually has the goal of eventual detoxification to a drug-free state, the use of medications to facilitate this transition is a clinically important treatment strategy. Anecdotal reports suggest that ibogaine has promise as an alternative medication approach for making this transition (*4*). Ibogaine has an added benefit to other detoxification strategies in that the treatment experience seems to bolster the patient's own motivational resources for change.

There have been few reports of the effects of ibogaine in humans. Anecdotal accounts of the acute and long-term effects of ibogaine have included only a small series of case reports from opiate and cocaine addicts with observations provided for only seven and four subjects, respectively (*2,3*). A retrospective case review of 33 ibogaine treatments for opioid detoxification in nonmedical settings under open label conditions has suggested further that the alkaloid has ameliorative effects in acute opioid withdrawal (*4*). However, objective investigations of ibogaine's effects on drug craving, and the signs and symptoms of opiate withdrawal, have not been done in either research or conventional treatment settings. Ibogaine is a drug with complex pharmacokinetics and an uncertain mechanism of action with regards to its alleged efficacy for the treatment of opiate dependence. Ibogaine is metabolized to noribogaine, which has a pharmacological profile that is different from that of the parent drug. We report here that ibogaine is effective in blocking opiate withdrawal, providing an alternative approach for opiate-dependent patients who have failed other conventional treatments. Identifying noribogaine's mechanism of action may explain how ibogaine promotes rapid detoxification from opiates after only a single dose.

II. Identification of a Primary Metabolite of Ibogaine

Our group developed an analytical method for quantifying ibogaine in blood samples from rats, primates, and humans (*7,8*). Using fullscan electron impact gas chromatography/mass spectrometry (GC/MS), a primary metabolite, 12-hydroxyibogamine (noribogaine) was detected for the first time in blood and urine samples. The analytical procedure involved a solvent extraction under basic conditions with D_3-ibogaine as an internal standard. Urines taken from dosed monkeys and humans were extracted under strongly basic conditions, extracts were evaporated, reconstituted, and analyzed by GC/MS in full scan electron impact ionization mode. Analysis of the resulting total ion chromatograms revealed a peak identified as ibogaine by comparison with an authentic standard. All samples were found to contain a second major component eluting after ibogaine. Similar spectral characteristics of this peak to ibogaine's spectrum defined it as an ibogaine metabolite, which is formed by the loss of a methyl group (Figure 1). The site for metabolic demethylation of ibogaine was the

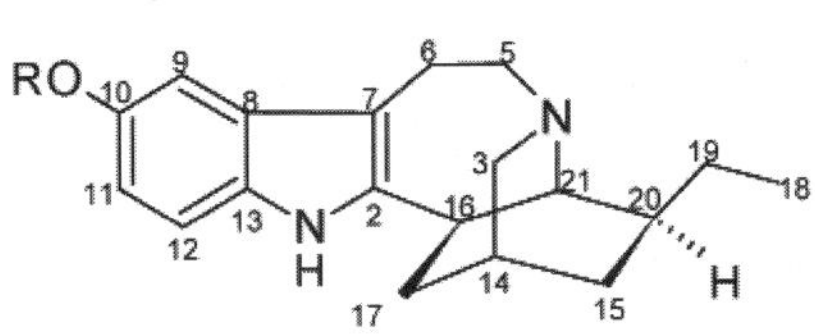

Ibogaine, R = CH_3 (Le Men-Taylor numbering)
Noribogaine (10-Hydroxyibogamine)*, R = H

*Noribogaine has frequently been referred to as 12-hydroxyibogamine in the biological and medical literature, based on the Chemical Abstracts numbering system for this alkaloid skeleton.

FIGURE 1. MOLECULAR STRUCTURES OF IBOGAINE AND NORIBOGAINE. Ibogaine undergoes *O*-demethylation to form 12-hydroxyibogamine (noribogaine).

methoxy group, resulting in the compound 12-hydroxyibogamine (noribogaine). The identity of the desmethyl metabolite was confirmed using an authentic standard of noribogaine (Omnichem S.A., Belgium) and gave a single peak at the same retention time and with the same electron impact fragmentation pattern as the endogenous compound isolated from monkey and human urine (*7*).

III. Cytochrome P450 Metabolism and Genetic Polymorphisms

Ibogaine, like most CNS drugs, is highly lipophilic and is subject to extensive biotransformation. Ibogaine is metabolized to noribogaine in the gut wall and liver (Figure 2, schematic). Ibogaine is *O*-demethylated to noribogaine primarily by cytochrome P4502D6 (CYP2D6). An enzyme kinetic examination of ibogaine *O*-demethylase activity in pooled human liver microsomes suggested that two (or more) enzymes are involved in this reaction (*8*). In this study, ibogaine was incubated with a set of microsomes derived from cell lines selectively expressing only one human cytochrome P450 enzyme and with a series of human liver microsome preparations, characterized with respect to their activities toward cytochrome P450 enzyme selective substrates to estimate the relative contributions of the various P450 enzymes to the metabolism of ibogaine *in vitro*. The enzyme CYP2D6 showed the highest activity toward the formation of noribogaine, followed by CYP2C9 and CYP3A4 (*9*).

Depending on whether a particular isoenzyme is present or absent, individuals are classified as extensive or poor metabolizers. The influence of genetic polymorphisms on the biotransformation of ibogaine under *in vivo* clinical conditions has been examined in recent studies (*9*). The results demonstrate that there are statistically significant differences in the two populations with regard to Cmax and $t_{1/2}$ (elim) and area under the curve (AUC) of the parent drug and metabolite, indicating that the disposition of ibogaine is dependent on polymorphic CYP2D6 distribution (Table 1). Since some of the CNS activity may be the result of noribogaine, the CYP2D6 phenotype may prove to be an important determinant in the clinical pharmacology of ibogaine. Many CYP2D6 substrates are subject to drug interactions. In considering the potential patient population who might benefit from ibogaine, many of these patients may have taken other medications (prescription and/or illicit), increasing the potential for serious adverse drug interactions.

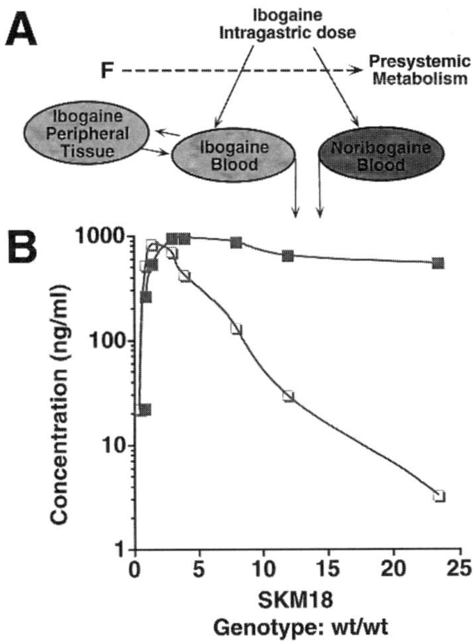

FIGURE 2. TIME COURSE OF WHOLE BLOOD CONCENTRATIONS OF IBOGAINE AND NORIBOGAINE AFTER ORAL ADMINISTRATION TO DRUG-DEPENDENT VOLUNTEER. Pharmacokinetics of ibogaine and noribogaine over the first 24 hours after oral dose in a human subject. Data shown are from a representative male subject (wt/wt, extensive metabolizer). Values for parent drug and desmethyl metabolite were measured in whole blood samples at the times indicated. Open squares indicate ibogaine concentrations and shaded squares indicate noribogaine concentrations. SK, St. Kitts, W.I., Subject Code.

TABLE 1.
PHARMACOKINETIC PARAMETERS OF IBOGAINE AND NORIBOGAINE IN HUMAN EXTENSIVE AND POOR METABOLIZERS (CYP2D6)

	*Extensive Metabolizers	**Poor Metabolizers
Ibogaine		
t_{max},hr	1.70 ± 0.15	2.50 ± 1.04
C_{max},ng/ml	737 ± 76	896 ± 166
$AUC_{0\text{-}24}$hr,ng • hr/ml	3936 ± 556	11471 ± 414
$t_{1/2}$,hr	7.45 ± 0.81	NQ
Noribogaine		
t_{max},hr	6.17 ± 0.85	3.17 ± 1.36
C_{max},ng/ml	949 ± 67	105 ± 30
$AUC_{0\text{-}24}$hr,ng • hr/ml	14705 ± 1024	3648 ± 435
$t_{1/2}$,hr	NQ	NQ

* N = 24 (10.0 mg/kg), 16 males and 8 females

** N = 3, 3 males (10.0 mg/kg)

IV. Ibogaine Pharmacokinetics

Pharmacokinetic measurements have been obtained from human drug-dependent patient volunteers who had received single oral doses of ibogaine (Table 1; Figure 2). Figure 2 illustrates the pharmacokinetic profile of ibogaine and the metabolite following oral doses of the drug in a representative male subject. Table 1 shows that CYP2D6 mediated metabolism of ibogaine resulted in high levels of noribogaine in blood, with Cmax values in the same range as the parent drug. The time required to eliminate the majority of absorbed ibogaine (>90%) was 24 hours post-dose (Figure 2). The pharmacokinetic profiles measured in whole blood demonstrate that the concentrations of noribogaine measured at 24 hours remained elevated, in agreement with previous findings (*10*). The still elevated concentrations of noribogaine in blood at 24 hours after drug administration limited the quantitation of the terminal half-life of the metabolite. Noribogaine was measured in CYP2D6 deficient subjects, but at concentrations that were markedly lower than for the extensive metabolizers. Conversion of the parent to noribogaine in CYP2D6 deficient subjects may reflect the metabolic contribution of other cytochromes (CYP2C9, CYP3A4). The concentration of noribogaine measured at 24 hours post-dose in the subject in Figure 2 was in the range of 800 ng/ml, similar to the peak concentration of ibogaine that was measured in this representative subject. Pharmacokinetic measurements in human volunteers administered oral doses of ibogaine showed that the area under the curve (AUC) for the parent compound was approximately three-fold less than for the active metabolite (Table 1). Thus, noribogaine reaches sustained high levels in blood after a single administration of the parent drug.

Since the metabolite has been shown in radioligand binding assays to have higher affinities for certain CNS targets, it can be estimated that the contribution of the metabolite to the total pharmacodynamic profile of ibogaine is significant. To display *in vivo* activity, it is necessary for CNS drugs to reach the brain. Since it is difficult to study these processes in humans, it is common to study the penetration of a CNS active drug into the brains of laboratory animals. The concentrations of ibogaine and noribogaine have been measured in rat brain following both oral and intraperitoneal (i.p.) administrations (*11,12*). The significance of micromolar interactions of ibogaine and noribogaine with various radioligand binding sites was related to the concentration of parent drug and metabolite in brain (Table 2). Regional brain levels of ibogaine and noribogaine were measured in rat cerebral cortex, striatum, brainstem, and cerebellum at 15 minutes, 1 and 2 hours postdrug administration. We have shown that ibogaine is rapidly detected in brain following oral administration. The metabolite was detected at the earliest time point (15 minutes), consistent with first pass metabolism of the parent drug (*11*). Administration of ibogaine (40 mg/kg i.p., 50

TABLE 2.
PHARMACOKINETIC PARAMETERS OF IBOGAINE AND NORIBOGAINE IN MALE RAT (SPRAGUE-DAWLEY)

	*Whole Blood 40 mg/kg i.p.	*Brain 40 mg/kg i.p.	**Brain 50 mg/kg p.o.
Ibogaine			
t_{max},hr	0.10 ± 0.03	1.00 ± 0.14	1.00 ± 0.21
C_{max},ng/ml or	3859 ± 789	3782 ± 418	5210 ± 480
ng/g [µM]	[11.2 ± 2.3]	[11.0 ± 1.2]	[15.1 ± 1.4]
AUC, ng • hr/ml or	10636 ± 341	22098 ± 922	NQ
ng/g [µM • hr]	[30.7 ± 1.0]	[63.9 ± 2.7]	
$t_{1/2}$,hr	2.38 ± 0.50	11.05 ± 1.15	NQ
Noribogaine			
t_{max},hr	2.40 ± 0.04	2.00 ± 0.16	2.00 ± 0.28
C_{max},ng/ml or	7265 ± 953	3236 ± 514	3741 ± 423
ng/g [µM]	[21.9 ± 2.9]	[9.8 ± 1.6]	[11.3 ± 1.3]
AUC, ng • hr/ml or	96920 ± 741	38797 ± 324	NQ
ng/g [µM • hr]	[292.0 ± 2.2]	[117.9 ± 1.0]	

NQ, not quantifiable
Noribogaine $t_{1/2}$ not quantifiable
* Noncompartmental pharmacokinetic analysis over a 24 hr. period
** Noncompartmental pharmacokinetic analysis over a 2 hr. period
Data represent the average values from individual animals (n = 4) assayed in duplicate.

mg/kg p.o.) in rodents resulted in levels of ibogaine and noribogaine that ranged from 10 to 15 µM and 10 to 12 µM, respectively. The results demonstrate that noribogaine reaches significant concentrations in brain following both routes of administration in rodents. Thus, the concentrations of noribogaine in brain may activate processes that cause the desired effects of suppressing opiate withdrawal signs and diminishing drug craving.

V. Setting and Study Design

We have had the opportunity to describe the clinical experience of a series of patients undergoing opiate detoxification with ibogaine. The study was conducted in a 12 bed freestanding facility in St. Kitts, West Indies. The treatment program had a planned duration of 12 to 14 days and stated goals of: (1) safe physical detoxification from opiates, (2) motivational counseling, and (3) referral to aftercare programs and community support groups (twelve-step programs). Subjects were self-referred for inpatient detoxification from opiates (heroin or methadone) and met inclusion/exclusion criteria. All individuals were deemed fit and underwent treatment following a physician's review of the history and physical examination. Participants did not have histories of stroke, epilepsy, or axis I psychotic disorders. Results of the electrocardiogram and clinical laboratory testing were within predetermined limits. All subjects signed an informed consent for ibogaine treatment. Overall, the sample of 32 patients was predominately male (69%) and white (82%), with a mean age of 33.6 years and a mean length of addiction of 11.1 years.

All participants met DSM-IV criteria for opioid dependence and had positive urine screens at entry to the study. Participants were assigned to fixed-dose (800 mg; 10 mg/kg) of ibogaine HCl under open-label conditions. Subjects were genotypyed for the CYP2D6 alleles (*2, *4, *5 and wt alleles), as described previously (*13*). On admission, participants were administered the Addiction Severity Index (*14*) and received structured psychiatric evaluations before and after ibogaine treatment (SCID I and II). In cases where the participant's responses were deemed questionable due to intoxication or withdrawal signs, portions of all interviews were repeated later, as necessary. Additional information about substance use history and past/current medical condition(s) was gathered and later cross-referenced for accuracy through a separate comprehensive psychosocial assessment.

VI. Physician Ratings of Withdrawal

Two physicians rated as present or absent 13 physical signs typically associated with opiate withdrawal, based on a 10-minute period of observation (*14,15*). The Objective Opiate Withdrawal Scale (OOWS) data were analyzed from three assessments performed during the period spent in the clinic under medical monitoring, given that those points in relation to ibogaine administration were highly comparable among all patients. The attending physician performed the first assessment following clinic admission an average of 1 hour before ibogaine administration and 12 hours after the last dose of opiate. A psychiatrist without knowledge of the admitting OOWS score performed the second assessment an average of 10 to 12 hours after ibogaine administration and 24 hours after the last opiate dose. The attending physician performed the third assessment 24 hours following ibogaine administration and 36 hours after the last opiate dose. Physician's ratings were subjected to repeated measures analysis of variance (ANOVA) with treatment phase (pre-ibogaine, post-ibogaine, and program discharge) as the within-subjects factor.

VII. Subjects' Self-Report of Withdrawal Symptoms

The Opiate-Symptom Checklist (OP-SCL) was developed for the present study as a subtle assessment of withdrawal symptoms, given that many subjects' verbal reports about withdrawal experience were generally exaggerated, both in number and severity of symptoms. Each of the 13 items that comprises the OP-SCL scale were taken from the Hopkins Symptom Checklist-90, with the criteria for selection based on whether it appeared in two other self-report withdrawal questionnaires, the Addiction Research Center Inventory (*16*) and the Subjective Opiate Withdrawal (*17*) scales. Subjects also completed a series of standardized self-report instruments relating to mood and craving at three different time points during the study within 7 to 10 days after the last dose of opiate. Subjects were asked to provide ratings of their current level of craving for opiates using questions from the Heroin Craving Questionnaire (HCQN-29) (*18*). Self-reported depressive symptoms were determined by the Beck Depression Inventory (BDI) (*19*). Subjects' scores were subjected to repeated measures analyses of variance across treatment phase (pre-ibogaine, post-ibogaine, and discharge) as the within-subjects factor for the total score from the OP-SCL, BDI, and the HCQN-29.

VIII. Acute Detoxification and Behavioral Outcomes

Physical dependence on opiates is characterized by a distinctive pattern of signs and symptoms that make up the naturalistic withdrawal syndrome. The physical dependence produced by an opiate is assessed usually by discontinuation of opioid treatment (spontaneous withdrawal) or by antagonist-precipitated withdrawal. All of the subjects identified opiates as one of the primary reasons for seeking ibogaine treatment and demonstrated active dependence by clinical evaluation, objective observations, and positive urine screen. Physician ratings demonstrate that ibogaine administration brings about a rapid detoxification from heroin and methadone (Figure 3A). The post-ibogaine OOWS rating obtained 10 to 12 hours after ibogaine administration and 24 hours following the last opiate dose was significantly lower than the rating obtained 1 hour prior to ibogaine administration and 12 hours after the last opiate dose. At 24 hours after ibogaine administration and 36 hours after the last opiate dose, the OOWS rating was significantly lower than the pre-ibogaine rating. The blinded post-ibogaine ratings between doctors agreed well item for item and were not significantly different from one another in terms of the mean total OOWS score (mean ± 1 SD, N = 32). These objective measures demonstrate the effects of ibogaine on opiate withdrawal assessed in this study. Objective signs of opiate withdrawal were rarely seen and none were exacerbated at later time points. The results suggest that ibogaine provided a safe and effective treatment for withdrawal from heroin and methadone. The acute withdrawal syndrome in addicts dependent on heroin begins approximately 8 hours after the last heroin dose, peaks in intensity at 1 to

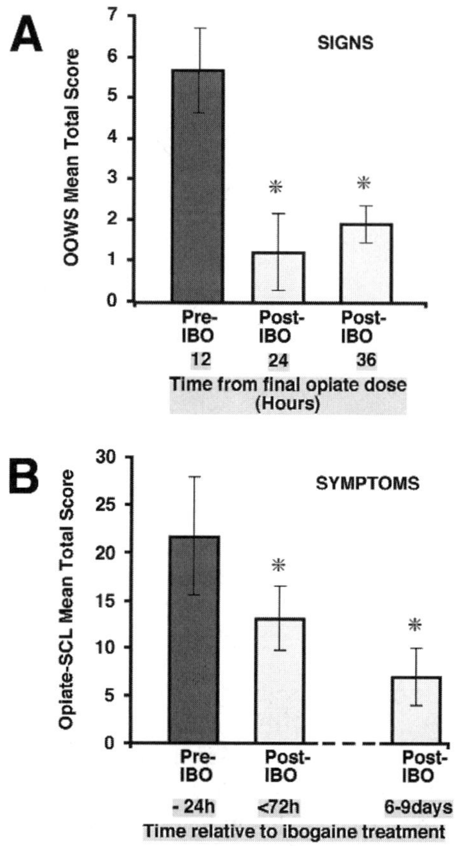

FIGURE 3. SCORES ON THE OBJECTIVE OPIATE WITHDRAWAL SCALE. (a) The effects of single-dose ibogaine treatment on opiate withdrawal signs at three physician-rated assessment times (12, 24, and 36 hours after the last dose of opiate). Average data are shown (mean ± 1 SD, N = 32). *$P < .05$. (b) The effects of single-dose ibogaine on patients self-report Opiate-Symptom Checklist (OP-SCL). The OP-SCL was developed for the present study as a subtle assessment of patients' subjective complaints based on 13 items selected from the Hopkins Symptom Checklist rated for intensity from 0 to 4. The maximum score attainable for the OP-SCL was 42.
* $p < .05$.

2 days, and subjective symptoms subside within 7 to 10 days. Self-reports of withdrawal symptoms shortly after recovery from ibogaine treatment (< 72 hours) were significantly decreased from the pre-ibogaine rating obtained 12 hours after the last use of opiates and were comparable to the level of discomfort reported at program discharge approximately one week later (Figure 3B). Thus, for subjects undergoing ibogaine detoxification, all of the subjects were successful during the detoxification process and many were able to maintain abstinence from illicit opiates and methadone over the months following detoxification (data not shown). Perhaps the most important observation was the ability of a single dose of ibogaine to promote a rapid detoxification from methadone without a gradual taper of the opiate. These preliminary observations of ibogaine treatment suggest that methadone withdrawal was not more difficult than heroin withdrawal following ibogaine detoxification. As discussed below, we suggest that the long-acting metabolite noribogaine may account for the efficacy of ibogaine treatment for both heroin and methadone withdrawal.

Craving is thought to be an important symptom contributing to continued drug use by addicts. Opiate-dependent subjects report increased drug craving during the early stages of withdrawal (*20*). We have previously reported that subjects undergoing opiate detoxification reported significantly decreased drug craving for opiates on five measures taken from the HCQN-29 scales at 36 hours posttreatment. These five measures inquired about specific aspects of drug craving, including urges, as well as thoughts about drug of choice or plans to use the drug. Questions are asked also about the positive reinforcing effects of the drug or the expectation of the outcome from using a drug of choice or the alleviation of withdrawal states. Perceived lack of control over drug use was included, since it is a common feature of substance-abuse disorders and is most operative under conditions of active use, relapse, or for subjects at high risk. The results demonstrated that across craving measures, the mean scores remained significantly decreased at program discharge (*10*). BDI scores were also significantly reduced both at program discharge and at 1-month follow-up assessments (*10*). Heroin craving is known to be dramatically reduced depending on the lack of availability of the abused drug in a controlled setting. Thus, more meaningful studies of ibogaine's ability to suppress heroin craving require further investigations done under naturalistic conditions.

IX. Cardiovascular Changes and Side Effects of Ibogaine

Ibogaine has a variety of dose-dependent pharmacological actions, which may not be relevant to its effectiveness for opiate detoxification and diminished drug cravings, but may influence considerations for safety. However, toxicological studies in primates have demonstrated previously that ibogaine administration at doses recommended for opiate detoxification is safe (*21*). The FDA Phase I Pharmacokinetic and Safety investigations by our group have not advanced in the United States due to a lack of funds to support clinical investigations of ibogaine in patient volunteers. However, we have had the opportunity to obtain additional safety data in drug-dependent subjects under controlled conditions in human studies conducted in St. Kitts, West Indies. For these subjects, baseline screening included a medical evaluation, physical examination, electrocardiogram, blood chemistries, and hematological workup, as well as psychiatric and chemical dependency evaluations. In some cases, more extensive evaluations were done to rule out cardiac risk factors and to exclude subjects for entry to the study. The recognition of the cardiovascular actions of ibogaine date back to the 1950s, when the CIBA Pharmaceutical Company investigated ibogaine as an antihypertensive agent. Ibogaine at doses used for opiate detoxification may lower blood pressure and heart rate when the drug reaches peak concentrations in blood. In contrast, the opiate withdrawal syndrome is associated with increases in pulse, systolic and diastolic blood pressures, and respiratory rate.

Our observations of the safety of ibogaine have not been limited to opiate-dependent subjects. To date, we have evaluated ibogaine's safety in more than 150 drug-dependent subjects that were assigned to one of three fixed-dose treatments under open label conditions: 8, 10, or 12 mg/kg ibogaine. Adverse effects were assessed by clinician side-effect ratings and open-ended query. To date, no significant adverse events were seen under these study conditions. The most frequent side effects observed were nausea and mild tremor and ataxia at early time points after drug administration. Random regression of vital signs (respiration rate, systolic and diastolic blood pressures, and pulse) revealed no significant changes across time or by treatment condition for opiate-dependent subjects. However, a hypotensive response to ibogaine was observed in some cocaine-dependent subjects, which required close monitoring of blood pressure and which was responsive to volume repletion. Comparison of pre- and postdrug effects demonstrated that blood cell count, neurotrophil levels, and sodium and potassium levels were in the normal range. There were no significant changes from baseline seen on liver function tests. No episodes of psychosis or major affective disorder were detected at posttreatment evaluations. Intensive cardiac monitoring demonstrated that no electrocardiographic abnormalities were produced or exaggerated following ibogaine administration in subjects that were not comorbid for any cardiovascular risk factors. These preliminary results demonstrate that single doses of ibogaine were well tolerated in drug-dependent subjects. These preliminary observations are encouraging, but they do not diminish the possibility that ibogaine may have other medical risks not ordinarily associated with opiate withdrawal or with the use of tapering doses of methadone.

However, we anticipate, based on our clinical experience from offshore studies, that any potential adverse cardiovascular responses can be well managed within routine clinical practice.

X. Mechanism of Action

While the precise mechanism(s) underlying the expression of opiate withdrawal signs and symptoms are not fully understood, and may be different between humans and laboratory animals, the cellular and behavioral changes resulting from withdrawal and that have motivational relevance to drug-seeking behavior may involve the same neural circuits as those that participate in opiate dependence. Ibogaine and its active metabolite noribogaine act on a number of different neurotransmitter systems in the brain that may contribute to ibogaine's ability to suppress the autonomic changes, objective signs, and subjective distress associated with opiate withdrawal. However, we have speculated that the actions of noribogaine at mu-opioid receptors may account in part for ibogaine's ability to reduce withdrawal symptoms in opiate-dependent humans (*22*). For example, the desmethyl metabolite noribogaine has been shown to be a full agonist at the mu-opioid receptor (Table 3). This pharmacological activity, coupled with the

TABLE 3.
INHIBITORY POTENCY OF IBOGAINE AND NORIBOGAINE

	Ibogaine		Noribogaine		Pharmacodynamic
	IC_{50}(μM)	nh	IC_{50}(μM)	nh	Action
Serotonergic					
5-HT Transporter (RTI-55)	0.59 ± 0.09	0.8	0.04 ± 0.01	0.76	Reuptake Blocker
Opioidergic					
Mu (DAMGO)	11.0 ± 0. 9	1.0	0.16 ± 0.01	0.99	Agonist
Kappa 1 (U69593)	25.0 ± 0.6	1.1	4.2 ± 0.3	1.05	Partial Agonist (?)
Kappa 2 (IOXY)	23.8 ± 7.1	1.0	92.3 ± 9.2	1.03	Partial Agonist (?)
Glutaminergic					
NMDA (MK-801)	5.2 ± 0.2	0.9	31.4 ± 5.4	1.1	Channel Blocker

The values represent the mean ± SE of the IC_{50} value (μM) from 3-4 independent experiments, each performed in triplicate. nh, Hill slope long duration of action may produce a self-taper effect in opiate-dependent patients.

The relative contributions of the parent and metabolite to the pharmacodynamic effects have yet to be established with precise certainty. Results from animal studies indicate that opiate withdrawal is associated with hyperactivity of the noradrenergic system and with changes in a variety of other neurotransmitter systems (*23*). Pharmacological agents may have differential effects on different components of opiate withdrawal. In addition to affecting mu-opioid receptors in the brain, noribogaine also has affinity at kappa-opioid receptors and the serotonin transporter (*8*). Indirect serotonergic agonists have been shown to attenuate neuronal opiate withdrawal (*24*). The 5-HT releaser d-fenfluramine and the 5-HT reuptake blockers fluoxetine and sertraline reduce the withdrawal-induced hyperactivity of locus ceruleus neurons. We have demonstrated previously that noribogaine elevates serotonin concentrations in brain by binding to the 5-HT transporter (Table 3) (*8*). Dysphoric mood states associated with opiate withdrawal may be a contributing factor for relapse, since addicts often experience drug craving in conjunction with dysphoric mood states (*20*). An action at the 5-HT transporter may explain the antidepressant effects seen following ibogaine administration in human opiate-dependent patients (*10*). Clinical studies have previously suggested that patients who abused opiates may have been self-medicating their mood disorders, indicating a possible role for endogenous opiates in major depression (*25*). Dysphoria and drug craving reportedly persist in opiate addicts even after detoxfication from opiates has been completed. Thus, noribogaine's effects at multiple opioid receptors and the 5-HT transporter may explain the easy transition following only a single dose of ibogaine in humans following abrupt discontinuation of opiates. These observations suggest that noribogaine may have potential efficacy for use as a rapid opiate detoxification treatment strategy. Recognition of the different components (autonomic changes and the objective signs versus subjective signs, dysphoric mood, and drug craving) may suggest the need for a medication strategy that targets multiple neurotransmitter systems for the treatment of opiate withdrawal and for relapse prevention. The identification of noribogaine's mix of neurotransmitter receptors and neurotransporter binding sites provides additional support for medications targeted to different aspects of the opiate withdrawal syndrome.

Opiate agonist pharmacotherapy with buprenorphine is a new alternative to methadone maintenance for the treatment of opiate dependence (*20*). Noribogaine has some pharmacologic similarities to the mixed agonist-antagonist analgesic buprenorphine. Buprenorphine and noribogaine both act as mu agonists. Compared to buprenorphine's high affinity partial agonist profile, noribogaine has lower receptor affinity, but increased intrinsic activity over buprenorphine as a mu agonist. Behavioral and physiological evidence suggest that buprenorphine has kappa antagonist effects in addition to its action as a partial mu agonist. Noribogaine binds to kappa receptors, but acts as a partial agonist (Table 3). Both drugs have a long duration of action due to the slow rates of dissociation from opiate receptor sites. Thus, ibogaine's ability to inhibit opiate craving may be accounted for by the mixed mu- and kappa-opioid profile of the active metabolite noribogaine.

XI. Conclusion and Future Directions

Pharmacological treatments for opiate dependence include detoxification agents and maintenance agents. New experimental approaches have also been tried to reduce the time it takes to complete the process of detoxification or to further reduce persisting subjective reports of dysphoria and opiate craving. Ibogaine treatment is a novel approach that has similarities with other detoxification pharmacotherapies, including substitution with a longer-acting opiate (e.g., methadone or buprenorphine). However, ibogaine appears to be a prodrug with the beneficial effects residing in the active metabolite noribogaine. Thus, it would be useful to demonstrate that noribogaine alone is effective in detoxification of heroin-dependent and methadone-maintained patients. If noribogaine alone is safe and effective in open label studies, a randomized, double-blind study comparing noribogaine to clonidine-naltrexone detoxification would be justified. This clinical study would demonstrate whether noribogaine is more effective and has fewer adverse hemodynamic effects. Based on its spectrum of pharmacological activities, we suggest that noribogaine should also be considered as an alternative to methadone maintenance.

A pharmacological approach for the compliance problem has been the development of depot formulations that might be injected as infrequently as once a month. The long-acting pharmacokinetics of noribogaine suggests that the drug may, in fact, persist in the body for weeks to months. Thus, future development of depot noribogaine preparations may provide an optimal therapeutic approach for treating intractable opiate abusers. Another approach would be to combine a noribogaine taper with naltrexone. This approach may provide a means to shorten the time needed to initiate opiate antagonist therapy. Previous studies have also suggested the need for combination pharmacotherapies, such as antidepressants with buprenorphine (*20*). Interestingly, noribogaine has a pharmacological profile that includes actions on both serotonin and opiate systems in the brain. Although not discussed in this report, ibogaine provides an approach for the treatment of abuse of multiple substances including alcohol and cocaine. Many opiate-dependent patients abuse multiple drugs and alcohol. Thus, ibogaine and its active metabolite noribogaine represent two additional pharmacological treatments for opiate dependence. However, clinical studies are needed to demonstrate whether they will become viable alternatives for treating opiate dependence in the future. It remains to be seen if the politics surrounding this controversial treatment approach will limit the promise for future development of either ibogaine or noribogaine.

Acknowledgments

This research was supported in part by the Addiction Research Fund. We are grateful to the staff of the Healing Visions Institute for Addiction Recovery, Ltd., St. Kitts, West Indies, for their support of the Ibogaine Research Project.

References

1. R. Goutarel, O. Gollnhofer, and R. Sillans, *Psychedel. Monographs and Essays* **6,** 71 (1991).
2. S.G. Shepard, *J. Subst. Abuse Treat.* **11**(4), 379 (1994).
3. B. Sisko, *Multidis. Assoc. Psyched. Stud.* **IV,** 15 (1993).
4. K.R. Alper, H.S. Lotsof, G.M. Frenken, D.J. Luciano, and J. Bastiaans, *Am. J. Addict.* **8,** 234 (1999).
5. E.D. Dzoljic, C.D. Kaplan, and M.R. Dzoljic, *Arch. Int. Pharacodyn. Ther.* **294,** 64 (1988).
6. S.D. Glick, K. Rossman, N.C. Rao, I.M. Maisonneuve, and J.N. Carlson, *Neuropharmacol.* **31,** 497 (1992).
7. W.L. Hearn, J. Pablo, G. Hime, and D.C. Mash, *Jour. Anal. Toxicology.* **19,** 427 (1995).
8. D.C. Mash, J.K. Staley, M.H. Baumann, R.B. Rothman, and W.L. Hearn, *Life Sci.* **57,** PL45-50 (1995).
9. R.S. Obach, J. Pablo, and D.C. Mash, *Drug Metab. Dispos.* **25**(12), 1359 (1998).
10. D.C. Mash, C.A. Kovera, J. Pablo, R. Tyndale, F.R. Ervin, I.C. Williams, E.G. Singleton, and M. Mayor, *Ann. N.Y. Acad. Sci.* **914,** 394 (2000).
11. J.K. Staley, Q. Ouyang, J. Pablo, W.L. Hearn, D.D. Flynn, R.B. Rothman, K.C. Rice, and D.C. Mash, *Psychopharmacology* **127,** 10 (1996).
12. C. Zubaran, M. Shoaib, I.P. Stolerman, J. Pablo, and D.C. Mash, *Neuropsychopharm.* **21,** 119 (1999).
13. M.H. Heim and U.A. Meyer, *Lancet* **336,** 529 (1990).
14. C.K. Himmelsbach, *Ann. Intern. Med.* **15,** 829 (1941).
15. W.R. Martin and D.R. Jasinski, *J. Psychiatr. Res.* **7,** 9 (1969).
16. C.A. Haertzen and M.J. Meketon, *Diseases of the Nervous System* **29,** 450 (1968).
17. L. Handelsman, K.J. Cochrane, M.J. Aronson, R. Ness, K.J. Rubinstein, and P.D. Kanoff, *Amer. J. Drug & Alcohol Abuse* **13,** 293 (1987).
18. E.G. Singleton, Unpublished research. Available from the Clinical Pharmacology and Therapeutics Branch, Intramural Research Program, NIDA, 55 Nathan Shock Drive, Baltimore, MD 21224, USA (1996).

19. A.T. Beck, C.H. Ward, M. Mendelson, J. Mock, and J. Erbaugh, *Arch. Gen. Psychiat.* **4,** 561 (1961).
20. E. Best, A.J. Olivieto, and T.R. Kosten, *CNS Drugs* **6,** 301 (1996).
21. D.C. Mash, Preclinical studies of ibogaine in the primate: Anatomical, neurochemical and behavioral observations. Presented to the NIDA-Sponsored Ibogaine Review Meeting. March (1995).
22. J. Pablo and D.C. Mash, *NeuroReport* **9,** 109 (1998).
23. T.R. Kosten, *J. Nerv. Ment. Dis.* **178,** 217 (1990).
24. H. Akaoka and G. Aston-Jones, *Neuroscience* **54,** 561 (1993).
25. E.J. Khantzian, *Rec. Dev. Alcohol* **8,** 255 (1990).

—— CHAPTER 9 ——

SIGMA RECEPTORS AND *IBOGA* ALKALOIDS

WAYNE D. BOWEN

Unit on Receptor Biochemistry and Pharmacology
Laboratory of Medicinal Chemistry
National Institute of Diabetes and Digestive and Kidney Diseases
National Institutes of Health
Bethesda, MD 20892

I. Introduction

Ibogaine, one of the naturally occurring indole alkaloids found in the shrub *Tabernanthe iboga* of central Africa, has been shown to have psychotropic effects, and was initially used for its hallucinogenic properties (*1,2*). Anecdotal reports of heroin and cocaine addicts suggested that taking ibogaine decreased drug craving, with the effects lasting for several months (*3,4*). This has been supported in several animal studies where ibogaine has been shown to reduce self-administration of both morphine and cocaine (*5-8*). On this basis, there has been interest in investigating ibogaine for its potential in treating drug abuse (*9*).

However, ibogaine has also been shown to have negative effects in animal studies that might potentially limit its clinical utility in humans. These effects include production of tremors and neurotoxicity (*1,2*). Specifically, treatment of rats with ibogaine at 100 mg/kg in one to three doses was found to cause

THE ALKALOIDS, Vol.56
0099-9598/01 $35.00

activation of microglia and astrocytes and loss of Purkinje cells in the parasagittal zones of the cerebellar vermis (*10,11*). Harmaline was found to have similar effects. The receptor sites through which ibogaine mediates its antiaddictive and neurotoxic effects are not known with certainty, since it interacts with low affinity at a number of neurotransmitter and transporter sites including NMDA-glutamatergic and kappa- opioid receptors (*1,2*). Current evidence indicates that ibogaine and other *iboga* alkaloids might produce some of their neurotoxic effects by interaction with sigma-2 receptors.

II. Sigma Receptors

A. General Characteristics and Functions

Sigma receptors are membrane proteins that bind several psychotropic drugs with high affinity (*12*). They were initially proposed to be related to opioid receptors (*13*) and then confused with the phencyclidine binding site on the NMDA-glutamatergic receptor ionophore. Sigma receptors, as defined today, are unique binding sites, with a pharmacological profile unlike any other known neurotransmitter or hormone receptor (*14*). Initial interest in sigma receptors came mainly from their high affinity for typical neuroleptic drugs, such as haloperidol, and their potential as alternative targets for antipsychotic agents (*15,16*).

Two major subclasses of sigma receptors have been identified. These have been termed sigma-1 and sigma-2, and they are differentiated by their pharmacological profile, function, and molecular size (*17,18*). Both subtypes have high to moderate affinity for typical neuroleptics, with haloperidol exhibiting the highest affinity for both sites. However, sigma-1 receptors exhibit high affinity for (+)-benzomorphans, such as (+)-pentazocine, whereas sigma-2 receptors have low affinity for the (+)-benzomorphans. The (-)-isomers of benzomorphans do not strongly differentiate the two sites. Photoaffinity labeling revealed a molecular weight of 25 kDa for sigma-1 receptors and of 18- 21.5 kDa for sigma-2 receptors (*17,19*).

Sigma receptors are widely distributed throughout the brain, but occur in particularly high density in the motor regions. These include cerebellum, brainstem, motor nuclei, and substantia nigra (*12*). Sigma receptors are also found in high density in many tissues outside of the nervous system. Sigma receptors are present in endocrine, immune, and reproductive tissues (*20*). Both subtypes are expressed in high density in the liver and kidney (*19*). In addition, both subtypes of sigma receptors are found to be expressed in very high density in tumor cell lines derived from various tissues (*21*). These include neuroblastomas, glioma, melanoma, and carcinoma cell lines of breast, prostate, and lung. Furthermore, the expression of sigma receptors in tumor cell lines increases when the cells are in a state of rapid proliferation (*22*), and tumor tissue has been found to express a higher density of sigma receptors than surrounding normal tissue (*23*). High sigma receptor expression in tumor cell lines and up regulation during rapid cell growth suggests a possible role of sigma receptors in cell growth and proliferation.

No endogenous functional ligand (agonist) for sigma receptors has been conclusively identified. There is evidence for the existence of sigma receptor binding substances in brain and tissue extracts (*24,25*), and for depolarization-induced release of a substance(s) from brain tissue slices that occupies sigma receptors (*26*). Progesterone has affinity for sigma-1 receptors (*27*) and certain neurosteroids have been shown to exhibit modulatory effects via sigma receptors (*28*). This has led to the proposal that certain steroids may be endogenous ligands for the sigma receptors.

The sigma-1 receptor has been cloned in guinea pig, mouse, rat, and human, and shown to be a novel protein with > 90% species homology (*29-32*). The sigma-1 protein is unrelated to any known receptor family. The protein sequence has substantial homology to the fungal sterol biosynthetic enzyme, $\Delta^{8,7}$-sterol isomerase (*29*). This has suggested a role of sigma-1 receptors in sterol metabolism, particularly in that of neurosteroids (*33*). However, the protein exhibits no enzymatic activity and is unrelated to the mammalian $\Delta^{8,7}$-sterol isomerase (*34*). Thus, the relevance of sigma-1 receptors to sterol metabolism is not yet clear. In light of the affinity of progesterone and some neurosteroids for sigma-1 receptors, it is possible that the homology represents a steroid binding activity. No information on the structure of the sigma-2 receptor is available at present.

Some of the functions attributed to sigma-1 receptors include: (1) modulation of synthesis and release of dopamine (*35,36*) and acetylcholine (*37*), (2) modulation of NMDA-type glutamatergic receptor electrophysiology (*38*), (3) modulation of NMDA-stimulated neurotransmitter release (*39,40*), (4) modulation of muscarinic receptor-stimulated phosphoinositide turnover (*41*), (5) neuroprotective and antiamnesic activity (*42*), (6) modulation of opioid analgesia (*43*), and (7) alteration of cocaine-induced locomotor activity and toxicity (*44*).

Less is known about the functions of sigma-2 receptors in the brain. As mentioned above, sigma receptors are highly expressed in regions of the brain that regulate posture or that are involved in motor control (*12*). Microinjection of sigma ligands into motor regions of the brain induces marked alterations in movement and posture. Microinjections of typical neuroleptics, as well as selective sigma ligands into the rat red nucleus, induces an acute dystonic reaction (*45*). Microinjection of sigma ligands into the facial nucleus, or spinal trigeminal nucleus oralis, produced orofacial dyskinesias (vacuous chewing and facial tremors) in rats (*46*). Unilateral microinjection of sigma ligands into the substantia nigra results in contralateral circling (*47*). These effects on motor behavior and posture were described by a pharmacological profile generally consistent with mediation by sigma-2 receptors (*47,48*). These results suggest that sigma-2 receptors might be involved in the regulation of motor behavior and may contribute to some of the motor side effects of typical antipsychotic drugs, particularly tardive dyskinesias and acute dystonias (*12,49*).

B. Sigma-2 Receptors and Cell Death

Results from some of the brain microinjection studies described above suggested that some sigma ligands might be neurotoxic. Reduced haloperidol (a major haloperidol metabolite and a potent sigma ligand) and the cyclohexane diamine, BD614, caused extensive gliosis and loss of magnocellular neurons in and around the injection site (*50,51*). Further investigation *in vitro* revealed that some ligands were cytotoxic to tumor cell lines of both neuronal and nonneuronal origin (e.g., SK-N-SH neuroblastoma and C6 glioma), as well as to primary cultures of rat central nervous system (e.g., cerebellar granule cells, cortical neurons, superior cervical ganglion cells) (*52-54*). Sigma ligands initially caused damage to cell processes, followed by a loss of processes, assumption of a spherical shape ("rounding"), and detachment from the surface. Continued exposure to sigma compounds ultimately resulted in cell death. The effect was dose dependent, with higher doses causing morphological changes and death at shorter time periods. In primary cultures, effects could be seen in relatively low doses (1 to 3 μM) for the most active compounds, with effects occurring over a course of up to 21 days with some cultures. This confirms the chronic nature of the effect, where the effective dose decreases as the period of exposure increases.

Detailed assessment of the pharmacology of this effect indicated the involvement of sigma-2 receptors. Compounds binding to both sigma-1 and sigma-2 sites, such as haloperidol, were active, whereas sigma-1-selective compounds such as (+)-pentazocine and compounds, which lack significant sigma affinity, but which are agonists or antagonists at other receptors, were inactive (*52-54*). Sigma-2 receptor specificity was confirmed using the sigma-2-selective ligands CB-64D and CB-184 (*55*), which were quite potent at producing cytotoxicity. Thus, chronic activation of sigma-2 receptors results in morphological changes and cell death.

Cell death may occur by either necrosis or apoptosis (*56-58*). Necrosis is thought to result from physical or chemical injury to the cell. It is typified by cell swelling, destruction of cytoplasmic organelles, and loss of membrane integrity, and is not controlled by a genetic program. Necrosis in tissues is accompanied by an inflammatory response. Apoptosis (or programmed cell death) can result from various and specific developmental or environmental stimuli. It is typified by cell shrinkage, membrane blebbing and cytoplasmic boiling, chromatin condensation, and nuclear DNA fragmentation, all with maintenance of membrane integrity (*58*). In tissues, apoptotic cells are removed by macrophages or adjacent epithelial cells, without generating an inflammatory response. Apoptosis is a highly regulated process, involving several signaling pathways, transcription factors, proteolytic enzymes (caspases), nucleases, and other intracellular molecules that both promote and prevent the death of the cell (*56,58*). Induction of apoptotic cell death or dysregulation of apoptosis plays a key role in several physiological and pathological processes (57). These include development, immune responses, carcinogenesis and tumor progression, hypoxia, viral infection, and degenerative disorders. Furthermore, many cytotoxic agents cause cell death via apoptosis.

The mode of cell death induced by sigma-2 ligands in various cell types was found to be apoptotic (*59,60*). Treatment of SK-N-SH neuroblastoma cells or breast tumor cell lines with sigma-2 agonists, including CB-64D and CB-184, caused inversion of phosphatidyl serine, DNA fragmentation, and nuclear condensation, as measured by annexin-V binding, TdT-mediated dUTP nick-end labeling (TUNEL), and bisbenzimide (Hoechst 33258) staining, respectively. All of these are known hallmarks of apoptosis (*58*). Similar results were observed using primary cultures of rat cerebellar granule cells (*59*). Treatment of cells with sigma-1 selective ligands (e.g. (+)-pentazocine) produced no change in the cells. Thus, activation of sigma-2 receptors subsequently activates the cellular machinery, which results in programmed cell death.

C. Sigma-2 Receptors and Calcium Signaling

The ability of sigma ligands to induce morphological changes and apoptosis led to an investigation of the signaling mechanisms that are utilized by sigma-2 receptors. It is well established that calcium plays a role in cytotoxicity and that alterations in cell calcium levels play a role in the induction of apoptosis in various cell types (*61-63*). Thus, the ability of sigma receptors to modulate intracellular calcium was investigated using indo-1-loaded human SK-N-SH neuroblastoma cells. Sigma receptor ligands from various structural classes produced two types of increases in intracellular (cytosolic) calcium concentration ($[Ca^{++}]i$) (*64,65*). Sigma receptor-inactive compounds structurally similar to the most active sigma ligands produced little or no effect. Mediation of the effect on $[Ca^{++}]i$ by sigma-2 receptors was strongly indicated by (1) the high activity of the sigma-2-selective ligand CB-64D, (2) the greater activity of CB-64D ((+)-isomer) over CB-64L ((-)-isomer), and (3) the very low activity of the sigma-1-selective (+)-benzomorphans, (+)-pentazocine, (±)-SKF-10,047, and dextrallorphan (*65*).

The two types of rise in $[Ca^{++}]i$ produced by sigma-2 receptor ligands were distinguishable both temporally and by source (*65*). The compounds all produced an immediate, dose-dependent, and transient rise in $[Ca^{++}]i$, which usually returned to near baseline within 7 to 10 minutes. This transient rise in $[Ca^{++}]i$ occurred in the absence of extracellular calcium and was virtually eliminated by pretreatment of cells with thapsigargin. Thus, sigma-2 receptors stimulate a transient release of calcium from the endoplasmic reticulum. Prolonged exposure of cells to sigma receptor ligands resulted in a latent and sustained rise in $[Ca^{++}]i$. This sustained rise in $[Ca^{++}]i$ was affected neither by removal of extracellular calcium nor by thapsigargin pretreatment. This indicates that sigma-2 receptor ligands also induce release of calcium from mitochondrial stores or from some other calcium store that is insensitive to thapsigargin, such as golgi apparatus. These findings indicate that sigma-2 receptors may utilize calcium signals in producing cellular effects.

The fact that production of a rise in $[Ca^{++}]i$, changes in cellular morphology, and induction of apoptosis all have the same pharmacological profile suggests that these processes are linked, and that sigma-2 receptors coordinate the events leading to apoptotic cell death. In view of the ability of sigma-2 receptors to induce cytotoxicity, and in light of the lack of information regarding the receptor sites(s) that might mediate ibogaine-induced neurotoxicity, we investigated whether ibogaine might interact with sigma receptors. *Iboga* alkaloids were found to interact selectively with sigma-2 receptors and to induce a rise in intracellular calcium levels, morphological changes, and apoptosis (*66-71*).

III. Binding of *Iboga* Alkaloids to Sigma Receptors

Table I shows the binding affinities of ibogaine and various related *iboga* alkaloids at sigma-2 receptors. Sigma-1 receptor affinities are given in the following text. Sigma-1 receptors were labeled with the sigma-1-selective probe, $[^3H]$(+)-pentazocine, in guinea pig brain membranes (*72*). Sigma-2 receptors were labeled with $[^3H]$DTG using rat liver membranes, in the presence of dextrallorphan to mask binding to sigma-1 sites (*19*). Ibogaine exhibited moderate affinity for sigma-2 sites ($K_i = 201 \pm 24$ nM), but had very low affinity for sigma-1 receptors ($K_i = 8{,}554 \pm 1{,}134$ nM), resulting in 43-fold selectivity for sigma-2 sites over sigma-1. Mach *et al.* (*67*) obtained similar results with ibogaine. Although the affinity of ibogaine for sigma-2 receptors is only moderate, this is none the less quite significant, since ibogaine generally has much lower affinity for other neurotransmitter receptors studied thus far (*73-78*). Although there is variation across studies, ibogaine is reported to bind with K_i values in the range of 1 - 15 μM to subtypes of muscarinic cholinergic, α-adrenergic, kappa-opioid, ionophore site of NMDA-glutamatergic receptor, as well as the dopamine and serotonin transporters. Ibogaine is reported to be inactive (K_i > 100 μM) at serotonergic, dopaminergic, metabotropic glutamatergic, benzodiazepine, γ-aminobutyric $acid_A$, and cannabinoid receptors. Furthermore, ibogaine turns out to be one of the rare sigma-2-selective ligands, since most compounds binding to sigma receptors either interact selectively with sigma-1 sites or bind to both sites with high affinity (*17-19, 65*). Interestingly, in addition to ibogaine, all of the ibogaine analogs shown in Table I also have a low affinity for sigma-1 receptors.

For discussion of the structure-activity relationships for affinity at sigma receptors, (±)-ibogamine will be considered as the parent compound for those shown in Table I. (±)-Ibogamine has an unsubstituted indole moiety, with a sigma-2 K_i = 137 ± 13 nM and sigma-1 K_i = 1,835 ± 131 nM. A methoxy group in the 10-position (ibogaine) did not markedly change the sigma-2 affinity, but decreased the sigma-1 affinity (K_i = 8,554 ± 1,134 nM). A methoxy group in the 11-position (tabernanthine) produced little change in sigma-2 affinity, and only a small decrease in sigma-1 affinity (K_i = 2,872 ± 37 nM), resulting in 14.8-fold selectivity for sigma-2 receptors. An *O-t*-butyl group in the 10-position also did not dramatically change the sigma-2 receptor affinity or the sigma-1 affinity (K_i = 4,859 ± 682 nM), resulting in 20-fold selectivity for sigma-2 sites. Thus, the

TABLE I.
AFFINITIES OF IBOGAINE AND RELATED INDOLE ALKALOIDS AT SIGMA-2 RECEPTORS

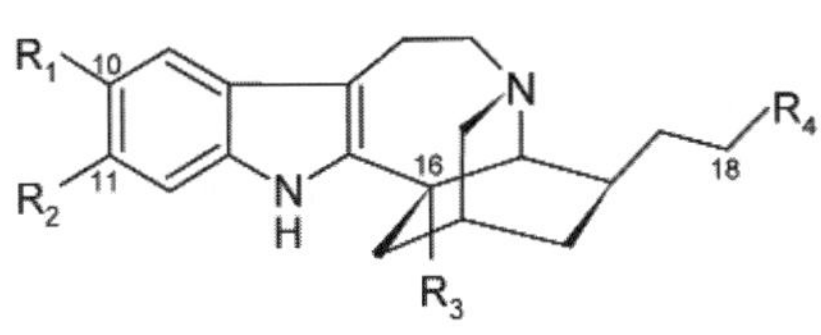

Alkaloid	R_1	R_2	R_3	R_4	Sigma-2 K_i (nM)
(±)-Ibogamine	H	H	H	H	137 ± 13
Ibogaine	OCH_3	H	H	H	201 ± 24
Tabernanthine	H	OCH_3	H	H	194 ± 10
10-*t*-Butoxy-ibogamine	*O-t*-Bu	H	H	H	247 ± 26
Noribogaine	OH	H	H	H	5,226 ± 1,426
(±)-Coronaridine	H	H	CO_2CH_3	H	>100,000
(±)-MC	H	H	CO_2CH_3	OCH_3	8,472 ± 1,237

Portions adapted from data in Bowen *et al.* (66). Sigma-1 receptor affinities are given in the text. Ibogaine was purchased from Sigma Chemicals (St. Louis, MO). See acknowledgments section for sources of other alkaloids. Alkaloids here and throughout the text without stereochemical designation are derived from natural ibogaine and are (-)-enantiomers.

presence or position of the methoxy group on the aromatic ring of the indole moiety is not critical for sigma-2 affinity. Furthermore, the size of the substituent appears not to be critical since the *O-t*-butyl group is just as well tolerated at the sigma-2 receptor as the methoxy group. However, a phenolic hydroxyl group in the 10-position (noribogaine) results in a 38-fold loss of binding affinity at sigma-2 receptors and an 8-fold loss of affinity at sigma-1 receptors (K_i = 15,006 ± 898 nM). Thus, a phenolic hydroxyl group appears not to be tolerated in the sigma-2 receptor binding site.

The effect of substitution in the saturated ring system was also examined. The presence of a carbomethoxy group in the 16-position ((±)-coronaridine) resulted in complete loss of sigma-2 receptor binding affinity and a 20-fold loss in sigma-1 affinity (K_i = 35,688 ± 2,858 nM) compared to (±)-ibogamine. Addition of a methoxy group at the 18-position of the 16-carbomethoxy analog, (±)-18-methoxycoronaridine ((±)-MC), led to a marked improvement of sigma-2 binding affinity compared to (±)-coronaridine, but was still of low affinity. Compared to (±)-ibogamine, (±)-MC had 62-fold lower sigma-2 binding affinity. (±)-MC had slightly improved sigma-1 binding affinity (K_i = 28,687 ± 283 nM) compared to (±)-coronaridine, but had 16-fold lower sigma-1 affinity compared to (±)-ibogamine. Thus, a carbomethoxy group at the 16-position is not tolerated in the sigma-2 receptor binding site. All of these analogs had a very low affinity at sigma-1 sites.

IV. Effect of *Iboga* Alkaloids on Intracellular Cytosolic Calcium

As described above, we have shown that sigma-2 receptors mediate a rise in cytosolic calcium levels (*64,65*). In view of the sigma-2 binding affinity of ibogaine and its analogs, we investigated whether *iboga* alkaloids could affect the levels of intracellular calcium in human SK-N-SH neuroblastoma cells. Human SK-N-SH neuroblastoma cells were loaded with Indo-1 calcium indicator dye, and $[Ca^{++}]i$ of individual cells was measured using the fluorescence ratio at 410 nm/485 nm (*65*).

The *iboga* alkaloid being tested was added to Indo-1-loaded SK-N-SH neuroblastoma cells, and the change in $[Ca^{++}]i$ was monitored for about 10 minutes. Ibogaine produced a dose-dependent rise in $[Ca^{++}]i$. The calcium levels began to rise almost immediately after addition of the alkaloid to the cells. Table II shows the effect of 100 μM of various *iboga* alkaloids on $[Ca^{++}]i$. The percent increase in $[Ca^{++}]i$ was calculated by determining the peak level of $[Ca^{++}]i$ relative to the starting basal level. In addition to ibogaine, (±)-ibogamine and 10-*t*-butoxy-ibogamine also produced a rise in $[Ca^{++}]i$. Noribogaine, (±)-coronaridine, and

TABLE II.
EFFECT OF IBOGAINE AND ITS ANALOGS ON $[Ca^{++}]i$

Alkaloid (100 μM)	Percentage increase in $[Ca^{++}]i$ above basal at 100 μM
Ibogaine	40.5 ± 2.0
(±)-Ibogamine	102 ± 14
10-*t*-Butoxy-ibogamine	100 ± 6.4
Noribogaine	0 ± 0
(±)-Coronaridine	0 ± 0
(±)-MC	5.0 ± 0.5
THAP (150 nM)/Ibogaine	0 ± 0
THAP (150 nM)/(±)-Ibogamine	0 ± 0

(±)-MC had little or no effect on $[Ca^{++}]i$. This pharmacological profile is consistent with mediation by sigma-2 receptors, since only those *iboga* alkaloids with significant sigma-2 affinity (Table I) are active at increasing $[Ca^{++}]i$.

To determine the source of calcium contributing to the *iboga* alkaloid-induced rise in $[Ca^{++}]i$, SK-N-SH neuroblastoma cells were pretreated for 10 minutes with 150 nM thapsigargin (THAP) to deplete the store of calcium in the endoplasmic reticulum. Table II shows that thapsigargin-pretreatment completely eliminated the rise in $[Ca^{++}]i$ produced by ibogaine and (±)-ibogamine. These results show that, like other sigma-2 receptor ligands, such as CB-64D and BD737 (*64,65*), ibogaine and related *iboga* alkaloids that have sigma-2 receptor affinity act as sigma-2 receptor agonists to gate calcium from the endoplasmic reticulum. Whether or not *iboga* alkaloids also produce a latent, sustained, and thapsigargin-insensitive rise in $[Ca^{++}]i$, like that produced by other sigma-2 agonists on long-term exposure, was not examined.

V. Effect of *Iboga* Alkaloids on Cellular Morphology and Induction of Apoptosis

As mentioned above, sigma-2 receptors were found to mediate morphological changes and apoptotic cell death in a number of cell types, including tumor cell lines and primary cultures of neuronal cells (*52-54,59,60*). The ability of *iboga* alkaloids to cause cytotoxicity was examined *in vitro* using rat C6 glioma cells and human SK-N-SH neuroblastoma cells. The cytotoxic effect of *iboga* alkaloids was also examined in primary cultures of rat cerebellar granule cells.

Cells were exposed to various concentrations (3 to 30 μM) of ibogaine or its analogs and the morphology of the cells examined by phase contrast microscopy.

The morphological state was given a score after the indicated time of exposure. Scoring of cell morphology was similar to that described previously (*52*): N, normal cells; A, loss or damage to cell processes; B, initial stages of cell rounding; C, complete rounding with or without detachment from substratum; D, cell death with presence of cell debris. Effects on rat C6 glioma cells and human SK-N-SH neuroblastoma cells are shown in Tables III and IV. The sigma-2 receptor-active compounds, ibogaine, (±)-ibogamine, and 10-*t*-butoxy-ibogamine produced dose- and time-dependent changes in cellular morphology. In C6 glioma cells, 30 μM ibogaine produced significant changes in cell morphology within 72 hours. 10-*t*-Butoxy-ibogamine was more potent, producing significant morphology changes within 24 hours and cell death within 72 hours of exposure. In SK-N-SH cells, 30 μM (±)-ibogamine and 10-*t*-butoxy-ibogamine induced cell death within 72 hours of exposure, with ibogaine producing significant cell rounding by this time point. Again, 10-*t*-butoxy-ibogamine was most potent, producing significant morphological change in as little as 6 hours at 30 μM, followed by (±)-ibogamine, and then ibogaine. Effects on rat cerebellar granule cells are shown in Table V. In cerebellar granule cells, 10-*t*-butoxy-ibogamine produced significant changes in cells within 72 hours at a concentration of 10 μM and induced cell death by 10 days at 30 μM. Ibogaine at a concentration of 30 μM induced cell rounding by 10 days.

Iboga alkaloids lacking sigma-2 affinity did not exhibit cytotoxic effects in these cells. Noribogaine and (±)-MC failed to produce any effect on cells. (±)-Coronaridine was inactive in C6 glioma cells at 30 μM, but did produce morphologic effects in SK-N-SH neuroblastoma cells at 30 μM. However, (±)-coronaridine-induced toxicity was distinct from that produced by the other *iboga* alkaloids and other sigma-2 receptor ligands. This alkaloid caused the appearance of abundant intracellular bodies with a granular appearance (indicated by "gran" in Table IV), which did not occur with the other *iboga* alkaloids or with other sigma-2 receptor agonists such as CB-64D and BD737. In addition, harmaline, an indole alkaloid that is also sigma receptor-inactive (*66*), caused morphological changes similar to those of (±)-coronaridine (not shown). Thus, these effects of (±)-coronaridine and harmaline on neuroblastoma cells appear not to be mediated by sigma-2 receptors and are due to some other mechanism.

DNA fragmentation is one hallmark of apoptotic cell death (*58*). DNA fragmentation occurring during apoptosis can be detected by incorporating fluorescein-12-dUTP at the 3'-OH DNA ends using the enzyme, terminal deoxynucleotidyl transferase (TdT). TUNEL (TdT-mediated dUTP Nick-End Labeling) was previously used to detect sigma-2 receptor-induced apoptotic cell death in both SK-N-SH neuroblastoma cells and cerebellar granule cells (*59*). SK-N-SH neuroblastoma cells were treated with a 100 μM concentration of various *iboga* alkaloids for 24 to 72 hours and then prepared for TUNEL staining and analysis by fluorescence microscopy. Treatment of SK-N-SH neuroblastoma

TABLE III.
EFFECT OF IBOGA ALKALOIDS ON RAT C6 GLIOMA CELLS

Alkaloid	Concentration	6 hours	24 hours	Time of exposure 48 hours	72 hours
Ibogaine	30 μM	N	N	N	A-B
10-*t*-Butoxy-ibogamine	30 μM	N	A-B	B-C	C-D
Noribogaine	30 μM	N	N	N	N
(±)-Coronaridine	30 μM	N	N	N	N

TABLE IV.
EFFECT OF IBOGA ALKALOIDS ON HUMAN SK-N-SH NEUROBLASTOMA CELLS

Alkaloid	Concentration	6 hours	24 hours	Time of exposure 48 hours	72 hours
Ibogaine	10 μM	N	N	N	A
	30 μM	N	N	A-B	B-C
(±)-Ibogamine	10 μM	N	N	N	A
	30 μM	N	A	B-C	C > D
10-*t*-Butoxy-ibogamine	10 μM	N	A	A	A-B
	30 μM	A-B	B-C	C	D
Noribogaine	30 μM	N	N	N	N
(±)-MC	30 μM	N	N	N	N
(±)-Coronaridine	10 μM	N	N	N-A (gran)	A (gran)
	30 μM	N	A-B (gran)	B-C (gran)	B-C (gran)

TABLE V.
EFFECT OF IBOGA ALKALOIDS ON RAT CEREBELLAR GRANULE CELLS

Alkaloid	Concentration	1 day	3 days	Time of exposure 7 days	10 days
Ibogaine	3 μM	N	N	N	N
	10 μM	N	N	N	A
	30 μM	N	A	A-B	B > C
10-*t*-Butoxy-ibogamine	3 μM	N	A	A	A-B
	10 μM	A	A > B	A-B	B-C
	30 μM	A-B	B-C	B < C	C-D
Noribogaine	3 μM	N	N	N	N
	10 μM	N	N	N	N
	30 μM	N	N	N	A-B

cell cultures with 100 μM ibogaine (48 hours), (±)-ibogamine (24 hours), and 10-*t*-butoxy-ibogamine (24 hours) resulted in TUNEL-positive cells, indicating apoptotic cell death. Treatment with 100 μM noribogaine for 72 hours failed to produce any TUNEL-staining cells, consistent with no change in morphology relative to untreated controls as observed above (Table IV). Similarly, TUNEL-positive cells were evident after treatment of rat cerebellar granule cells with 30 μM ibogaine (72 hours), (±)-ibogamine (48 hours), and 10-*t*-butoxy-ibogamine (48 hours). No TUNEL-positive cells were present after treatment with 30 μM noribogaine for up to 7 days. Thus, consistent with the profile for production of morphological changes, only those *iboga* alkaloids with affinity for sigma-2 receptors produced DNA fragmentation and apoptotic cell death.

VI. Summary and Discussion

The specific receptor sites at which ibogaine interacts to produce neurotoxicity *in vivo* have not yet been delineated with certainty, and the exact relevance of the cytotoxicity of ibogaine as demonstrated *in vitro* with regard to administration of the drug *in vivo* is not clear. O'Hearn and Molliver (*79*) have proposed an indirect toxicity model for ibogaine-induced cerebellar toxicity whereby acute administration of ibogaine (100 mg/kg, i.p., once) activates neurons in the inferior olive, resulting in sustained release of glutamate from climbing fiber synapses onto the Purkinje cells. This results in excitotoxic degeneration of the Purkinje cells in the cerebellum. This notion is strongly supported by the observation that ablation of the inferior olive abolishes the neurotoxic effect of an acute dose of ibogaine (*79*). Furthermore, ibogaine can potentiate neuronal glutamatergic activity, as evidenced by its ability to slightly increase the electrophysiological response to NMDA in the CA_3 region of the rat dorsal hippocampus (*80*). This enhancing effect was proposed to be mediated *via* a sigma-2 receptor-related site (*80*). Interestingly, an effect of ibogaine involving glutamate might appear paradoxical, since ibogaine has been shown to be a noncompetitive antagonist at the NMDA-glutamatergic receptor (*75,81*) and thus would be expected to have neuroprotective activity in models of glutamate-induced excitotoxicity. It is possible, however, that glutamatergic receptors other than the NMDA-type contribute to the cerebellar excitotoxicity. Also, the redundancy of the synaptic input onto Purkinje cells could make them exquisitely sensitive to glutamate-induced neurotoxicity (*79*).

It at first appears unlikely that sigma-2 receptors are solely responsible for the highly selective Purkinje cell toxicity produced by ibogaine, since harmaline, which lacks sigma-2 affinity (*66*), produces the same effect (*11*). The most parsimonious explanation for this is that ibogaine and harmaline both act at some other site to activate the olivocerebellar projection. However, it remains possible that ibogaine and harmaline act through different mechanisms to activate the same pathway, with ibogaine acting at sigma-2 receptors and harmaline acting through a different site (see below).

Based on the *in vitro* results currently described, an additional model to consider is one where ibogaine causes activation of sigma-2 receptors and results in a direct cytotoxic effect on neuronal and/or glial cells through an apoptotic mechanism. It is possible that this direct neurotoxicity combines with excitotoxicity due to enhanced response to glutamate, both effects being mediated by sigma-2 receptors. In conjunction with the greater vulnerability of Purkinje cells to excitotoxic injury, this could result in the cerebellar degeneration caused by ibogaine. This would also explain the apparent paradox of ibogaine-induced excitotoxicity, despite ibogaine's properties as an NMDA-glutamatergic antagonist. Furthermore, it was observed in the *in vitro* model that harmaline also caused cell morphology changes, but these effects were clearly distinct from the effects produced by ibogaine and other sigma-2 receptor agonists. This suggests that harmaline and ibogaine act *via* different mechanisms *in vitro*, and might do so *in vivo*.

Whereas the climbing fiber model accounts for the specificity of ibogaine toxicity for cerebellar Purkinje cells, the direct toxicity model would apply to any ibogaine-induced cytotoxicity that might be observed in other brain regions or in peripheral tissues due to the wide tissue distribution of sigma-2 receptors (*19-21*). Such widespread cytotoxicity of ibogaine has not yet been reported in the brain or the periphery. No significant pathological effects were observed in liver, kidney, heart, or brain following chronic treatment of rats with ibogaine (10 mg/kg for 30 days or 40 mg/kg for 12 days, i.p.) (*82*). However, it should be noted that the neurotoxic effect of ibogaine is reported to be highly dependent on dose, whereby a single dose that is effective at reducing morphine and cocaine self-administration (40 mg/kg, i.p.) does not produce cerebellar neurotoxicity in the rat (*83*). Also, chronic administration of a behaviorally active dose of ibogaine (10 mg/kg, i.p., every other day for 60 days) failed to produce loss of cerebellar Purkinje cells in rats (*84*). Thus, it is conceivable that an acute dose of ibogaine higher than that used by O'Hearn and Molliver (*79*), a different route of administration, or a chronic paradigm at a dose greater than 40 mg/kg might produce widespread, direct toxicity to rat brain neurons as well as to peripheral tissues expressing high densities of sigma-2 receptors such as rat liver and kidney (*19*).

Noribogaine has been shown to be the major ibogaine metabolite in humans and results from *O*-demethylation (*85, 86*). Interestingly, noribogaine lacks affinity for sigma-2 receptors (Table I), produces no effects on $[Ca^{++}]i$ (Table II), and is devoid of cytotoxicity *in vitro* (Tables III-V). Therefore, after administration of a dose of ibogaine, *O*-demethylation to noribogaine would eliminate the sigma-2 receptor binding affinity and therefore would abolish its potential cytotoxicity. This could have important implications for the treatment of drug abusers with ibogaine, since subjects with a low level of hepatic *O*-demethylase activity ("slow metabolizers") might be more susceptible to the potential cytotoxic effects of ibogaine than "rapid metabolizers." Differences in the rate of ibogaine demethylation could also explain the observed species differences in sensitivity to the neurotoxic effects of ibogaine. For example, ibogaine clearly produces neurotoxicity in rats at a dose of 100 mg/kg (*10,11,79*), but no neurotoxicity was observed in African green monkeys after treatment for 5 days with repeated doses of either 25 mg/kg (p.o.) or 100 mg/kg (s.c.) of ibogaine (*9*). Furthermore, no cerebellar degeneration or degeneration in any other brain area was observed on postmortem neuropathological examination of a female patient who had received four doses of ibogaine ranging from 10 to 30 mg/kg over a 15-month period (*9*). Thus, ibogaine may be neurotoxic in rodents, but not in primates, and this could conceivably be due to differences in its rate of conversion to the much less cytotoxic metabolite, noribogaine. This notion deserves further study.

Another implication of these findings is that it appears possible to dissociate the neurotoxic effects from the beneficial effects of *iboga* alkaloids. In rats, noribogaine (40 mg/kg) has effects similar to ibogaine in suppressing morphine and cocaine self-administration, but does not have the tremorigenic effects of an equal dose of ibogaine (also, see below) (*87*). 18-Methoxycoronaridine (MC) is a synthetic analog of ibogaine (*88*). MC suppresses morphine and cocaine self-administration. However, rats treated with up to 100 mg/kg MC showed no evidence of cerebellar neurotoxicity (*88*). This absence of *in vivo* neurotoxicity with MC is consistent with the lack of sigma-2 receptor binding affinity, lack of effect on $[Ca^{++}]i$, and lack of cytotoxicity *in vitro* (Tables I, II, and IV). Thus, sigma-2 receptors appear not to be involved in the positive effects of ibogaine and may specifically contribute to the neurotoxic effects. It should be possible to develop synthetic ibogaine analogs that have low sigma-2 receptor affinity and low neurotoxicity, but that remain potent at blocking drug self-administration. This could be accomplished by incorporating hydroxyl groups on the aromatic ring of the indole moiety, as in noribogaine, or by making substitutions at the 16-position of the saturated ring system, as in the case of MC.

Sigma-2 receptors may contribute to other toxic effects of *iboga* alkaloids. Ibogaine and some of its congeners are known to cause tremors with marked ataxia in both mice and rats (*89-91*). Singbartl and colleagues (*89,90*) have examined the structure-activity relationships for the tremorigenic effect of a number of *iboga* alkaloids. They found that a carbomethoxy group had a clear negative effect on tremorigenic activity, and that an aromatic methoxy group enhanced, whereas a hydroxyl group decreased, tremorigenic activity. They concluded that due to this defined structure-activity relationship, indole

TABLE VI.
TREMORIGENIC STRUCTURE-ACTIVITY RELATIONSHIP AND SIGMA BINDING AFFINITIES OF IBOGA ALKALOIDS

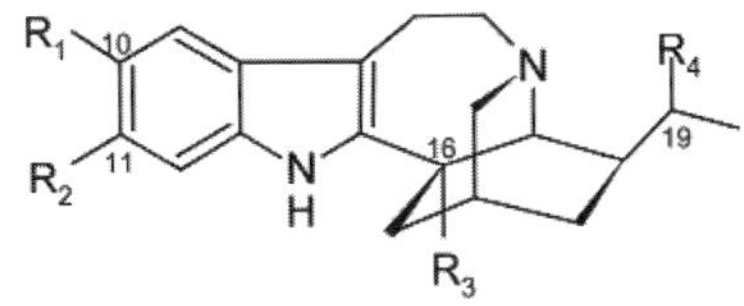

Alkaloid	R_1	R_2	R_3	R_4	Tremors (ED_{50}, μmol/kg s.c.)	Sigma-2 K_i (nM) or *predicted affinity
Ibogaine	OCH_3	H	H	H	34.8	201
Tabernanthine	H	OCH_3	H	H	4.5	194
Ibogaline	OCH_3	OCH_3	H	H	7.6	High*
Iboxygaine	OCH_3	H	H	OH	80.4	High*
Noribogaine	OH	H	H	H	176	5,226
Voacangine	OCH_3	H	CO_2CH_3	H	Inactive	Low*
Conopharyngine	OCH_3	OCH_3	CO_2CH_3	H	Inactive	Low*
Voacristine	OCH_3	H	CO_2CH_3	OH	Inactive	Low*

Adapted from data in Singbartl and colleagues (*89,90*) and Bowen *et al.* (*66*).

derivatives must interact with a specific receptor site for the generation of tremors (*90*).

In view of the high density of sigma receptors in brain motor control regions, and the effects of sigma-2 receptor ligands on movement and posture (*12,45-49*), it is interesting to note that the pharmacological profile for the tremorigenic effect of *iboga* alkaloids is also consistent with mediation by sigma-2 receptors. Table VI shows the structure-activity relationship for tremors described by Singbartl and colleagues (*89,90*), along with the observed sigma-2 binding K_i value, or a prediction of whether or not the alkaloid would exhibit high or low sigma-2 binding activity based on the structure-activity relationship described in Table I. The sigma-2 receptor-active alkaloids, ibogaine and tabernanthine, both produced tremors. The *iboga* alkaloids iboxygaine and ibogaline are predicted to have good sigma-2 affinity, since the position of the aromatic methoxy group does not affect sigma-2 binding activity. Both of these alkaloids had tremorigenic activity. Noribogaine, which has very weak sigma-2 binding affinity due to the presence of a phenolic hydroxyl group, also had relatively weak tremorigenic activity. Table I shows that a carbomethoxy group at the 16-position, greatly reduces or eliminates sigma-2 receptor binding affinity. All of the *iboga* alkaloids that have a carbomethoxy group at the 16-position (voacangine, voacristine, and conopharyngine) were all inactive at producing tremors. Furthermore, Glick and colleagues have shown that MC is devoid of tremorigenic activity (*88*). Thus, the tremorigenic activity of *iboga* alkaloids, like the neurotoxic effect, is consistent with binding to sigma-2 receptors.

Further study will be needed in order to determine whether sigma-2 receptors contribute to the neurotoxic and/or tremorigenic effects of ibogaine and other *iboga* alkaloids observed *in vivo*. As pointed out earlier, harmaline, a β-carboline indole alkaloid structurally related to ibogaine, but devoid of sigma-2 binding affinity (*66*), also causes cerebellar neurotoxicity and tremors (*11, 79*). This suggests that sigma-2 receptors do not explain all of the neurotoxic actions of these indole alkaloids and that other receptor sites may also be involved. However, as relatively selective sigma-2 receptor ligands, *iboga* alkaloids may serve as templates on which to design selective agonists and antagonists for further study of sigma-2 receptor function. Designing ibogaine derivatives that lack sigma-2 receptor affinity may result in effective and nontoxic agents for the treatment of drug abuse.

Acknowledgments

The author wishes to acknowledge Ms. Wanda Williams, for generating the receptor binding data, and Dr. Bertold J. Vilner, for carrying out experiments on the effects of *iboga* alkaloids on cultured cells. Both are members of this laboratory. Noribogaine and 10-*t*-butoxy-ibogamine were synthesized by Dr. Craig Bertha (NIDDK, Bethesda, MD). All of the other ibogaine analogs were provided by Dr. Upul Bandarage and Dr. Martin Kuehne (Department of Chemistry, University of Vermont, Burlington, VT).

References

1. P. Popik, R. T. Layer, and P. Skolnick, *Pharmacol. Rev.* **47,** 235 (1995).
2. P. Popik and S.D. Glick, *Drugs of the Future* **21,** 1109 (1996).
3. H.S. Lotsof, *U. S. Pat.* **4,499,096**; *Chem. Abstr.* **102,** 160426w (1985).
4. H.S. Lotsof, *U. S. Pat.* **4,587,243**; *Chem. Abstr.* **106,** 12967r (1986).
5. S.D. Glick, K. Rossman, S. Steindorf, I.M. Maisonneuve, and J.N. Carlson, *Eur. J. Pharmacol.* **195,** 341 (1991).
6. S.D. Glick, M.E. Kuehne, J. Raucci, T.E. Wilson, D. Larson, R.W. Keller, Jr., and J.N. Carlson, *Brain Res.* **657,** 14 (1994).
7. S.L.T. Cappendijk and M.R. Dzoljic, *Eur. J. Pharmacol.* **241,** 261 (1993).
8. H. Sershen, A. Hashim, and A. Lajtha, *Pharmacol. Biochem. Behav.* **47,** 13 (1994).
9. D.C. Mash, C.A. Kovera, B.E. Buck, M.D. Norenberg, P. Shapshak, W.L. Hearn, and J. Sanchez-Ramos, *Ann. N.Y. Acad. Sci.* **844,** 274 (1998).

10. E. O'Hearn, D.B. Long, and M.E. Molliver, *NeuroReport* **4,** 299 (1993).
11. E. O'Hearn and M.E. Molliver, *Neurosci.* **55,** 303 (1993).
12. J.M. Walker, W.D. Bowen, F.O. Walker, R.R. Matsumoto, B.R. de Costa, and K.C. Rice, *Pharmacological Rev.* **42,** 355 (1990).
13. W.R. Martin, C.G. Eades, J.A. Thompson, R.E. Huppler, and P.E. Gilbert, *J. Pharmacol. Exp. Ther.* **197,** 517 (1976).
14. R. Quirion, R. Chicheportiche, P.C. Contreras, K.M. Johnson, D. Lodge, S.W. Tam, J.H. Woods, and S.R. Zukin, *Trends Neurosci.* **10,** 444 (1987).
15. T.-P. Su, *J. Pharmacol. Exp. Ther.* **223,** 284 (1982).
16. S.W. Tam and L. Cook, *Proc. Natl. Acad. Sci. USA* **81,** 5618 (1984).
17. S.B. Hellewell and W.D. Bowen, *Brain Res.* **527,** 244 (1990).
18. R. Quirion, W.D. Bowen, Y. Itzhak, J.L. Junien, J.M. Musacchio, R.B. Rothman, T.-P. Su, S.W. Tam, and D.P. Taylor, *Trends Pharmacol. Sci.* **13,** 85 (1992).
19. S.B. Hellewell, A.Bruce, G. Feinstein, J. Orringer, W. Williams, and W.D. Bowen, *Eur. J. Pharmacol. - Mol. Pharmacol. Sect.* **268,** 9 (1994).
20. S.A. Wolfe, Jr. and E.B. De Souza, *in* "Multiple Sigma and PCP Receptor Ligands: Mechanisms for Neuromodulation and Neuroprotection?" (J.M. Kamenka and E.F. Domino, eds.), p. 927. NPP Books, Ann Arbor, MI, 1992.
21. B.J. Vilner, C.S. John, and W.D. Bowen, *Cancer Res.* **55,** 408 (1995).
22. R.H. Mach, C.R. Smith, I. al-Nabulsi, B.R. Whirrett, S.R. Childers, and K.T. Wheeler, *Cancer Res.* **57,** 156 (1997).
23. W.T. Bem, G.E. Thomas, J.Y. Mamone, S.M. Homan, B.K. Levy, F.E. Johnson, and C.J. Coscia, *Cancer Res.* **51,** 6558 (1991).
24. T.-P. Su, A.D. Weissman, and S.-Y. Yeh, *Life Sci.* **38,** 2199 (1986).
25. P.C. Contreras, D.A. DiMaggio, and T.L. O'Donohue, *Synapse* **1,** 57 (1987).
26. J.F. Neumaier and C.Chavkin, *Brain Res.* **500,** 215 (1989).
27. T.-P. Su, E.D. London, and J.H. Jaffe, *Science* **240,** 219 (1988).
28. F.P. Monnet, V. Mahe, P. Robel, and E.E. Baulieu, *Proc. Natl. Acad. Sci. USA* **92,** 3774 (1995).
29. M. Hanner, F.F. Moebius, A. Flandorfer, H.G. Knaus, J. Striessnig, E. Kempner, and H. Glossmann, *Proc. Natl. Acad. Sci. USA* **93,** 8072 (1996).
30. R. Kekuda, P.D. Prasad, Y.-J. Fei, F.H. Leibach, and V. Ganapathy, *Biochem. Biophys. Res. Commun.* **229,** 553 (1996).
31. P. Seth, F.H. Leibach, and V. Ganapathy, *Biochem. Biophys. Res. Commun.* **241,** 535 (1997).
32. P. Seth, Y.-J. Fei, H.W. Li, W. Huang, F.H. Leibach, and V. Ganapathy, J. *Neurochem.* **70,** 922 (1998).
33. F.F. Moebius, J. Striessnig, and H. Glossmann, *Trends Pharmacol. Sci.* **18,** 67 (1997).
34. S. Silve, P.H. Dupuy, C. Labit-Lebouteiller, M. Kaghad, P. Chalon, A. Rahier, M. Taton, J. Lupker, D. Shire, and G. Loison, *J. Biol. Chem.* **271,** 22434 (1996).
35. R.G. Booth and R.J. Baldessarini, *Brain Res.* **557,** 349 (1991).
36. S.L. Patrick, J.M. Walker, J.M. Perkel, M. Lockwood, and R.L. Patrick, *Eur. J. Pharmacol.* **231,** 243 (1993).
37. K. Matsuno, T. Senda, T. Kobayashi, and S. Mita, *Brain Res.* **690,** 200 (1995).
38. F.P. Monnet, G. Debonnel, and C. De Montigny, *J. Pharmacol. Exp. Ther.* **261,** 123 (1992).
39. G.M. Gonzalez-Alvear and L.L. Werling, *Brain Res.* **673,** 61 (1995).
40. F.P. Monnet, B.R. de Costa, and W.D. Bowen, *Brit. J. Pharmacol.* **119,** 65 (1996).
41. W.D. Bowen, P.J. Tolentino, K.K. Hsu, J.M. Cutts, and S.S. Naidu, *in* "Multiple Sigma and PCP Receptor Ligands: Mechanisms for Neuromodulation and Neuroprotection?" (J.M. Kamenka and E.F. Domino, eds.), p. 155. NPP Books, Ann Arbor, MI, 1992.
42. T. Maurice and B.P. Lockhart, *Prog. Neuro-Psychopharmacol. Biol. Psychiat.* **21,** 69 (1997).
43. M. King, Y.-X. Pan, J. Mei, A. Chang, J. Xu, and G.W. Pasternak, *Eur. J. Pharmacol.* **331,** R5 (1997).
44. K.A. McCracken, W.D. Bowen, B.R. de Costa, and R.R. Matsumoto, *Eur. J. Pharmacol.* **370,** 225 (1999).

45. J.M. Walker, R.R. Matsumoto, W.D. Bowen, D.L. Gans, K.D. Jones, and F.O. Walker, *Neurology* **38,** 961 (1988).
46. T.T. Tran, B.R. de Costa, and R.R. Matsumoto, *Psychopharmacol.* **137,** 191 (1998).
47. J.M. Walker, W.D. Bowen, S.L. Patrick, W.E. Williams, S.W. Mascarella, X. Bai, and F.I. Carroll, *Eur. J. Pharmacol.* **231,** 61 (1993).
48. R.R. Matsumoto, M.K. Hemstreet, N.L. Lai, A. Thurkauf, B.R. de Costa, K.C. Rice, S.B. Hellewell, W.D. Bowen, and J.M. Walker, *Pharmacol. Biochem. Behav.* **36,** 151 (1990).
49. J.M. Walker, W.J. Martin, A.G. Hohmann, M.K. Hemstreet, J.S. Roth, M.L. Leitner, S.D. Weiser, S.L. Patrick, R.L. Patrick, and R.R. Matsumoto, *in* "Sigma Receptors, Neuroscience Perspectives Series" (Y. Itzhak, ed.), p. 205. Academic Press, London, 1994.
50. W.D. Bowen, E.L. Moses, P.J. Tolentino, and J.M. Walker, *Eur. J. Pharmacol.* **177,** 111 (1990).
51. W.D. Bowen, J.M. Walker, B.R. de Costa, R. Wu, P.J. Tolentino, D. Finn, R.B. Rothman, and K.C. Rice, *J. Pharmacol. Exp. Ther.* **262,** 32 (1992).
52. B.J. Vilner, B.R. de Costa, and W.D. Bowen, *J. Neurosci.* **15,** 117 (1995).
53. W.D. Bowen and B.J. Vilner, *Soc. Neurosci. Abstr.* **20,** 747, #314.10 (1994).
54. B.J. Vilner and W.D. Bowen, *Soc. Neurosci. Abstr.* **22,** 2006, #787.6 (1996).
55. W.D. Bowen, C.M. Bertha, B.J. Vilner, and K.C. Rice, *Eur. J. Pharmacol.* **278**, 257 (1995).
56. R.A. Schwartzman and J.A. Cidlowski, *Endocrine Rev.* **14,** 133 (1993).
57. I. Vermes and C. Haanan, *Adv. Clin. Chem.* **31,** 177 (1994).
58. A.H. Wyllie, *Brit. Med. Bull.* **53,** 451 (1997).
59. B.J. Vilner and W.D. Bowen, *Soc. Neurosci. Abstr.* **23,** 2319, #905.6 (1997).
60. K.W. Crawford, B.J. Vilner, and W.D. Bowen, *Proc. Am. Assoc. Cancer Res.* **40,** 166, #1104 (1999).
61. M.J. Berridge, M.D. Bootman, and P. Lipp, *Nature* **395,** 645 (1998).
62. P. Nicotera, B. Zhivotovsky, and S. Orrenius, *Cell Calcium* **16,** 279 (1994).
63. D.J. McConkey and S. Orrenius, *J. Leuk. Biol.* **59,** 775 (1996).
64. B.J. Vilner and W.D. Bowen, *Soc. Neurosci. Abstr.* **21,** 1608, #631.3 (1995).
65. B.J. Vilner and W.D. Bowen, *J. Pharmacol. Exp. Ther*. **292,** 900 (2000).
66. W.D. Bowen, B.J. Vilner, W. Williams, C.M. Bertha, M.E. Kuehne, and A.E. Jacobson, *Eur. J. Pharmacol.* **279,** R1 (1995).
67. R.H. Mach, C.R. Smith, and S.R. Childers, *Life Sci.* **57,** PL57 (1995).
68. W. Williams, U.K. Bandarage, M.E. Kuehne, C.M. Bertha, and W.D. Bowen, *in* "Problems of Drug Dependence, 1997: Proceedings of the 59th Annual Scientific Meeting, National Institute on Drug Abuse Research Monograph 178" (L.S. Harris, ed.), p. 236. U.S. Government Printing Office, Washington, DC, 1998.
69. W.D. Bowen, B.J. Vilner, U.K. Bandarage, and M.E. Kuehne, *Soc. Neurosci. Abstr.* **22,** 2006, #787.5 (1996).
70. B.J. Vilner, U.K. Bandarage, M.E. Kuehne, C.M. Bertha, and W.D. Bowen, *in* "Problems of Drug Dependence, 1997: Proceedings of the 59th Annual Scientific Meeting, National Institute on Drug Abuse Research Monograph 178" (L.S. Harris, ed.), p. 235. U.S. Government Printing Office, Washington, DC, 1998.
71. W.D. Bowen, B.J. Vilner, W. Williams, U.K. Bandarage, and M.E. Kuehne, *Soc. Neurosci. Abstr.* **23,** 2319, #905.7 (1997).
72. W.D. Bowen, B.R. de Costa, S.B. Hellewell, J.M. Walker, and K.C. Rice, *Mol. Neuropharmacol.* **3,** 117 (1993).
73. D.C. Deecher, M. Teitler, D.M. Soderlund, W.G. Bornmann, M.E. Kuehne, and S.D. Glick, *Brain Res.* **571,** 242 (1992).
74. H. Sershen, A. Hashim, L. Harsing, and A. Lajtha, *Life Sci.* **50,** 1079 (1992).
75. P. Popik, R.T. Layer, and P. Skolnick, *Psychopharmacol.* **114,** 672 (1994).
76. D.B. Repke, D.R. Artis, J.T. Nelson, and E.H.F. Wong, *J. Org. Chem.* **59,** 2164 (1994).
77. P.M. Sweetnam, J. Lancaster, A. Snowman, J. Collins, S. Perschke, C. Bauer, and J. Ferkany, *Psychopharmacol.* **118,** 369 (1995).
78. M.E.M. Benwell, P.E. Holtom, R.J. Moran, and D.J.K. Balfour, *Brit. J. Pharmacol.* **117,** 743

(1996).
79. E. O'Hearn and M.E. Molliver, *J. Neurosci.* **17,** 8828 (1997).
80. S. Couture and G. Debonnel, *Synapse* **29,** 62 (1998).
81. D.C. Mash, J.K. Staley, J.P. Pablo, A.M. Holohean, J.C. Hackman, and R.A. Davidoff, *Neurosci. Lett.* **192,** 53 (1995).
82. H.I. Dhahir, "A Comparative Study of the Toxicity of Ibogaine and Serotonin," *Diss. Abstr. Int.* 32/04-B, 2311, 1971.
83. H.H. Molinari, I.M. Maisonneuve, and S.D. Glick, *Brain Res.* **737,** 255 (1996).
84. S. Helsley, C.A. Dlugos, R.J. Pentney, R.A. Rabin, and J.C. Winter, *Brain Res.* **759,** 306 (1997).
85. D.C. Mash, J.K. Staley, M.H. Baumann, R.B. Rothman, and W.L. Hearn, *Life Sci.* **57,** PL 45 (1995).
86. J.K. Staley, Q. Ouyang, J. Pablo, W.L. Hearn, D.D. Flynn, R.B. Rothman, K.C. Rice, and D.C. Mash, *Psychopharmacol.* **127,** 10 (1996).
87. S.D. Glick, S.M. Pearl, J. Cai, and I.M. Maisonneuve, *Brain Res.* **713,** 294 (1996).
88. S.D. Glick, M.E. Kuehne, I.M. Maisonneuve, U.K. Bandarage, and H.H. Molinari, *Brain Res.* **719,** 29 (1996).
89. G. Zetler, G. Singbartl, and L. Schlosser, *Pharmacol.* **7,** 237 (1972).
90. G. Singbartl, G. Zetler, and L. Schlosser, *Neuropharmacol.* **12,** 239 (1973).
91. S.D. Glick, K. Rossman, N.C. Rao, I.M. Maisonneuve, and J.N. Carlson, *Neuropharmacol.* **31,** 497 (1992).

—CHAPTER 10—

IBOGAINE NEUROTOXICITY ASSESSMENT: ELECTROPHYSIOLOGICAL, NEUROCHEMICAL, AND NEUROHISTOLOGICAL METHODS

ZBIGNIEW K. BINIENDA,
ANDREW C. SCALLET, LARRY C. SCHMUED,
AND SYED F. ALI

Division of Neurotoxicology
FDA/National Center for Toxicological Research
Jefferson, AR 72079-9502

I. Introduction

Ibogaine, a psychoactive indole alkaloid, is derived from the root bark of the tropical shrub *Tabernanthe iboga*. The powdered root bark of *T. iboga* is used for medicinal and religious purposes in the Bwiti cult in Gabon (*1*). Anecdotal reports and published studies in laboratory animals have indicated that ibogaine may reduce the craving for cocaine (*2,3*). Ibogaine is also reportedly effective in the blockade of morphine self-administration and decreasing the signs of opiate withdrawal (*4*). Worldwide social and medical problems of substance abuse make evaluating the efficacy of potential compounds exhibiting antiaddictive properties of prime importance. However, in animal studies, ibogaine administration has been associated with neurotoxic side effects. Observations from several labora-

0099-9598/01 $35.00

tories, including our own, of ibogaine's neuronal cytotoxicity in rats, have raised the question of whether treatment of substance dependence with ibogaine may also lead to ibogaine-induced neurotoxicity (*5-7*)

Interactions have been reported between ibogaine and many neurotransmitter systems, (i.e., dopaminergic, serotonergic, opioid, glutamate, nicotine, noradrenergic, and cholinergic, reviewed by Popik and Skolnick [*8*]). Thus far, it is not completely understood how those interactions contribute to ibogaine's putative antiaddictive effects. The involvement of the dopaminergic system is described in publications from multiple laboratories (*9-11*). Acute *in vivo* response to ibogaine has been reported to involve a decrease in striatal and cortical dopamine concomitant with an increase in dopamine metabolites, 3,4-dihydroxyphenylacetic acid (DOPAC) and homovanillic acid (HVA), and dopamine turnover (*11*). Increased or decreased dopamine levels in specific brain regions have been seen, together with increased or decreased motor activity after amphetamine or cocaine administration and ibogaine pretreatment (*12,13*).

A complex ibogaine interaction with other neurotransmitter receptor sites has been suggested to have modulatory effects on the dopamine system. For example, it has been speculated that ibogaine's action as an NMDA antagonist (*14*), together with kappa-opioid agonist and nicotinic antagonist effects, underlies the ibogaine modulatory effect (*15*). Other complex interactions have also been proposed to explain the mechanisms of ibogaine's therapeutic actions. Recently, neurotensin, a neuromodulator peptide, has been suggested to be an important intermediary in ibogaine's apparent antiaddictive actions against cocaine's stimulation of the dopaminergic system (*16*).

The chemical structure of ibogaine is similar to serotonin (5-HT) and melatonin. Several *in vitro* and *in vivo* studies indicated that the serotonergic system plays a role in ibogaine actions (*10,13,19*). In fact, acute behavioral responses (tremor, ataxia) in rats to ibogaine, particularly at high doses (*17*), resemble a stereotyped behavioral response observed after central serotonergic stimulation (*18*). Ibogaine was also reported to increase extracellular 5-HT concentration in rat striatum and nucleus accumbens (*19,20*). The 5-HT-like response to ibogaine may result from a direct action of ibogaine on 5-HT receptors and/or ibogaine-induced increase in 5-HT level.

II. Electroencephalography and Ibogaine

Electroencephalography (EEG) is a technique applied in the assessment of spontaneous electrocerebral activity using either scalp (surface) electrodes, or in the case of the electrocorticogram (ECoG), from electrodes implanted in specific brain regions. Electrocerebral activity represents local action potentials and widespread excitatory and inhibitory postsynaptic potentials. The ECoG records an average of synchronous, widespread postsynaptic potentials arising in vertically oriented pyramidal cells of the upper layers of the cerebral cortex (*21*). EEG synchrony is reduced (desynchronization) by arousal and cognitive activity. On the other hand, reduced vigilance (drowsiness, sleep) increases synchrony. Transitory hypersynchronous cortical activity may also be elicited by afferent stimuli (evoked potentials), and pathological epileptiform discharges.

Rapid advances in computer technology during the past 20 years have allowed expansion of quantitative EEG analysis in neuroscience, as well as clinical neurology. Among the variety of techniques in this field, frequency (spectral) analysis provides a sensitive tool for time-course studies of different compounds acting on particular neurotransmitter systems. Frequency data are often analyzed as the power spectrum, measured as total power in microvolts-squared divided by frequency or over a particular power band.

The complex nature of ibogaine actions on neurotransmitters and neuromodulators in the cerebral cortex may have both an inhibitory and stimulatory effect on the neuronal firing reflected as the bioelectric neuronal activity and recorded as the EEG (*22*). We previously applied electroencephalography and spectral analysis to characterize the ECoG profiles in rats anesthetized with isoflurane and exposed to ibogaine or to one of two other NMDA receptor antagonists, MK-801 or phencyclidine (PCP). While some features of the neurochemical response to all three compounds were similar, a distinctly different EEG response to each treatment was observed (*23*). Recently, to extend our research on ibogaine neurotoxicity assessment, we aimed to analyze the effects of ibogaine/cocaine interaction on electrocerebral activity in conscious rats (*24*).

A. EEG Study

Three-month-old, male, Sprague-Dawley rats of the Charles River cesarean delivered (CD) strain were used in this study. Bipolar stainless steel electrodes were implanted above the somatosensory cortex, 3 mm laterally from the sagittal fissure, 1 and 4 mm posterior to the bregma. They were referenced to a ground electrode placed in the dorsal neck. The ECoG was recorded via a tether and swivel system at least one week after implantation. During recording, the animals remained in a microdialysis bowl placed inside a Faraday cage. Amplified signals were rectified to pass frequencies of 1-40 Hz and processed with LabView software (National Instruments, Austin, Texas). The power spectra obtained by use of Fast Fourier Transformations were divided into 1.25-4.50 Hz (delta), 4.75-6.75 Hz (theta), 7.00-9.50 Hz (alpha$_1$), 9.75-12.50 Hz (alpha$_2$), 12.75-18.50 Hz (beta$_1$), and 18.75-35.00 Hz (beta$_2$) frequency bands. Following the recording of the 30-minute baseline ECoG in the morning, rats were either injected intraperitoneally (i.p.) with cocaine alone (20 mg/kg.) or pretreated i.p. with ibogaine (50 mg/kg), followed an hour later by cocaine (20 mg/kg).

B. Results and Comments

Administration of cocaine was accompanied within 10 to 15 minutes after the injection by increased stereotypical behavior (hyperactive sniffing, chewing) and locomotor stimulation that lasted throughout the 60 minute recording. On the other hand, treatment with ibogaine alone produced tremors and ataxia. Administration of cocaine following ibogaine led to locomotor activity, but less than that observed after only cocaine.

Analysis of the ECoG in rats injected with ibogaine revealed a significant increase in total power (1-40 Hz) during first 30 minutes postinjection (Figure 1). A power increase in the theta frequency band lasting for approximately 10 minutes was observed. The total power was again significantly activated throughout the 60 minute recording when cocaine was injected after ibogaine pretreatment (Figure 1). Administration of cocaine alone was associated only with a significant power increase in the alpha1 frequency band during the first 30 min. postinjection (Figure 2). However, when cocaine was injected after ibogaine pretreatment, the alpha1 increase was maintained throughout recording. In addition, ibogaine/cocaine treatment resulted in a significant power increase in the delta and theta bands (Figure 3).

Studies have indicated that the alteration of ECoG patterns observed after cocaine administration appear to be related to increased release of dopamine in the striatum and prefrontal cortex (*25,26*). However, besides the dopaminergic effect of cocaine, (i.e., inhibition of dopamine reuptake), serotonergic effects of cocaine administration have also been reported (reviewed by Sershen *et al.* [*10*]). Ibogaine administered intraperitoneally is reported to markedly increase extracellular 5-HT in the nucleus accumbens and striatum (*19,20*). Activation of 5-HT receptors has been shown to the increase power in the $alpha_1$ band (*27*). The spectral patterns obtained after ibogaine/cocaine treatment in our study, mainly showing increased power in the low frequency bands and enhancement of power in the $alpha_1$ band, appear to indicate the contribution of the serotonergic system in the ibogaine-mediated response to cocaine.

Although no behavioral convulsive effects of cocaine injected after ibogaine were found, the enhancement of power observed in low frequency bands after the ibogaine/cocaine treatment may suggest that ibogaine at high dose decreases the threshold for cocaine-induced seizures. This effect seems to be contradictory to the fact that ibogaine was shown to be a neuroprotectant due to its NMDA noncompetitive antagonist action, suggesting that ibogaine should suppress seizures. However, a similar effect exerted by two other NMDA noncompetitive antagonists was reported earlier by other investigators. Ketamine and MK-801,

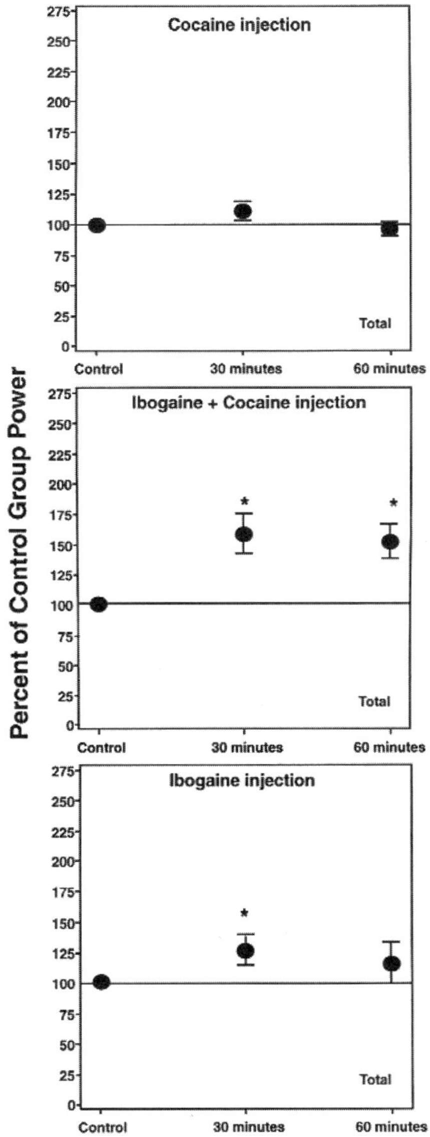

FIGURE 1. EFFECTS PRODUCED BY COCAINE (20 MG/KG), IBOGAINE (50 MG/KG) I.P., AND IBOGAINE PRETREATMENT 1 HR PRIOR TO COCAINE ON ELECTROENCEPHALOGRAPHIC ACTIVITY. Total = total power 1-40 Hz. Power values calculated as percent of the 30 min baseline power recorded after saline injection (assigned as a value of 100%). Mean ± SEM; n=3 rats
*$p<0.05$ significantly different from baseline.

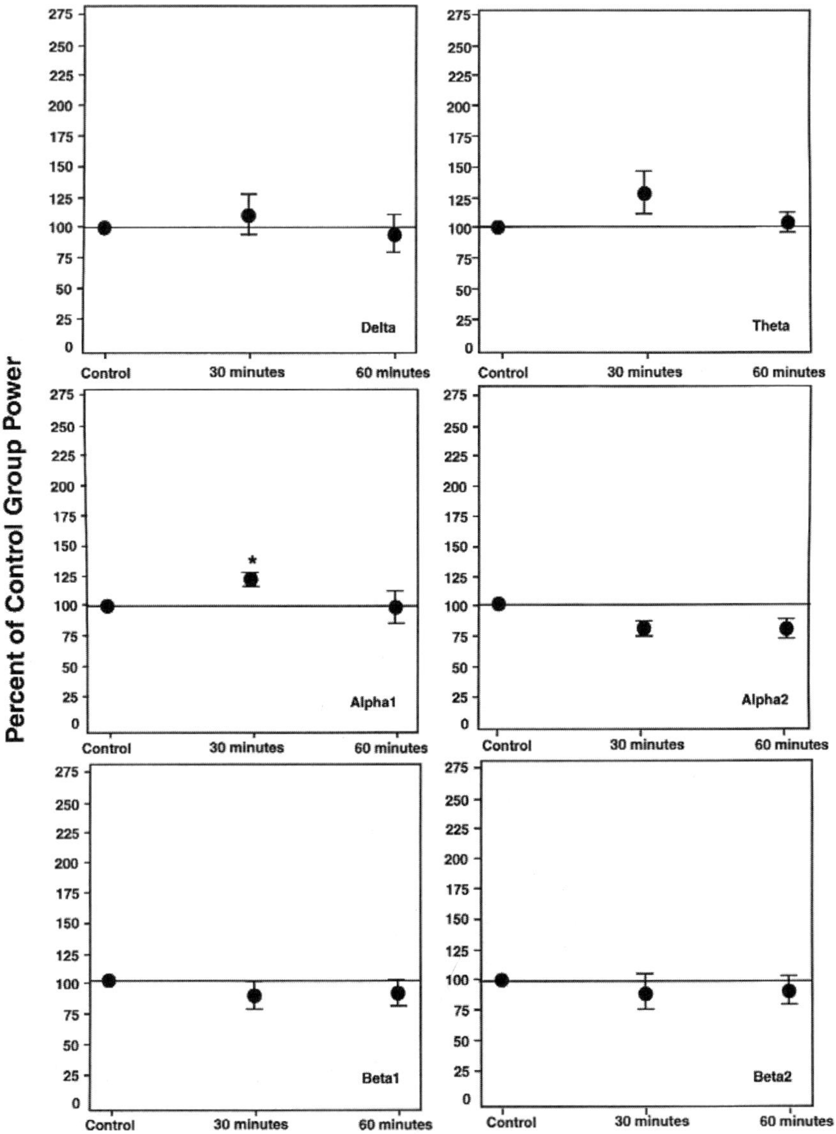

FIGURE 2. EFFECTS PRODUCED BY COCAINE ON THE CORTICAL POWER SPECTRA. Cocaine was injected at 20 mg/kg i.p. Power values calculated as percent of the 30 min baseline power recorded after saline injection (assigned as a value of 100% in each band). Mean ± SEM; n=3 rats. $*p < 0.05$ significantly different from baseline.

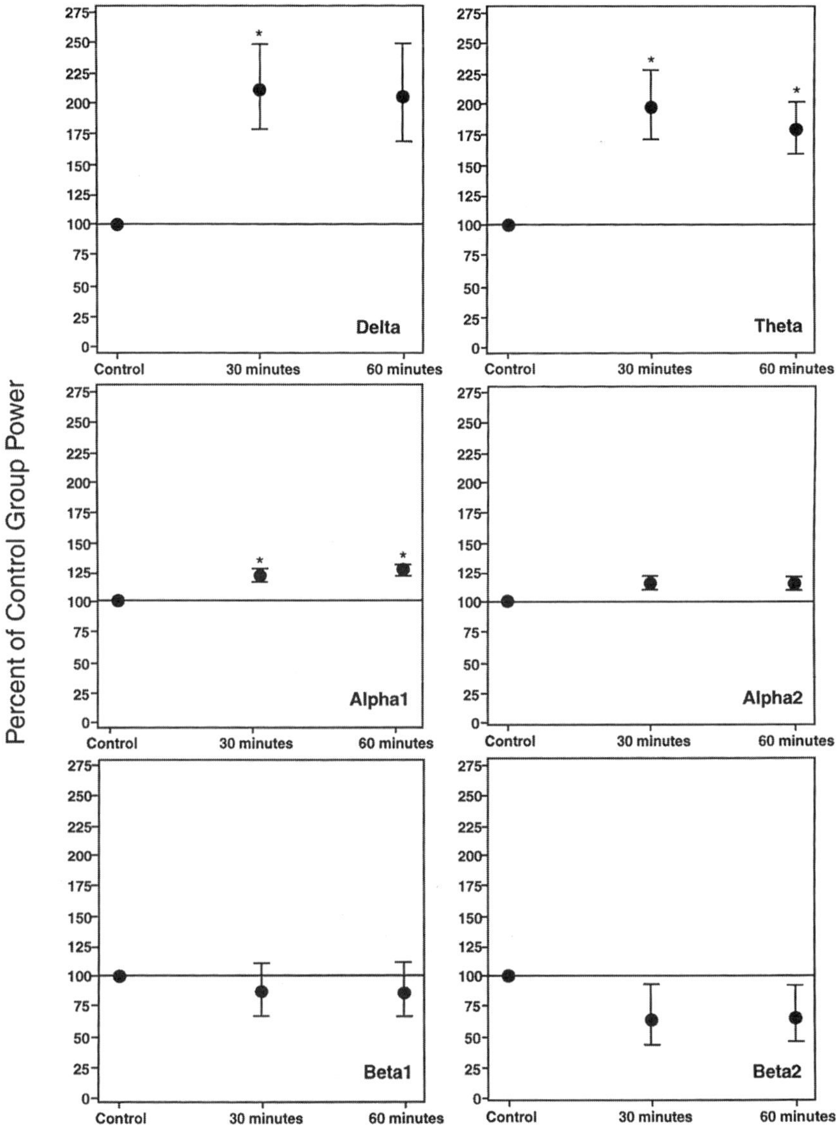

FIGURE 3. EFFECTS PRODUCED BY COCAINE INJECTED AT 20 MG/KG I.P. AND IBOGAINE PRETREATMENT AT 50 MG/KG I.P. 1 HR PRIOR TO COCAINE. Power values calculated as percent of the 30 min baseline power recorded after saline injection (assigned as a value of 100% in each band). Mean ± SEM; n=3 rats. *$p<0.05$ significantly different from baseline.

tested for their antiepileptic activity, induced a paradoxical enhancement of electrographic seizures that preceded suppression of status epilepticus (*28*). IBO, like MK-801, stimulates corticosterone release (*29*) and corticosterone has been shown to increase susceptibility to seizures (*30*).

III. Other Studies on Ibogaine Neurotoxicity at FDA/NCTR

A. Neurochemistry

Although ibogaine has been known to produce effects on multiple neurotransmitter systems, the neurochemical basis of ibogaine's effects is still poorly understood. Several reports have suggested that acute administration of ibogaine alters the extracellular concentration of dopamine and its metabolites in different regions of the rat and mouse brain (9,12). However, we have reported that pretreatment with ibogaine failed to alter either the spontaneous activity of ventral tegmental dopamine neurons, or the response of these dopamine neurons to morphine or cocaine (31). The excitatory effects of ibogaine on ventral tegmental dopamine neurons are not long lasting, nor does ibogaine persistently alter cocaine- or morphine-induced changes in dopamine neuronal impulse activity.

In our collaborative time course study reported earlier (*11*), adult, male, CD strain Sprague-Dawley rats were treated with a single injection of ibogaine (50 mg/kg, i.p.). They were sacrificed at 15, 30, 60, 120 minutes, and 24 hours later by decapitation. Trunk blood was collected and brains were dissected into different regions. We have shown that acute injection of ibogaine produced a significant increase in blood plasma prolactin levels within 15 and 30 minutes. While prolactin was observed to return to the control level by 120 minutes (Figure 4a), the corticosterone concentration that increased within 15 minutes returned to the control level by 24 hours after ibogaine treatment (Figure 4b). Besides neuroendocrine alterations, ibogaine produced significant changes in monoamine neurotransmitter systems. A single injection of ibogaine produced a significant reduction in the dopamine concentration in the striatum after 30, 60, and 120 minutes. Dopamine levels returned to control values after 24 hours. The dopamine metabolites (DOPAC and HVA) increased significantly within 15 minutes after ibogaine administration and remained elevated up to 120 minutes. While HVA returned to the control level, DOPAC concentration decreased to below control values 24 hours after ibogaine administration. In the frontal cortex, the concentration of dopamine decreased 30 minutes after ibogaine injection and returned to control values within 60 minutes (Figures 5a and b).

The endocrine profile observed in our ibogaine study resembles those obtained with the administration of other 5-HT releasing agents, such as fenfluramine (*32*). Our data suggest that ibogaine effects, like fenfluramine, might be mediated via stimulation of the serotonergic system. Ibogaine administration elicits a serotonergic-like syndrome, such as tremors and forepaw treading, and interactions between ibogaine and serotonergic system have been reported (*11,13*). In addition, the affinity of ibogaine for the 5-HT transporter is higher than for the dopamine transporter (*10*). Ibogaine produced time-dependent changes in the dopamine system, which also have been reported by several laboratories, including ours (*9-12,15*). However, these effects do not involve ibogaine binding to dopamine receptors (*13,19*). Ibogaine displays different dopamine transporter

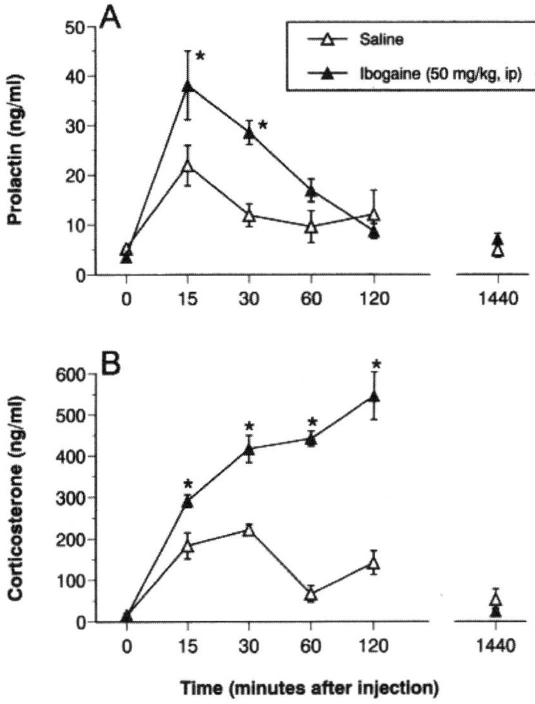

FIG. 4. EFFECTS OF SALINE (1 ML/KG, I.P.) OR IBOGAINE (50 MG/KG, I.P.) ON PLASMA PROLACTIN (A) AND CORTICOSTERONE (B) IN ADULT MALE RATS. Trunk blood was collected immediately before and at 15, 30, 60, 120 and 1440 minutes (24 h) after ibogaine administration. Data represent mean ± S.E.M. of n=4-8 rats/group.
*$p<0.05$ compared to saline treated group (Adopted from Ali *et al.*[*11*]).

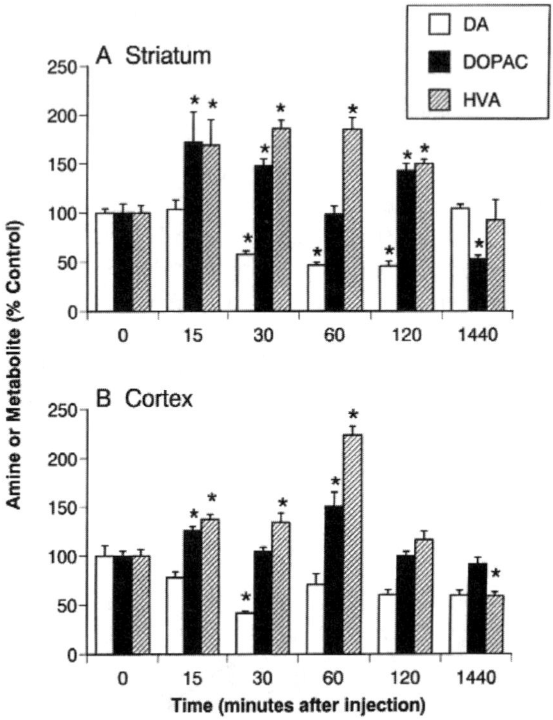

FIG. 5. EFFECTS OF IBOGAINE (50 MG/KG, I.P.) ON DA, DOPAC AND HVA CONCENTRATION IN THE STRIATUM (A) AND FRONTAL CORTEX (B) OF ADULT MALE RATS. Rats were sacrificed (n=4-8 rats/group) immediately before and at 15, 30, 60, 120 and 1440 min (24 hr) after ibogaine administration. Values are expressed as % of control of the data pooled from saline-treated rats at all time points (n=20).
*p<0.05 with respect to time zero control (Adopted from Ali *et al.*[*11*]).

binding affinity depending on the radioligand used to label these sites. Therefore, different domains may be present on the dopamine transporter protein that binds to ibogaine.

The neurochemistry/neurobiology of ibogaine is complex, and the binding of ibogaine to the multiple target sites in the central nervous system, and the coactivation of multiple transmitter systems, probably accounts for the diverse actions of this alkaloid, including its putatively antiaddictive effects.

B. NEUROHISTOLOGY

In addition to a structural resemblance to 5-HT, ibogaine is closely related structurally to harmaline, a tremorigenic agent known to produce neurotoxic damage to the cerebellum. This observation led O'Hearn and Molliver (*17*) to evaluate the neurohistology of the rat cerebellum following acute exposure to 100 mg/kg ibogaine, i.p. As with harmaline, they observed a loss of Purkinje neurons in the cerebellar vermis, as indicated by several neurohistological biomarkers: argyrophilic degeneration, loss of calbindin immunoreactivity, astrocytosis, and microgliosis. Efforts by other laboratories failed to obtain any evidence for the neurotoxicity of ibogaine in nonhuman primates (*33*). However, the methods used in those studies were primarily conventional hematoxylin and eosin (H and E) staining of paraffin sections, rather than the more specialized techniques of O'Hearn and Molliver.

Both the nature and the extent of ibogaine neurotoxicity, as well as its efficacy, must be understood in order that the risks and benefits can be appropriately balanced to provide the necessary information for regulatory decisions regarding the therapeutic use of ibogaine in humans. Therefore, our research group at FDA/NCTR replicated the initial observations of O'Hearn and Molliver (*6,17*), using their specialized neurohistological methods, which included degeneration-selective silver-staining of dead (argyrophilic) neurons, as well as several immunohistochemical approaches. We sought to eliminate, as much as possible, the controversy that had been generated during the early 1990s regarding their initial observations of ibogaine neurotoxicity. Just as O'Hearn and Molliver had reported, we also observed that a single i.p. dose of 100 mg/kg ibogaine produced "patches" of dead cerebellar Purkinje neurons (*6*). These "patches" comprised clusters of perhaps five to eight adjacent, or nearly adjacent, neurons that had died and become argyrophilic within a week after the ibogaine injection (*6*). Similar sized "patches" were observed by using antisera to reveal the enhanced presence of glial fibrillary acidic protein (GFAP; an astrocyte-specific protein) (*6,17*). As a third method to identify neuropathology, we highlighted the appearance of normal cerebellar Purkinje neurons by immunostaining the dense deposits of calbindin contained in each cell body. IBO treatment (100 mg/kg) resulted in similar "patches," each again about five to eight neurons long, where no calbindin-immunoreactive neurons could be observed (*6,17*). Our data thus strongly supported the initial report of ibogaine neurotoxicity (*17*), using essentially the same treatment and evaluation approaches (*6*). A third independent evaluation by Molinari *et al.* (*7*), using degeneration-selective silver-staining, has also confirmed the occurrence and character of ibogaine neurotoxicity in the rat cerebellum following 100 mg/kg, i.p., but not after a lower dose of 40 mg/kg, i.p.

Finally, our own recent dose-response study once again replicated the several previous observations of ibogaine neurotoxicity one week following doses of 100 mg/kg i.p. and additionally evaluated doses of 75, 50, and 25 mg/kg in female rats. This investigation also demonstrated the dose-response relationship, for each of the three different neuropathological techniques, by which ibogaine produced signs of Purkinje cell damage. A dose of 25 mg/kg was the highest level at which no observable adverse effects (NOAEL) of ibogaine occurred in any of the rats evaluated by any of the techniques in our study. The most sensitive procedures seemed to be immunohistochemistry for GFAP in the cerebellar cortex and the silver stain for degenerating axons in the deep cerebellar nuclei. Both of these methods detected the effects of 50 mg/kg ibogaine in the same two rats (out of a total of six) that were tested at this dose. Clearly neurotoxic effects of ibogaine were apparent in all six rats dosed with either 75 mg/kg or 100 mg/kg of ibogaine. However, the degenerating "patches" of Purkinje neurons were narrower, and fewer of their degenerating axons (as projections terminating in the deep cerebellar nuclei) could be observed in the 75 mg/kg compared to 100 mg/kg rats (*34*).

As mentioned previously, ibogaine shows a close structural resemblance to melatonin and 5-HT, whose receptors are widely distributed in the cerebellum, and throughout the entire brain. We were interested in exploring other histological biomarkers, such as c-fos, to comprehensively demonstrate the localization of brain cells activated by ibogaine (*35,39*). These data on regional c-fos responses may be compared to the effects of ibogaine on EEG described above. Previously, localization of c-fos activation has been compared to EEG findings for the convulsant neurotoxicants such as kainic acid and domoic acid (*36,37*). Under control conditions, only scattered and occasional neuronal nuclei express immunoreactive c-fos, an early-immediate gene product, located throughout the brain. However, stimuli resulting in the generation of neuronal action potentials have been shown to effectively initiate c-fos expression (*38*). Indeed, in our studies, exposure to 100 mg/kg of IBO evoked a widespread pattern of c-fos expression that served to indicate the specific regions of the brain that were most affected by ibogaine (*39*).

We believed that mapping the locations of c-fos activation might afford further insight into both the therapeutic and neurotoxic actions of ibogaine, so that the two might be dissociated. Intact excitatory input to the Purkinje neurons is required for the neurotoxic action of either harmaline or ibogaine (*17*). This may be demonstrated by using systemic injections of the neurotoxicant 3-aminopyridine to lesion the inferior olive, which provides the climbing fibers that ascend from the brainstem and innervate the Purkinje neurons. Under these circumstances, neither harmaline nor ibogaine can effectively produce cerebellar neurotoxicity (*17*). It was interesting to note that c-fos in the nuclei of the inferior olivary neurons was greatly increased following ibogaine exposure (*39*, and see Figure 6). Patches of cerebellar Purkinje neurons and their nearby granule cells also were strongly stimulated to express c-fos (*39*). Thus, it is likely that ibogaine's excitation of this pathway, which contains endogenous glutamate and/or aspartate, each capable of causing "excitotoxic" neurotoxicity, is sufficient to explain the loss of Purkinje neurons that was observed.

However, many other regions of the rat brain, where no neurotoxicity can be observed, are also induced into increased c-fos expression by ibogaine (*39*, and see Figures 6 and 7). These especially include neurons located throughout the rat neocortex, as well as the granule cells of the dorsal blade of the hippocampal dentate gyrus, and the pyramidal neurons of the hippocampal CA1 region. Ibogaine's strong activation of c-fos in the hippocampus may well relate to its induction of the EEG theta rhythm, as we previously observed (*24*), since theta rhythms are thought to arise from the hippocampal CA1 region in rats (*40*).

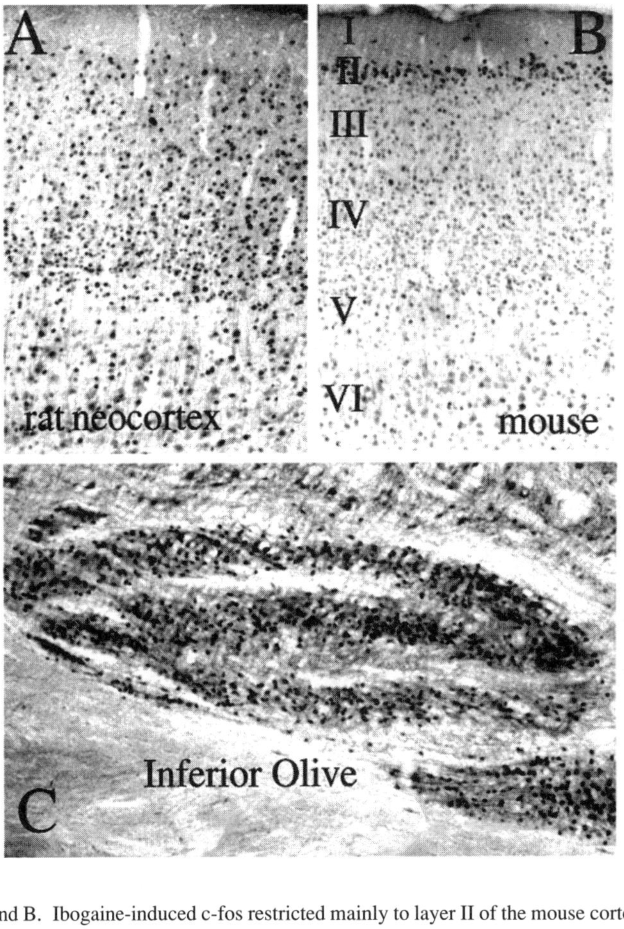

FIG. 6. A. and B. Ibogaine-induced c-fos restricted mainly to layer II of the mouse cortex, but, in the rat, considerable c-fos activation occurs throughout the deeper cortical layers, as well. C. Ibogaine induces many c-fos-immunoreactive neuronal nuclei in the inferior olive. These neurons project excitatory climbing fibers to innervate the Purkinje neurons of the cerebellum.

The paraventricular nucleus (PVN) of the hypothalamus is also highly activated by ibogaine (*39*, and see Figure 7b). The PVN is an important neurosecretory nucleus and regulator of the pituitary. Since its parvocellular neurons contain nearly all of the hypothalamic neuropeptide corticotrophin-releasing hormone (CRH), the effects of ibogaine on neuroendocrine functions, such as corticosterone release, may thus be explained. These additional effects of ibogaine outside the cerebellum may also be relevant to its psychoactive and

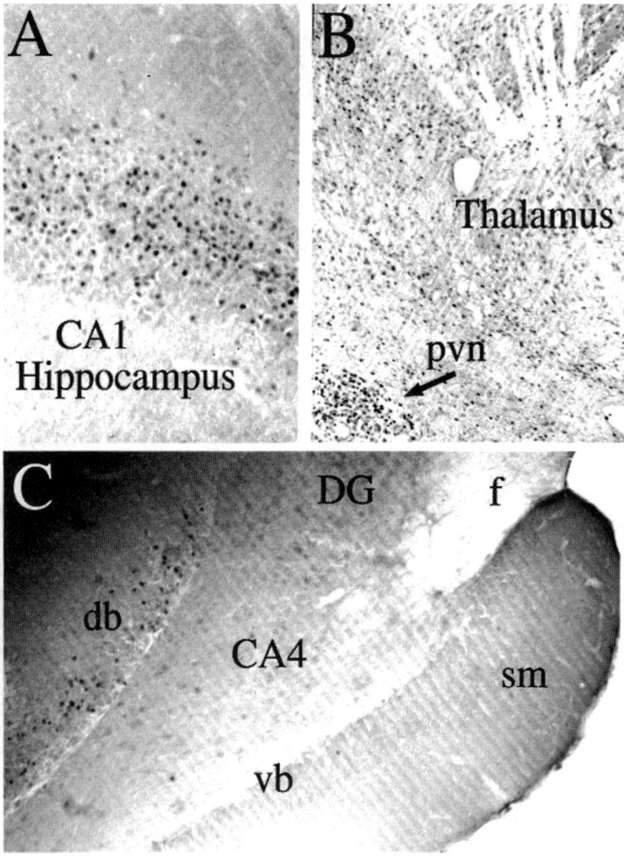

FIG. 7. Activation of c-fos occurs within the hippocampus, primarily in CA1 pyramidal neurons (A) and in neurons of the dorsal blade (db) of the dentate granule cells (C). Fig. 7B. Demonstrates that c-fos is strongly activated in the thalamus and in the hypothalamic paraventricular nucleus (pvn) as well. Abbreviations: CA, cornu ammonis; f, fornix; sm, stratum moleculare; vb, ventral blade; DG, dentate gyms. (Ibogaine treated).

therapeutic actions.

As we have argued elsewhere (*39*), it appears likely that, in rats, an excitatory projection from the deep layers of the neocortex to the neurons of the inferior olive activate their climbing fibers sufficiently to cause excitotoxic damage to the Purkinje neurons that they innervate. This contention is based on the observation that, in mice, ibogaine at 100 mg/kg, i.p. was a completely ineffective neurotoxicant. The only obvious difference in the intensity and pattern of c-fos activation in the mouse, compared to rat, was the striking lack of activation of the

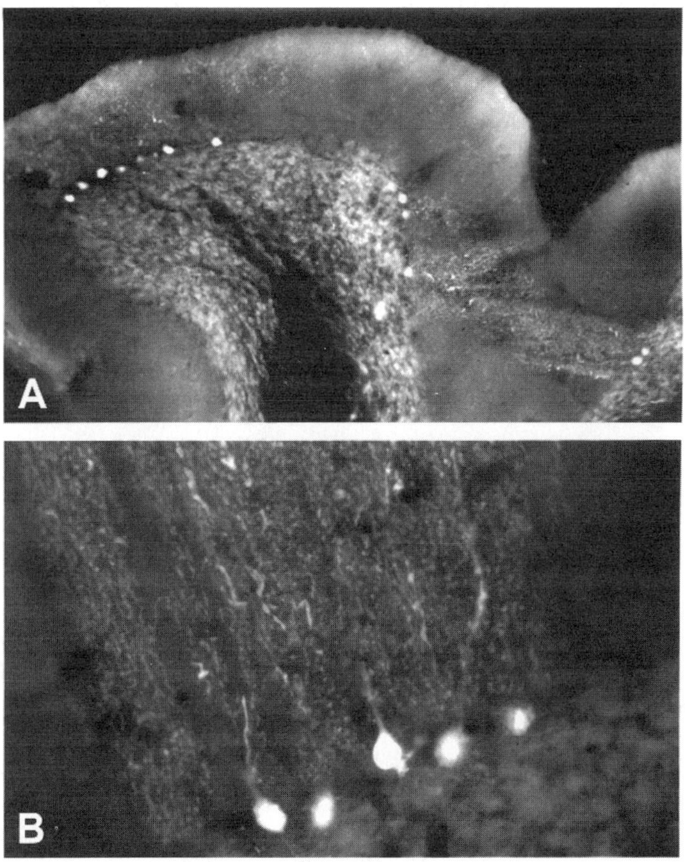

FIG. 8. A. Survey view of the paravermal region of the cerebellum of a rat exposed to ibogaine reveals the patchy distribution of Fluoro-Jade positive degenerating Purkinje cells. B. High magnification view of Fluoro-Jade positive Purkinje cells reveals both cellular and dendritic degeneration while granule cells (lower right) are not labeled.

deeper cortical layers, despite a prominent band demarking a strong excitation of layer 2. These differences between mice and rats may relate to different concentrations of ibogaine-related receptors in their deeper cortical neurons. For more optimal prediction of potential human neurotoxic responses to ibogaine, it might be informative to know if they are more "rat-like" or "mouse-like" in this regard.

In addition to verifying the cerebellar neuropathology using the aforementioned methods of Molliver and O'Hearn, a recently developed marker of neuronal degeneration was also used to validate the previous findings. This marker was Fluoro-Jade, which has been shown to localize neuronal degeneration following a wide variety of insults (*41*). This fluorescent tracer confirmed the existence of small patches of degenerating Purkinje cells. A survey view reveals the patchy appearance of Fluoro-Jade positive cells of the paravermal region of the cerebellum (Figure 8a), while a higher magnification view of these regions reveals the shrunken cytoplasm and extensive dendritic labeling (Figure 8b).

One of the more surprising aspects of ibogaine pathology is the relatively restricted pattern of neuronal degeneration observed. This pattern seen with ibogaine does not obviously correlate with that of neurotoxicants known to act via a specific transmitter system. For example, it is not similar to the distribution of neuropathology commonly associated with either NMDA agonists, which typically involves limbic system degeneration, or NMDA antagonists, which typically involves retrosplenial cortex degeneration (*42,43*). Likewise, there is little similarity to the pattern of degeneration that is observed following dopamine agonists, such as the degeneration in the parietal cortex and midline thalamus seen with methamphetamine, or the pattern that is observed following dopamine toxicants, such as degeneration of neurons of the substantia nigra and dorsal medial thalamus induced by 1-methyl-4-phenyl-1,2,3,6-tetrahydropyridine (MPTP)(*44-47*). Inhibitors of oxidative respiration also resulted in a differential pattern of neuronal degeneration. For instance, 3-nitropropionic acid (3-NPA) results in neuronal degeneration within the basal ganglia, medial thalamus, and deep nuclei of the cerebellum (*48,49*). 5-HT agonists may result in a pattern of degeneration most similar to that produced by ibogaine. For example, the 5-HT agonist d-fenfluramine is capable of inducing degeneration of cerebellar Purkinje neurons, as well as neuronal degeneration within frontal cortex and medial thalamus (*44*). This raises the question as to why ibogaine treatment does not also result in degeneration of forebrain structures with a robust serotonergic innervation. One possible explanation is that, like *d*-fenfluramine, hyperthermia may be necessary to potentiate forebrain degeneration. Another possible explanation would be that serotonergic input to glutaminergic forebrain nuclei was not as damaging as the serotonergic input to the aspartate-containing neurons of the brainstem inferior olive.

IV. Conclusions

Anecdotal reports and published studies in laboratory animals have suggested antiaddictive properties of ibogaine. Ibogaine, like many other indole alkaloids, has hallucinogenic as well as stimulant properties. So the question arose whether treatment of substance addiction with ibogaine may also lead to ibogaine-induced neurotoxicity.

We used electrophysiological, neurochemical, and neurohistological tools to evaluate neurotoxicity of ibogaine. Electrophysiological studies suggest that ibogaine stimulates monoaminergic neurons and may lower the threshold for cocaine induced electrographic seizures. Ibogaine interacts with several neurotransmitter-binding sites, produces significant alterations in neurotransmitter concentrations in different regions of the brain, and also induces immediate early genes (c-fos and erg-1). A single injection of ibogaine produces a spectrum of effects that includes elevation of plasma prolactin and corticosterone, short and long-term effects on dopamine neurotransmission, and modest transient effects on 5-HT. Neuropathological studies reveal that ibogaine administered at high doses produces selective neuronal degeneration. Therefore, we conclude that ibogaine might have potential utility for the treatment of drug addiction, but may also be neurotoxic at high doses, and that more studies are needed to elucidate the apparently complex mechanism of action of this drug.

References

1. H. G. Pope, Jr., *Econ. Bot.* **23,** 174 (1969).
2. S. L. Cappenduk and M. R. Dzojic, *Eur. J. Pharmacol.* **241,** 261 (1993).
3. H. Sershen, A. Hashim, and A. Lajtha, *Pharmacol. Biochem. Behav.* **47,** 13 (1994).
4. S. D. Glick, K. Rossman, S. Steindorf, I. M. Maisonneuve, J. N., and Carlson, *Eur. J. Pharmacol.* **195,** 341 (1991).
5. E. O'Hearn and M.E. Molliver, *Neurosci.* **55,** 303 (1993).
6. C. Scallet, X. Ye, R. Rountree, P. Nony, and S. F. Ali, *Ann. N.Y. Acad. Sci.* **801,** 217 (1996).
7. H. H. Molinari, I. M. Maisonneuve, and S. D. Glick, *Brain Res.* **737,** 255 (1996).
8. P. Popik and P. Skolnick, *in* "The Alkaloids" (G.A. Cordell, ed.), Volume 52, p. 197. Academic Press, San Diego, CA, 1999.
9. M. Maisonneuve, K. L. Rossman, R. W. Keller, Jr., and S.D. Glick, *Brain Res.* **575,** 69 (1992).
10. H. Sershen, A. Hashim, and A. Laitha, *Brain Res. Bull.* **42,** 161 (1997).
11. S. F. Ali, G. D. Newport, W. Slikker, Jr., R. B. Rothman, and M. H. Baumann, *Brain Res.* **737,** 215 (1996).
12. S. D. Glick, K. Rossman, S. Wang, N. Dong, and R. W. Keller, Jr., *Brain Res.* **628,** 201 (1993).
13. P. A. Broderick, F. T. Phelan, F. Eng, and R. T. Wechsler, *Pharmacol. Biochem. Behav.* **49,** 711 (1994).
14. P. Popik, R. T. Layer, L. H. Fossom, M. Benveniste, B. Geter-Douglass, J. M. Witkin, and P.

Skolnick, *J. Pharmacol. Exp. Ther.* **275,** 753 (1995).
15. S. D. Glick and I. M. Maisonneuve, *Ann. N.Y. Acad. Sci.* **844,** 214 (1998).
16. M. E. Alburgens and G. R. Hanson, *Brain Res.* **818,** 96 (1999).
17. E. O'Hearn and M. E. Molliver, *Neuroscience* **55,** 303 (1993).
18. L. Jacobs, *Life Sci.* **19,** 777 (1976).
19. C. Mash, J. K. Staley, M. H. Baumann, R. B. Rothman, and S. L. Hearn, *Life Sci.* **57,** 45 (1995).
20. Wei, I. M. Maisonneuve, M. E. Kuehne, and S. D. Glick, *Brain Res.* **800,** 260 (1998).
21. C. D. Binnie and P. F. Prior, *J. Neurol. Neurosurg. Psych.* **57,** 1308 (1994).
22. J. S. Barlow, *in* "The Electroencephalogram: Its Patterns and Origins" (J. Barlow, ed.), p. 169. The MIT Press, Cambridge, MA, 1993.
23. Z. Binienda, J. Johnson, B. Thorn, and S. F. Ali, *Abst. Soc. Neurosci.* **23,** 1227 (1997).
24. Z. Binienda, M. A. Beaudoin, B. T. Thorn, N. V. Sadovova, W. Slikker, Jr., and S. F. Ali, *Ann. N.Y. Acad. Sci.* **914,** 387 (2000).
25. H-F. Luoh, T. B. J. Kuo, S. H. H. Chan, and W. H. T. Pan, *Synapse* **16,** 29 (1994).
26. B. Ferger, W. Kropf, and K. Kuschinsky, *Psychopharmacol.* **114,** 297 (1994).
27. W. Dimpfel, M. Spüler, and D. E. Nichols, *Psychopharmacol.* **98,** 297 (1989).
28. H. Bertram and E. W. Lothman, *Epilepsy Res.* **5,** 177, 1990.
29. M. H. Baumann, R. B. Rothman, and S. F. Ali, *Ann. N.Y. Acad. Sci.* **844,** 252 (1998).
30. A. Cottrell, C. Nyakas, E. R. De Kloet, and B. Bohus, *Brain Res.* **309,** 377 (1984).
31. E. D. French, K. Dillon, and S. F. Ali, *Life Sci.* **59,** 199 (1996).
32. L. D. Van de Kar, *Ann. Rev. Pharmacol. Toxicol.* **31,** 289 (1991).
33. D. C. Mash, C. A. Kovera, B. E. Buch, M. D. Norenberg, P. Shapshak, W. L. Hearn, and J. Sanchez, *Ann. N.Y. Acad. Sci.* **844,** 274 (1998).
34. Z. Xu, L. W. Chang, W. Slikker, Jr., S. A. Ali, R. L. Rountree, and A. C. Scallet, *Tox. Sci.* **57,** 95 (2000).
35. S. F. Ali, N. Thiriet, and J. Zwiller, *Mol. Brain Res.* **74,** 237 (1999).
36. R.C. Collins, E. W. Lothman, and J. W. Olney, *Adv. Neurol.* **34,** 277 (1983).
37. Robertson, K. Renton, J. Kohn, and T. White, *Ann. N.Y. Acad. Sci.* **648,** 330 (1992).
38. M. Kasof, A. Mandelzys, S.D. Maika, R. E. Hammer, T. Curran, and I. L. Morgan, *J. Neurosci.* **15,** 4238 (1995).
39. A.C. Scallet, X.Ye, and S.F. Ali, *Ann. NY Acad. Sci.* **801,** 227 (1996).
40. L. S. Leung, *Neurosci. Biobehav. Rev.* **22,** 275 (1998).
41. L. C. Schmued, C. Albertson, and W. Slikker, Jr., *Brain Res.* **751,** 37 (1997).
42. W. Olney, V. Rhee, and O. L. Ho, *Brain Res.* **77,** 507 (1974).
43. W. Olney, J. Labruyere, and M. T. Price, *Science* **244,** 1360 (1989).
44. C. Schmued and J. F. Bowyer, *Brain Res.* **759,** 135 (1997).
45. J. Eisch, L. C. Schmued, and J. F. Marshall, *Synapse* **30,** 329 (1998).
46. R. E. Heikkila, A. Hess, and R. C. Duvoisin, *Science* **224,** 1451 (1983).
47. T. E. Freyaldenhoven, S. F. Ali, and L. C. Schmued, *Brain Res.* **759,** 9 (1997).
48. F. Hamilton and D. H. Gould, *Acta Neuropathol. (Berl.)* **72,** 286 (1987).
49. Z. Binienda, D. L. Frederick, S. A. Ferguson, R. L. Rountree, M. G. Paule, L. Schmued, S. F. Ali, W. Slikker, Jr., and A. C. Scallet, *Metab. Brain Dis.* **10,** 269 (1995).

—CHAPTER 11—

MODULATION OF THE EFFECTS OF REWARDING DRUGS BY IBOGAINE

LINDA A. PARKER

Department of Psychology
Wilfrid Laurier University
Waterloo, Ontario N2L 3C5

SHEPARD SIEGEL

Department of Psychology
McMaster University
Hamilton, Ontario L8S 4L8

THE ALKALOIDS, Vol.56
0099-9598/01 $35.00

I. Introduction

Ibogaine (NIH 10567, Endabuse™) is an alkaloid obtained from the root of *Tabernanthe iboga*, a shrub indigenous to equatorial Africa. There are anecdotal reports that ibogaine is effective in the treatment of a variety of addictive disorders (*1,2*). Indeed, Howard Lotsof (NDA International) has patented ibogaine treatments for dependencies on heroin (U.S. Patent 4,499,091), amphetamine (U.S. Patent 4,587,243), cocaine (U.S. Patent 4,587,243), alcohol (U.S. Patent 4,857,523), and nicotine (U.S. Patent 5,026,697). Results of research with ibogaine indicate that the alkaloid may modulate the rewarding properties of drugs (e.g. *3,4*) and interfere with withdrawal reactions (*5-7*) in animals. This chapter reviews some of the evidence that ibogaine modulates drug effects in animals.

II. Ibogaine-Morphine Interactions

A. Ibogaine Potentiates Morphine Analgesia

If ibogaine is to be used as a treatment for opiate addiction, it is important to understand the interaction between ibogaine and opiate drugs. Although Matwyshyn and Bhargava (*8*) reported that ibogaine does not modulate some effects of morphine (analgesia and hypothermia) in mice, others have reported that the alkaloid augments the effects of morphine in rats (*9*). An early report by Schneider and McArthur (*10*) reported that ibogaine (at doses ranging from 50-100 mg/kg) potentiated the lethal effect of morphine and (at doses ranging from 3-24 mg/kg) enhanced the analgesic effect of morphine in rats. Such potentiation of morphine's effects by ibogaine has profound implications for the manner in which ibogaine is used clinically in treating opiate dependence. All efforts must be made to ensure that the opiates are not available to the patient during treatment.

B. Ibogaine Interferes with Morphine Tolerance

Ibogaine has been reported to modulate the development of morphine tolerance (*11*). Rats were injected with morphine (5 mg/kg) at 48-hour intervals, and, 30 minutes after each injection, analgesia was assessed with the "hot-plate" procedure and the latency to respond (paw lick or jump) was measured. The effect of ibogaine injected prior to morphine on the development of analgesic tolerance was evaluated. Various doses of ibogaine (10-40 mg/kg), and intervals between ibogaine and morphine administration (30 min-24 hr), were assessed. The results indicated that although ibogaine (at all doses) by itself had no effect on hot-plate response latency, 20 mg/kg ibogaine administered 30 minutes prior to morphine clearly retarded the development of morphine analgesic tolerance. A lower dose of ibogaine, or longer pretreatment intervals (e.g., 40 mg/kg ibogaine injected 24 hours prior to each morphine injection), did not affect the development of tolerance.

III. Ibogaine and Morphine Reward

A. Animal Models of Drug Reward

There are two commonly used methods for evaluating drug reward in animals: Drug self-administration and place preference conditioning. In drug self-administration, an animal must make some response, such as a lever press, which is followed by a presentation of the drug reward. This approach evaluates the ability of a drug to directly reinforce behavior. In essence, this method determines whether the animal will "work for" or "seek" the drug. In fact, rats will "work for" most drugs that humans tend to abuse (see ref. *12*), suggesting that these drugs are also rewarding for rats.

The place-conditioning paradigm is an alternative measure of drug reward (for review, see ref. *13*). With this procedure, during a training period, rats are confined to one distinctive compartment following injection of a drug and an alternative compartment following injection of an inert substance. Their preference for one or the other compartment is subsequently evaluated during a test in which they are undrugged, and have access to both compartments. It is inferred that the drug is rewarding if the rat displays a preference for the drug-associated compartment (and aversive if the rat displays a preference for the alternative compartment). Rats form a preference for a place paired with the same drugs that they tend to self-administer (*13*).

B. Ibogaine and Morphine Self-Administration

The results of several experiments, using rats and mice, suggest that ibogaine reduces the self-administration of drugs of abuse (see ref. *14*). Glick *et al.* (*3*) reported that ibogaine (in doses ranging from 2.5 to 40 mg/kg) decreased intravenous self-administration of morphine in a dose-dependent manner. The 40 mg/kg dose of ibogaine, administered 5 minutes prior to a self-administration session, interfered with water-reinforced responding (as well as morphine-reinforced responding), and this acute action may be attributable to the motoric effects of ibogaine. However, the alkaloid continued to suppress responding for morphine reinforcement, but not water reinforcement, long after the tremorigenic effect of ibogaine was no longer apparent. Glick *et al.* (*3*) reported that a single 40 mg/kg ibogaine dose attenuated morphine self-administration for a period of several days (see also ref. *15*).

In contrast with findings suggesting that ibogaine has a long-lasting effect on drug reinforcement is a report that the alkaloid has a short-term effect, but not a persistent (24-hour) effect, on responding for heroin (*16*). The reasons for the discrepant findings are not clear. However, as discussed by Glick *et al.* (*15*), there are many procedural differences between the experiments that have reported various effects of ibogaine on drug self-administration.

C. Ibogaine and Morphine-Induced Place Preference

1. Ibogaine Attenuates the Establishment of a Morphine-Induced Place Preference

Although self-administration of drugs provides an intuitively appealing measure of drug reward, the effect of pharmacological pretreatment on operant responding for drug reinforcement cannot be interpreted unambiguously. As discussed by Glick *et al.* (*3*), there are several reasons why a pretreatment may decrease drug self-administration: (1) the pretreatment agent reduces drug reward, (2) the pretreatment agent interferes with responding for reward, or (3) the pretreatment agent increases the rewarding properties of the drug, rendering each infusion more potent. The place preference paradigm provides an unambiguous measure of the rewarding properties of morphine, because the strength of a place preference is proportional to the dose of morphine (*13,17,18*). Therefore, if ibogaine attenuates the rewarding effect of morphine, it should attenuate a morphine-induced place preference.

Using the place preference paradigm, Parker, Moroz, and Siegel (*4*) found that ibogaine reduces the ability of a single injection of morphine to produce a preference in rats. This attenuation of morphine's rewarding properties is seen if ibogaine (40 mg/kg) is administered either immediately (i.e., 10 min) or 24 hours before the opiate. Ibogaine-induced interference with morphine place preference does not appear to be the result of the summation of the independent hedonic properties of ibogaine and morphine, because ibogaine alone produced neither a place preference nor place aversion. Although the effect of ibogaine was apparent with a one-trial, morphine-induced place preference, it was not maintained after four training trials. That is, rats receiving ibogaine pretreatment prior to each of four place preference trials responded with a preference for the morphine side that was similar in strength to that of rats not pretreated with ibogaine. These results suggest that: (1) ibogaine may only modulate a weak (i.e., one-trial) place preference, but not a strong (i.e., four-trial) place preference, or (2) repeated exposure to ibogaine reduces its efficacy in attenuating the rewarding effect of morphine. Moroz, Parker, and Siegel (*19*) later found evidence that supports the latter alternative, as discussed subsequently.

Additional experiments by Parker *et al.* (*4*) were designed to determine if ibogaine attenuated morphine-induced place preference because it interfered with morphine reward or because it generally interfered with drug-place associations. If ibogaine generally affected drug-place associations, it should not only interfere with place preference learning, but it should also interfere with place-aversion learning. Ibogaine, administered either 10 minutes or 24 hours prior to several doses of drugs known to induce place aversion (naloxone and lithium chloride) did not affect the magnitude of the place aversions. Ibogaine selectively modulated the rewarding properties of morphine, rather than generally interfering with place conditioning.

2. *Ibogaine and the Expression of Morphine-Induced Place Preference Learning*

The finding that ibogaine interferes with morphine place preference learning (*4*) is consistent with suggestions that ibogaine is an N-methyl-D-aspartate (NMDA) antagonist (*14*), inasmuch as other NMDA antagonists also interfere with this type of learning (*20-22*). However, NMDA antagonists are known to interfere selectively with the acquisition, but not with the expression, of a variety of previously learned tasks (e.g., *23,24*). Thus, it might be expected that ibogaine would not interfere with a previously established morphine place preference. We recently have reported such a finding (*25*). A one-trial morphine place preference was induced (using procedures similar to those of Parker *et al.* [*4*]). A single injection of 40 mg/kg ibogaine, either 24 hours, 12 hours, or 4 hours prior to a preference, test did not interfere with the expression of the morphine place preference. A variety of other ibogaine pretreatment regimens (involving higher doses or multiple injections of ibogaine and various intervals between ibogaine administration and place preference testing) were similarly ineffective. In summary, the results of Luxton *et al.* (*25*), together with the previous results of Parker *et al.* (*4*), indicate that ibogaine, like other NMDA antagonists, interferes with the establishment, but not the expression, of a morphine place preference.

D. Ibogaine and Morphine Withdrawal Symptoms

Advocates for the use of ibogaine as a treatment for addiction emphasize its efficacy in reducing opiate withdrawal symptoms (*26,27*). Indeed, laboratory experiments with animals have demonstrated that ibogaine interferes with somatic symptoms of naloxone-precipitated withdrawal (*5-7,14*), although there are some conflicting results (*28,29*).

In morphine-dependent animals, administration of an opioid antagonist drug such as naloxone or naltrexone, produces somatic signs of withdrawal that include rearing, grooming, jumping, wet-dog shakes, teeth chattering, salivation, and diarrhea. Glick *et al.* (*7*) reported that ibogaine (40 mg/kg, i.p.) attenuated some signs of withdrawal (wet-dog shakes, grooming, teeth chattering, and diarrhea) when administered either 4 hours or 30 minutes prior to naltrexone (1 mg/kg, i.p.) in rats that had received 5 days of exposure to slow-release morphine pellets. Therefore, ibogaine may reverse somatic signs of opiate withdrawal.

Opiate dependence is typically thought to occur only after prolonged opiate exposure. However, human and animal research has shown that naloxone-precipitated withdrawal can be observed even when naloxone is administered up to several hours after a single administration of morphine (*30-34*). This phenomenon has been termed "acute opiate dependence" and may represent the early developmental states of the dependence process. In nondependent opiate users, June *et al.* (*33*) examined opioid agonist effects, morphine plasma levels, and withdrawal effects precipitated by naloxone (10 mg/70 kg, administered intramuscularly) at 1, 3, 6, 12, 18, 24, 30, 36, and 42 hours after a single dose of morphine (18 mg/70 kg, administered intramuscularly). The intensity of subjectively reported precipitated withdrawal effects was greatest when testing was conducted at 6 hours after morphine administration and persisted for up to 24 hours after morphine administration, whereas, peak intensity of agonist effects (pupil constriction and subjective ratings) and highest plasma morphine concentrations were observed at the shortest test interval (1 hour) after morphine. Therefore, acute opiate dependence produced by a single dose of morphine peaks later and persists over a longer duration after morphine administration than do other agonist effects. The discrepancy between peak agonist effects and peak withdrawal effects suggests that neural adaptations underlying acute morphine dependence develop over a 6-hour time period and gradually decay over time. The discrepancy also suggests that the agonist effect of acute morphine administration is mediated by a different mechanism than the effect of naloxone-precipitated morphine withdrawal (*33*).

In the rat, acute opiate dependence can be demonstrated when naloxone is administered between 30 min and 48 hr (*30,31*) after a single morphine exposure. Acute dependence is evidenced as specific somatic withdrawal reactions (e.g., wet dog shakes) or as aversive stimulus properties which modify behavior. For instance, Gellert and Sparber (*31*) reported that administration of a low dose of naloxone 48 hours after a single morphine exposure significantly decreased operant responding for food, whereas it was without effect in opiate naïve rats. The aversive properties of withdrawal apparently suppressed responding. The aversive properties of withdrawal have also been evaluated using the place-conditioning paradigm (*35-39*); rats learn to avoid a place previously paired with naloxone-precipitated withdrawal (*37*). In fact, a naloxone-precipitated, withdrawal-induced place aversion can be produced 24 to 48 hours after a single injection of morphine (*34*).

Ibogaine interferes with acute opioid dependence. Parker and collaborators (*40*) recently reported that ibogaine interferes with naloxone-precipitated withdrawal in rats treated with morphine 24 hours prior to the conditioning trial. On each of two conditioning trial cycles, rats were administered morphine (20 mg/kg, s.c.) or saline 24 hours prior to an injection of naloxone (1 mg/kg, s.c.) and placement in a chamber. Half of the rats were injected with ibogaine (40 mg/kg, i.p.) and half were injected with saline 4 hours before the naloxone injection. Ninety-six hours later, the rats received a drug-free place preference test.

Ibogaine attenuated naloxone-precipitated morphine withdrawal. Figure 1 presents the mean number of seconds that the rats spent on the naloxone-paired minus the saline-paired floor during the place preference test. The groups displayed on the abscissa include those injected with morphine 24 hours prior to

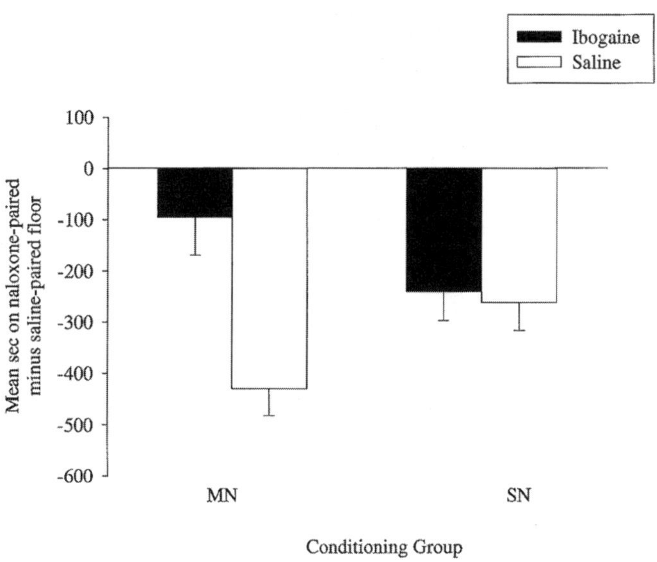

FIGURE 1. Mean (±sem) seconds spent on the naloxone-paired minus the saline-paired floor during the place conditioning test by groups MN and SN pretreated with ibogaine or saline during conditioning.

naloxone (MN) and those injected with saline 24 hours prior to naloxone (SN). Among each conditioning group, the black bars represent the rats pretreated with ibogaine 4 hours prior to naloxone and the white bars represent the rats pretreated with saline 4 hours prior to naloxone. Ibogaine pretreatment interfered with the naloxone-induced place aversion displayed by Group MN, without modulating the weaker aversion displayed by Group SN. This latter finding is consistent with those of Parker, Luxton, and Siegel (*4*) who reported that ibogaine did not interfere with the establishment of a naloxone-induced place aversion, but did interfere with the establishment of a morphine-induced place preference.

In a second experiment, Parker *et al.* (*40*) evaluated the potential of ibogaine to interfere with the expression of naloxone-precipitated somatic withdrawal symptoms. As in the previously described place-conditioning experiment, on each of two conditioning trial cycles, rats were administered morphine (20 mg/kg, s.c.) or saline 24 hours prior to an injection of naloxone (1 mg/kg, s.c.) and placement in a chamber. Half of the rats in each group were injected with ibogaine (40 mg/kg, i.p.) and the other half of the rats were injected with saline, 4 hours before the naloxone injection. The rats were then placed in the observation chamber for 1 hour. During the second repetition of the cycle, all subjects were videotaped for a period of 10 minutes following the naloxone injection. Videotapes were later scored for instances of wet dog shakes, mouth movements, teeth chattering, and genital licks. The frequency of each withdrawal behavior displayed by each rat was scored by two observers (one who was aware of testing conditions, and one who was unaware of testing conditions).

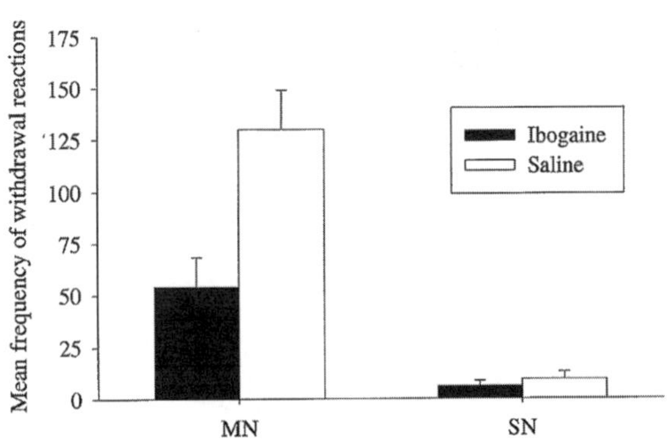

FIGURE 2. Mean (±sem) frequency of somatic withdrawal reactions (mouth movements + wet-dog shakes + teeth chattering + genital licking) by groups MN and SN pretreated with ibogaine or saline solution.

Ibogaine attenuated naloxone-precipitated morphine withdrawal behavior as it had attenuated the withdrawal-induced place aversion. Figure 2 presents the mean frequency of withdrawal behaviors (the summed frequencies of wet dog shakes, teeth chattering, mouth movements, and genital licks) for the various groups. Rats in Group MN displayed fewer somatic withdrawal behaviors when pretreated with ibogaine than they did when pretreated with saline. Consistent with the effect of ibogaine on the motivational effects of withdrawal, ibogaine did not modulate somatic withdrawal behaviors in Group SN.

Ibogaine, therefore, not only interferes with withdrawal from chronically administered morphine (e.g. *5-7,14*), but it also interferes with withdrawal from acutely administered morphine. It interferes with both the aversive motivational properties of withdrawal, as revealed by the place conditioning paradigm, and the somatic properties of withdrawal, as revealed by direct observation of somatic withdrawal symptoms.

IV. Ibogaine and Other Drugs of Abuse

A. Ibogaine and Self-Administration of Other Drugs of Abuse

Anecdotal reports indicate ibogaine decreases dependence on a wide variety of drugs of abuse, including alcohol, amphetamine, cocaine, nicotine, and opiates (*27*). Indeed, there is evidence that ibogaine interferes with self-administration of drugs other than morphine. For example, Rezvani, Overstreet, and Lee (*41*) reported that ibogaine reduced ethanol consumption in several strains of ethanol-preferring rats. There are also reports that ibogaine has a long-lasting suppressive effect on intravenous (*42*) and oral self-administration (*43*) of cocaine.

B. Ibogaine and Amphetamine-Induced Place Preference Learning

Ibogaine also interferes with the establishment of amphetamine-induced place preference learning (*19*). Since ibogaine produces prolonged (24-hour) interference with the establishment of a morphine-induced place preference (*4*), rats were injected with ibogaine or saline solution 24 hours prior to an amphetamine injection that preceded placement in a conditioning chamber. Rats were tested drug-free following one, and again following four, conditioning trials. The preference test results revealed that ibogaine interfered with the establishment of an amphetamine-induced place preference following one, but not following four, conditioning trials. This effect paralleled that found with morphine place preference learning (*4*).

In order to determine whether the reduced effectiveness after four trials was the result of tolerance to ibogaine, rats were given weekly injections of ibogaine (group ibogaine-experienced) or saline (group ibogaine-naïve) in their home cage over four weeks prior to place conditioning trials. On place-conditioning trials, rats in groups ibogaine-experienced and ibogaine-naïve were injected with ibogaine 24 hours before amphetamine conditioning trials on each of two weekly cycles. In a subsequent place preference test, ibogaine blocked an amphetamine place preference only in the group ibogaine-naïve—that is, prior experience with ibogaine eliminated its ability to interfere with amphetamine place preference learning. Therefore, ibogaine's reduced effectiveness across multiple conditioning trials may be the result of the development of tolerance to ibogaine.

V. Ibogaine and Craving: Future Directions

A. The Reinstatement Paradigm

Anecdotal reports by abstinent individuals suggest that re-exposure to a formerly self-administered substance induces a strong motivational state or "craving" for the drug, and is a frequent precursor to relapse (e.g., *44*). Several investigators have developed animal models of relapse to account for this phenomenon. One of these models is "reinstatement."

In the reinstatement model, rats are first trained to self-administer a rewarding drug by pressing a lever. Following such training, the response is extinguished—that is, an inert substance is substituted for the drug, resulting in a decrease in responding. Then, to "whet the rat's appetite," the drug is administered and the rat is provided the opportunity to lever press. In the reinstatement paradigm, a single administration of the reinforcing drug can reinstate previously extinguished self-administration behavior (e.g., *45,46*). In fact, following extinction training, a priming injection of morphine can also reinstate a conditioned place preference (*47*). Future experiments might examine the ability of ibogaine administered 24 hours prior to a priming injection of morphine to interfere with the reinstatement of a conditioned place preference. Such a finding would provide evidence that ibogaine, indeed, can interfere with drug craving in animals.

Most research evaluating the motivational effects of drugs has used either self-administration or place preference procedures. Ettenberg and colleagues (e.g., *48-50*) have demonstrated the utility of the runway procedure in investigations of drug reward. Rats are trained to traverse an alley for intravenous drug administration. With some rewarding drugs, the well-trained rat exhibits both "approach" and "retreat" behavior. That is, it quickly runs toward the goal, then reverses direction and runs away from the goal, then again runs toward the goal. There may be several cycles of this retreat behavior each trial, apparently indicating conflict: "retreat behavior might be reflective of a conflict resulting from concurrent positive and negative associations with the goal box" (*51*).

There is considerable evidence that some commonly abused drugs have, simultaneously, both rewarding and aversive effects (*52-55*). One possible mechanism by which ibogaine may attenuate drug reward is by increasing the aversive effect of drugs (rather than by decreasing the reinforcing effect of drugs). As suggested by Ettenberg and Geist (*48*), the spatiotemporal record of behavior in the straight alley permits evaluation of the aversive effect of the drug (e.g., number of retreats), as well as the rewarding effect of the drug. If ibogaine augments the aversive effects of drugs, this should be manifest as an increase in retreat behavior by ibogaine pretreated rats.

Ettenberg, MacConell, and Geist (*56*) described evaluation of reinstatement in the straight alley. Rats were first trained to run down the alley for heroin reinforcement. When the response was established, it was extinguished—rats received an IV infusion of saline, rather than heroin, in the goal box. When rats reached an extinction criterion, the effect of a single priming infusion of heroin was evaluated. Rats received a single "treatment trial"—they received a single i.v. infusion of either heroin or saline in the goal box. The effect of the prime was evaluated 24 hours later, on the "test day". Rats were undrugged on this test day. Rats primed with heroin on the post-extinction treatment trial ran significantly faster on the test day than did rats primed with saline, demonstrating the reinstatement effect. It would be of interest to determine if ibogaine would interfere with this reinstatement effect in the runway situation.

B. Ibogaine and Morphine Reward in Dependent Rats

Although our previously described studies employed nondependent rats, the anecdotal reports of ibogaine's effectiveness in humans are based on reports from drug-dependent humans (e.g., *26,27*). Furthermore, some current animal models of craving (*57-59*) emphasize the role of drug experience in establishing the motivational state necessary for craving.

If ibogaine interferes with craving, then one might expect that it would more effectively interfere with the establishment or the expression of morphine place preference learning and runway behavior in drug-dependent than in drug-naïve rats. In fact, Pearl, Johnson, and Glick (*60*) reported that ibogaine modified morphine-induced motoric effects and morphine-induced dopamine release in morphine-experienced rats more effectively than in morphine-inexperienced rats.

Future studies will evaluate the efficacy of ibogaine to interfere with morphine reward in rats that are maintained on morphine over a period of one month and rats that are morphine naïve. It is conceivable that ibogaine may more effectively modulate drug reward in drug-experienced rats than in naïve rats.

C. Ibogaine and Nonpharmacological Reward

Interest in the effects of ibogaine have focused on the alkaloid's potential in attenuating drug reward, but it is possible that ibogaine generally affects reward processes. There are few studies evaluating the effects of ibogaine on nonpharmacological reinforcement, and the available data are inconsistent. Glick *et al.* (*3*) reported that ibogaine does not have a long-lasting effect on operant responding for water reinforcement, but Dworkin *et al.* (*16*) reported that ibogaine has a long-lasting effect on responding for food.

Rats show a conditioned preference for a chamber in which they were previously given access to food (e.g., *61*), even when they are sated. Furthermore, both sucrose (*62*) and saccharin solution (*63*) have been reported to produce a place preference in sated rats. In pilot experiments, we have also found that nondeprived rats will readily learn to traverse a runway for highly palatable sweet solution (a mixture of 0.16% saccharin and 3% glucose in water). As discussed by others (e.g., *12*) such instrumental responding for a palatable solution by non-deprived subjects would seem to approximate the motivational properties of responding for a drug reward. It would be of interest to examine the effects of ibogaine on non-pharmacological reward using the place conditioning and runway paradigms.

D. Future Studies: Summary

These proposed directions for research should provide greater insight into the mechanism by which ibogaine modulates the rewarding properties of drugs as well as withdrawal effects. Considerable evidence is accruing that future research on the putative antiaddictive properties of ibogaine would be fruitful.

V. Conclusions

Considerable evidence indicates that ibogaine modulates a variety of opiate effects in rats. Ibogaine potentiates opiate-induced analgesia and lethality (*10*) and interferes with morphine tolerance (*11*). It also interferes with the rewarding properties of morphine when assessed in self-administration (*3*) and in place preference learning (*4*). Finally, ibogaine also interferes with the aversive properties of opiate withdrawal (*7,40*).

Advocates for ibogaine as an antiaddictive medication argue that ibogaine modulates a variety of addictive disorders, not only opiate addiction. Animal evidence reported above also indicates that ibogaine modulates the rewarding properties of stimulants when evaluated in the self-administration paradigm (*42*) or in the place preference paradigm (*19*).

The encouraging results of animal work with ibogaine suggest that further work with this agent is warranted to evaluate its potential antiaddictive properties. Yet any enthusiasm for ibogaine as an antiaddictive drug must be tempered by a report that, at high doses, it may produce cerebellar damage in rats (*64*). However, there are reports that ibogaine does not produce such damage at doses that effectively modulate drug reward in animals (*65,66*).

Acknowledgments

This research was supported by grants from the Natural Sciences and Engineering Research Council of Canada to Linda Parker and to Shepard Siegel. We would like to thank Marion Corrick and Doreen Mitchell for assistance with animal care throughout the course of these experiments. All procedures were approved by the respective institutional animal care committee in accordance with the guidelines of the Canadian Council on Animal Care.

References

1. S. Sheppard, J. *Subs. Abuse Treat.* **11,** 379 (1994).
2. B. Sisko, *Bull. Multidiscipl. Assoc. Psychedelic Stud.* **2,**15 (1993).
3. S.D. Glick, K. Rossman, S. Steindorf, I.M. Maisonneuve, and J.N. Carlson. *Eur. J. Pharmacol.* **195,** 341 (1991).
4. L.A. Parker, S. Siegel, and T. Luxton, *Exp. Clin. Psychopharmacol.* **3,** 344 (1995).
5. M.D. Aceto, E.R. Bowman, and L.S. Harris, *NIDA Research Monograph* **95,** 578 (1990).
6. E.D. Dzoljic, C.D. Kaplan, and M.R. Dzoljic, *Arch. Int. Pharmacodyn.* **294,** 64 (1988).
7. S.D. Glick, K. Rossman, N.C. Rao, I.M. Maisonneuve, and J.N. Carlson *Neuropharmacol.* **31,** 497 (1992).
8. G.A. Matwyshyn and H.N. Bhargava, *FASEB J.* **10,** 2618 (1996).
9. I.M. Maisonneuve, K.L. Rossman, R.W. Keller, and S.D. Glick, *Brain Res.* **575,** 69 (1992).
10. J.A. Schneider and M. McArthur, *Experientia* **8,** 323 (1956).
11. S. Siegel, J. Kim, L. Weise-Kelly, and L.A. Parker, *Exp. Clin. Psychopharmacol.* **4,** 258 (1996).
12. R.A. Wise, in "The Neuropharmacological Basis of Reward" (J.M. Liebman and S.J. Cooper, eds.), p. 377. Clarendon Press, Oxford, 1989.

13. D. van der Kooy, in "Methods of Assessing the Reinforcing Properties of Abused Drugs" (M. Bozarth, ed.), p. 229. New York: Springer-Verlag, 1987.
14. P. Popik, R.T. Layer, and P. Skolnick, *Pharmacol. Rev.* **47,** 235 (1995).
15. S.D. Glick, M.E. Kuehne, J. Raucci, T.E. Wilson, D. Larson, R.W. Keller, and J.N. Carlson, *Brain Res.* **657,** 14 (1994).
16. S.I. Dworkin, S. Gleeson, D. Meloni, T.R. Koves, T.H. Martin, *Psychopharmacology* **117,** 257 (1995).
17. R.F. Mucha, D. van der Kooy, M. O'Shaugnessy, and P. Bucenieks, *Brain Res.* **243,** 91 (1982).
18. M. Sala, D. Braida, P. Calcaterra, M.P. Leone, and E. Gori, *Eur. J. Pharmacol.* **217**, 37 (1992).
19. I. Moroz, L.A. Parker, and S. Siegel, *Exp. Clin. Psychopharmacol.* **5,** 119 (1997).
20. L. Cervo and R. Samanin *Brain Res.* **673,** 242 (1995).
21. T.M. Tzschentke and W.J. Schmidt, *Neurosci. Let.* **193,** 37 (1995).
22. J.F. Toth and L.A. Parker, *Ann. Learn Behav.* **27,** 481 (1999).
23. M.L. Shapiro and Z. Carmanos, *Psychobiology* **18,** 231 (1990).
24. U. Staubli, T. Olivier, M. DiLorenzo, and G. Lynch, *Behav. Neurosci.* **103,** 54 (1989).
25. T. Luxton, L.A. Parker, and S. Siegel, *Prog. Neuro-Psychopharmacol. Biol. Psychiat.* **20,** 857 (1996).
26. K.R. Alper, H.S. Lotsof, G.M.N. Frenken, D.J. Luciano, and J. Bastiaans, *Am. J. Addict.* **8,** 234 (1999).
27. H. Lotsof, *Bull. Multidiscipl. Assoc. Psychedelic Stud.* **5** (1995).
28. B. Frances, R. Gout, J. Cros, and J.M. Zajac, *Fundament. Clin. Pharmacol.* **6,** 327 (1992).
29. L.G. Sharpe and J.H. Jaffe, *NeuroReport* **1,** 5 (1990).
30. R.M. Eisenberg, *Life Sci.* **31,** 1531 (1982).
31. V.F. Gellert and S.B. Sparber, *J. Pharm. Exper. Ther.* **201,** 44 (1977).
32. S.J. Heishman, M.L. Stitzer, G.E. Bigelow, and I.A. Liebson, *Pharmacol. Biochem. Behav.* **36,** 393 (1990).
33. H.L. June, M.L. Stitzer, and E.C. Cone, *Clin. Pharmacol. Ther.* **57,** 270 (1995).
34. L.A. Parker, J.A. Cyr, A.N. Santi, and P.D. Burton, *Pharmacol., Bioch. and Behav.* (submitted).
35. H.A. Baldwin and G.F. Koob, *Neuropsychopharmacology* **8,** 15 (1993).
36. J.J. Jacob and G.M. Michaud, *Psychol. Med.* **4,** 270 (1974).
37. R.F. Mucha, in "Animal Models of Drug Addiction" (A.A. Boulton, G.B. Baker, and P.H. Wu, eds.), Humana Press, NJ, 1992.
38. R.F. Mucha, M.D. Gritti, and C. Kim, *NIDA Res. Monogr.* **75,** 567(1986).
39. L.A. Parker and A. Joshi, *Pharm. Biochem. Behav.* **61,** 331 (1998).
40. L.A. Parker, S. Siegel, P. Burton, R.V. McDonald, and J. Kim, *Prog. Neuro-Psychopharmacol. Biol. Psychiat.* (in press).
41. A.H. Rezvani, D.H. Overstreet, and W.Y. Lee, *Pharmacol. Biochem. Behav.* **52,** 615 (1995).
42. S.L.T. Cappendjik and M.R. Dzoljic, *Eur. J. Pharmacol.* **241,** 261 (1993).
43. H. Sershen, A. Hashim, and A. Lajtha, *Pharmacol. Biochem. Behav.* **47,** 13 (1994).
44. Anonymous, "Alcoholics Anonymous World Services," New York, 1976.
45. R. Stretch and G.J. Gerber, *Can. J. Psychol.* **27,** 168 (1973).
46. H. deWitt and J. Stewart, *Psychopharmacology* **75,** 134 (1981).
47. L.A. Parker and R.V. McDonald, *Pharmacol. Biochem. Behav.* **66,** 559 (2000).
48. A. Ettenberg and T.D. Geist, *Pharmacol. Biochem. Behav.* **44,** 191 (1993).
49. T.D. Geist and A. Ettenberg, *Pharmacol. Biochem. Behav.* **36,** 703 (1990).
50. K. McFarland and A. Ettenberg, *Psychophamacology* **122,** 346 (1995).
51. R.A. Wise, R. Yokel, and H. deWitt, *Science* **191,** 1273 (1976).
52. M.A. Reicher and E.W. Holman, *Anim. Learn. Behav.* **5,** 343 (1977).
53. A. Bechara and D. van der Kooy, *Nature* **314,** 533 (1985).
54. S. Siegel, L.A. Parker, and I. Moroz, *Pharmacol. Biochem. Behav.* **50,** 299 (1995).
55. A. Ettenberg, L.A. MacConell, and T.D. Geist, *Psychopharmacology* **124,** 205 (1996).
56. A. Bechara, F. Harrington, K. Nader, and D. van der Kooy, *Behav. Neurosci.* **106,** 798 (1992).

57. A. Markou, F. Weiss, L.H. Gold, B. Caine, G. Schulteis, G. and Koob, *Psychopharmacology* **112,** 163 (1995).
58. T.E. Robinson and K.C. Berridge, *Brain Res. Rev.* **18,** 247 (1993).
59. S.M. Pearl, D.W. Johnson, and S.D. Glick, *Psychopharmacology* **121,** 470 (1995).
60. N.R. Swerdlow, D. van der Kooy, G.F. Koob, and J.R. Wenger, *Life Sci.* **32,** 2087 (1983).
61. N.M. White and G.D. Carr, *Pharmacol. Biochem. Behav.* **23,** 37 (1985).
62. T. Stefurak and D. van der Kooy, *Behav. Neurosci.* **108,** 972 (1994).
63. E. O'Hearn and M.E. Molliver, *Neurosci.* **55,** 303 (1993).
64. L.J. Larson-Prior, and P.L. Hawkins, *Soc. Neurosci. Abstr.* (318.6), 799 (1996).
65. H.H. Molinari, I.M. Maisonneuve, and S.D. Glick, *Brain Res.* **737,** 255 (1996).
66. Z. Xu, L.W. Chang, W. Slikker, Jr., S.F. Ali, R.L. Roontree, and A.C. Scalllet, *Toxicol. Sci.* **57,** 95 (2000).

——CHAPTER 12——

ANXIOGENIC ACTION OF IBOGAINE

PIOTR POPIK
MALGORZATA WRÓBEL

Institute of Pharmacology
Polish Academy of Sciences
31-343 Kraków, Poland

I. Introduction

Almost 100 years ago, Dybowsky and Landrin (*1*), as well as Haller and Heckel (*2*), isolated a crystalline alkaloid from the root of the West African shrub, *Tabernanthe iboga* and named it "ibogaine." Ibogaine was "rediscovered" during the past decade because of the anecdotal (*3*) and preliminary clinical (*4,5*) observations suggesting that it may offer a novel means of treating drug addictions. Preclinical studies are, in general, consistent with these claims. Thus, ibogaine reduces self-administration of cocaine (*6*) and morphine (*7*), attenuates the severity of morphine withdrawal (*8,9*), and inhibits conditioned place preference produced by morphine and amphetamine (*10,11*).

The neurochemical actions of ibogaine and related alkaloids were intensively investigated during the past decade, which has witnessed increasing growth of information attempting to explain the "antiaddictive" effects of ibogaine. The majority of work concentrated on its neurochemical effects, in the hope of identifying neuronal receptors and systems affected by this molecule. These findings, described in detail in the present and former volumes of *The Alkaloids*

THE ALKALOIDS, Vol.56
0099-9598/01 $35.00

(*12-14*), have profound implications for both understanding the mechanism of action of ibogaine and the molecular mechanism of drug addiction and dependence. Ibogaine affects a substantial number of neurotransmitter pathways, including *N*-methyl-D-aspartate (NMDA) and κ-opioid receptors, as well as dopamine and serotonin uptake sites and σ sites [for a review, see (*14*)]. However, the "antiaddictive" effects of ibogaine are still poorly understood, particularly at the psychological level.

In the current paradigm of drug development, findings gained in behavioral research ultimately direct our understanding of the drug's mechanism of action, provide evidence of its effectiveness in an animal model, and ultimately decide whether a substance becomes a therapeutic agent (*15*). Findings gathered in behavioral studies often raise questions concerning whether the given substance may or may not be effectively used in a given clinical application, and also about its "side effects." In addition, the effects found in behavioral studies that are not directly related to the purported mechanism of action often raise new questions such as the relevance of a "side effect" to a therapeutic action. This appears to be the case for ibogaine that, besides diminishing self-administration of drugs of abuse and alleviating the severity of opioid withdrawal, produces psychotomimetic effects, affects learning and memory processes (*16-18*), and has anxiogenic actions (*14*).

It is beyond the scope of this overview to provide direct evidence linking these "miscellaneous" or "side-effect"-like actions of ibogaine to its purported "antiaddictive" effects. Nonetheless, these "miscellaneous" or "side-effect"-like actions of ibogaine cannot be neglected, since they may indeed have some therapeutic implications, and their understanding may facilitate the discovery of its molecular mechanism(s) of action.

Operant and classical conditioning, habituation, and sensitization play important roles in determining the level of drug tolerance, addiction, dependence, and craving (*19*). Thus, inhibitory effects of ibogaine on drug addiction could be linked to a general interference with learning and memory processes. Often, drug-addicted individuals do not realize the roots of their habits—why they are, indeed, drug addicted. There are numerous possibilities, including traumatic experiences, such as events that took place in childhood. It can be hypothesized that a therapy, able to uncover these experiences, would be helpful in localizing and eliminating the cause of the disorder. This is the role of the psychotherapist, but perhaps with the help of ibogaine, this role could be made easier.

It is known that ibogaine produces a variety of psychotomimetic effects. It is debatable whether these are important from the clinical perspective. It would be hard to argue that the "antiaddictive" effects in laboratory animals are due to the hallucinogenic actions. It is also worth noting that there are many psychotomimetic compounds that are not "antiaddictive." Nonetheless, Naranjo, who explored the possibility of using ibogaine to facilitate psychotherapy, concluded that ibogaine could act as a psychological catalyst, which could compress a long psychotherapeutic process into a shorter time (*20*).

II. Ibogaine and Anxiety

A. Experimental Rationale

In humans, ibogaine has been reported to produce anxiety, fear, and apprehension (*21*). These effects are important, as they may influence the process of psychotherapy and/or in some cases, can be regarded as a confounding factor. Recently, Benwell and colleagues (*22*), while studying the possible effects of ibogaine on the neurochemical actions of nicotine in rats, found that ibogaine administered at the dose of 40 mg/kg, 22 hours before the test, produced an anxiogenic effect. It must be considered that such an anxiogenic effect of ibogaine may confound several measures of drug-seeking and taking behavior in animal models of drug addiction. The present work was designed to determine if the same effects on anxiety could be identified in mice following the administration of lower doses of ibogaine given after a shorter time interval.

B. Experimental Methodology

1. Animals

Male Albino Swiss mice (26-32 g) obtained from the Institute of Pharmacology breeding facility, were housed under standard laboratory conditions (lights on at 0600 hours, lights off at 1800 hours; room temperature 23 ± 1°) with pelleted food and tap water available *ad libitum*. They were kept in 43 x 27 x 15 cm plastic cages (eight mice per cage). All animals were used only once.

2. Drugs

Hydrochloride salt of ibogaine (Sigma) and the reference compound picrotoxin were dissolved in the physiological saline that served as a placebo. Injections were done i.p. in the volume of 10 ml/kg. The selection of doses was based on previous studies with ibogaine (*22*) and picrotoxin (*23,24*).

3. Apparatus and Procedure

An elevated plus maze (*23*) was used to study the anxiety in mice. It was made of plywood and consisted of two open arms (30 x 5 cm) and two enclosed arms (30 x 5 x 15 cm). The arms extended from a central 5 x 5 cm platform. The maze was painted black and mounted on a wooden base, raising it 50 cm above the floor. The apparatus was lit with two 15-watt lightbulbs placed 60 cm above the open arms.

The experiments were carried out between 0900 and 1700 hours. Drugs were administered 30 minutes before the test that lasted for 5 minutes. Mice were tested in an order counterbalanced for the treatment condition. There were 8 to 10 mice in each group. Mice were placed on the central platform, and the time spent on each of the four arms as well as the number of entries into arms was manually recorded using PLUS MAZE 2.0 program on a PC-compatible computer. After each mouse, the apparatus was cleaned and dried. The experimenter was blind to the treatment condition. The percentage of open/total time ([open arm time/open arm + closed arm time] x 100) as well as the percentage of open/total arm entries ([open arm entries/open arm + closed arm entries] x 100) served as the measures of anxiety. The number of closed arm entries was used as the measure of locomotor activity.

The experiments were carried out according to the National Institutes of Health Guide for Care and Use of Laboratory Animals (publication No. 85-23, revised 1985) and were approved by the internal Bioethics Commission.

TABLE 1.
ANXIOGENIC EFFECTS OF IBOGAINE IN MICE

Treatment (mg/kg)	% Open/ total time	% Open/total arm entries	Closed arm entries [n]
Placebo [10]	6.21 ± 1.33	18.6 ± 1.36	13.1 ± 0.81
Ibogaine 10 [10]	3.07 ± 1.33	10.3 ± 3.12	12.3 ± 1.34
Ibogaine 20 [10]	2.03 ± 0.90*	10.7 ± 2.96	11.2 ± 0.73
Ibogaine 40 [9]	1.56 ± 0.73*	5.77 ± 2.50**	10.4 ± 1.27
ANOVA	F(3,35) = 3.49, P<0.05	F(3,35) = 4.21, P<0.05	F(3,35) = 1.19, NS
Placebo [10]	7.53 ± 0.86	20.8 ± 1.66	12.6 ± 0.68
PIC 0.25 [8]	5.03 ± 1.10	16.0 ± 1.85	12.0 ± 1.39
PIC 0.5 [8]	3.64 ± 1.19*	13.2 ± 2.66	11.4 ± 1.29
PIC 1 [10]	0.56 ± 0.35***	9.15 ± 3.83*	6.20 ± 0.83***
ANOVA	F(3,32) = 11.9, P<0.001	F(3,32) = 3.53, P<0.05	F(3,32) = 8.81, P<0.0001

Presented are means ± S.E.M. of: (1) percentage of open/total arm time, (2) percentage of open/total arm entries, and (3) the number of closed arm entries in the elevated plus maze. Results of (1) and (2) reflect the measures related to the anxiety. The number of closed arm entries is regarded as the measure of general locomotor activity. Mice were treated with placebo (physiological saline), ibogaine, and picrotoxin [PIC] 30 minutes before the test.

NS not significant

* (P < 0.05)

** (P < 0.01)

*** (P < 0.001)

Statistically significant difference compared to placebo treatment (Newman-Keul's Multiple Comparison Test). The number of mice used for each dose is indicated in the brackets.

4. Statistical Analysis

For the statistics, one way between subjects ANOVA was performed for each treatment condition, followed by Newman-Keul's multiple comparison post-hoc test.

C. Results

Ibogaine reduced the percentage of open/total arm time (20 and 40 mg/kg) and the percentage of open/total arm entries (40 mg/kg). The decrease of the percentage of open/total time and the percentage of open/total arm entries produced by ibogaine was comparable to that observed in mice treated with picrotoxin (0.5 and 1 mg/kg) (Table I). Mean locomotor activity (closed arm entries) was not affected by the drugs tested, with the exception of picrotoxin (1.0 mg/kg) (Table I).

III. Discussion

The elevated plus maze (*25*) offers a reliable means of investigating anxiety in rodents. Ibogaine and picrotoxin reduced the percentage of open/total time as well as percentage of open/total arm entries, indicating that these compounds increased the anxiety. The anxiogenic effect of picrotoxin is well known (*23,25*) and therefore supports the use of this compound as reference. As in our study, Dalvi and Rodgers (*24*) reported that the anxiogenic effect of picrotoxin in mice at doses higher than or equal to 1 mg/kg was confounded by the behavioral suppression.

Perhaps the first documented notion that ibogaine produces a subjective state of anxiety was made by Sigg, who, after ingestion of 200 mg of ibogaine, reported: "Subjectively, the most unpleasant symptoms were the anxiety, the extreme apprehension, and the unheimliche Grundstimmung (~unfamiliar mood) associated with visual and bodily hallucinations" (*26*), p. 94. More recent studies indicate that ibogaine reduced the number of open arm entries in the elevated plus-maze test in rats tested 22 hours after pretreatment with ibogaine (40 mg/kg, i.p.). Such a long time between the drug administration and the test may suggest the involvement of a long-lasting ibogaine metabolite, since ibogaine's plasma half-time in rodents is about 1 hour (*27*).

Ibogaine affects a substantial number of neurotransmitter pathways, including NMDA and κ-opioid receptors, as well as dopamine and serotonin uptake sites and σ sites [for a review, see (*14*)]. Ibogaine antagonistic activity at NMDA receptors cannot explain its anxiogenic effects, because NMDA receptor antagonists reduce, rather than increase, anxiety in rodents (*28*). The high affinity of ibogaine at the k-opioid receptors is supported by its ability to block the effects of a κ-opioid agonist to inhibit dopamine and serotonin release (*29*), suggesting that ibogaine may act as a κ-opioid antagonist. This is, in turn, supported by findings indicating that κ-opioid agonists produce anxiolytic effects in the elevated plus maze in rats (*30*). However, in light of other reports suggesting agonist activity of ibogaine at κ-opioid receptors (*31*), it remains to be assessed whether ibogaine's activity at κ-opioid receptors plays a role in its anxiogenic effect.

In conclusion, it remains to be established whether the acute (present study) and delayed (*22*) anxiogenic effects of ibogaine are related to its inhibitory effects on drug-seeking and taking behavior or if it may be regarded only as the confounding factor. Alternatively, it is possible that the anxiogenic actions of ibogaine revealed in mice may have nothing to do with their putative antiaddictive actions in humans.

Acknowledgments

This work was supported by the statutory activity of the Institute of Pharmacology Polish Academy of Sciences.

References

1. J. Dybowsky and E. Landrin, C. *R. Acad. Sci. (Paris)* **133,** 748 (1901).
2. A. Haller and E. Heckel, *Compt. Rend. Soc. Biol.* **133,** 850 (1901).
3. H.S. Lotsof, *U.S. Patent* 4,499,096 (1985).
4. S.G. Sheppard, *J. Subst. Abuse Treat.* **11,** 379 (1994).
5. K.R. Alper, H.S. Lotsof, G.M.N. Frenken, D. Luciano, and J. Bastiaans, *Am. J. Addict.* **8,** 234 (1999).
6. S.L.T. Cappendijk and M.R. Dzoljic, *Eur. J. Pharmacol.* **241,** 261 (1993).
7. S.D. Glick, M.E. Kuehne, J. Raucci, T.E. Wilson, D. Larson, R.W. Keller, and J.N. Carlson, *Brain Res.* **657,** 14 (1994).
8. P. Popik, R.T. Layer, L. Fossom, M. Benveniste, B. Getter-Douglas, J.M. Witkin, and P. Skolnick, *J. Pharmacol. Exp. Ther.* **275,** 753 (1995).
9. R.T. Layer, P. Skolnick, C.M. Bertha, M.E. Kuehne, and P. Popik, *Eur. J. Pharmacol.* **309,** 159 (1996).
10. I. Moroz, L.A. Parker, and S. Siegel, *Exp. Clin. Psychopharm.* **5,** 119 (1997).
11. L.A. Parker, S. Siegel, and T. Luxton, *Exp. Clin. Psychopharm.* **3,** 344 (1995).
12. W.I. Taylor, *in* "The Alkaloids: Chemistry and Physiology" (R.H.F. Manske, ed.), Volume 8, p. 203. Academic Press, New York, NY, 1965.

13. W.I. Taylor, *in* "The Alkaloids: Chemistry and Physiology" (R.H.F. Manske, ed.), Volume 11, p. 79. Academic Press, New York, NY, 1968.
14. P. Popik and P. Skolnick, *in* "The Alkaloids: Chemistry and Biology" (G.A. Cordell, ed.), Volume 52, p. 197. Academic Press, San Diego, CA, 1999.
15. P. Willner *in* "Behavioural Models in Psychopharmacology: Theoretical, Industrial and Clinical Perspectives" (P. Willner, ed.), p. 3. Cambridge University Press, Cambridge, 1991.
16. R.P. Kesner, P. Jackson-Smith, C. Henry, and K. Amann, *Pharmacol. Biochem. Behav.* **51,** 103 (1995).
17. P. Popik, *Life Sci.* **59,** 379 (1996).
18. S. Helsley, D. Fiorella, R.A.Rabin, and J.C.Winter, *Pharmacol. Biochem. Behav.* **58,** 37 (1997).
19. S. Siegel, *in* "Psychopathology in Animals, Research and Clinical Implications" (J.D. Keehn, ed.), p. 143. Academic Press, New York, NY, 1979.
20. C. Naranjo, *Clin. Toxicol.* **2,** 209 (1969).
21. J.A. Schneider and E.B. Sigg, *Ann. N.Y. Acad. Sci.* **66,** 765 (1957).
22. M.E.M. Benwell, P.E. Holtom, R.J. Moran, and D.J.K. Balfour, *Br. J. Pharmacol.* **117,** 743 (1996).
23. R.G. Lister, *Psychopharmacology* **92,** 180 (1987).
24. A. Dalvi and R.J.Rodgers, *Psychopharmacology* **128,** 380 (1996).
25. S. Pellow, P. Chopin, S.E. File, and M. Briley, *J. Neurosci. Meth.* **14,** 149 (1985).
26. J.A. Schneider and E.B. Sigg, *in* "Progress in Neurobiology. Psychopharmacology; Pharmacologic Effects on Behavior" (H.H. Pennes, ed.), p. 75. New York, NY, 1958.
27. H.I. Dhahir, *Diss. Abstr. Int.* **32/04-B,** 2311 (1971).
28. J.L. Wiley, A.F. Cristello, and R.L. Balster, *Eur. J. Pharmacol.* **294,** 101 (1995).
29. H. Sershen, A. Hashim, and A. Lajtha, *Brain Res. Bull.* **36,** 587 (1995).
30. T.H. Privette and D.M. Terrian, *Psychopharmacology* **118,** 444 (1995).
31. S.D. Glick, I.M. Maisonneuve, and S.M. Pearl, *Brain Res.* **749,** 340 (1997).

—CHAPTER 13—

"RETURNING TO THE PATH": THE USE OF IBOGA[INE] IN AN EQUATORIAL AFRICAN RITUAL CONTEXT AND THE BINDING OF TIME, SPACE, AND SOCIAL RELATIONSHIPS

JAMES W. FERNANDEZ
RENATE L. FERNANDEZ

Department of Anthropology
The University of Chicago
Chicago, IL 60637

I. Introduction

In this chapter the authors, a cultural (JWF) and a biocultural anthropologist (RLF), reexamine their late 1950s ethnographic work among iboga(ine) users in Equatorial Africa in light of Western societies' current questions regarding the craving for, addiction to, withdrawal from, and relapse back into the hard drugs (*1,2*). The possibility of the addiction interruption obtained by this alkaloid for harder drugs in the West relates to its use in Africa as transitional to the rediscovery of a fully consecrated and dedicated existence. We may also learn by comparing the therapeutic measures or psychosocial interventions attempted in

THE ALKALOIDS, Vol.56
0099-9598/01 $35.00

the West with those understood in the African context. Initiates and members of Bwiti, the religious movement that regularly used iboga(ine) to obtain its ritual effects, were what is now called consonant addicts, if we can talk about iboga addiction at all (*3*). They ingested the alkaloid willingly and without noticeable regret or remorse, and almost exclusively as a ritually circumscribed routine. They regarded iboga(ine) as a sacred substance capable of both prolonging and enhancing their weekly religious experience and confirming and consecrating them in a revitalized relation to their ancestors and to their descendants. This consonance causes us to reflect on the words "addiction," "craving," "withdrawal," or "relapse" so central to the "drug wars" of the West, and prompts us to reexamine the ritual context and social consequences of iboga(ine)'s use in Bwiti. In this autochthonous context, the words hardly have equivalent meanings to those of the West, for iboga is there regarded as crucial to the management of space and time and generational relationships. It effects a time and space binding of social experience in the broadest sense. Despite the disparities of drug use in these contrasting cultural settings, there are also commonalities, and thus comparative observations on the use of iboga in the ethnographically observed situation may hold some implications for Western understanding and management of addiction in our contemporary drug-saturated world.

II. Time Binding in Professional and Religious Worlds

It was very far from the authors' minds during the several years spent among Fang and Fang-Bwiti adepts in Equatorial Africa that some 40 years later we would be invited to return and reflect on those intense ethnographic years in the context of present and ongoing research, research at once important and hopeful into the role of iboga(ine) in relation to addiction (*1,2*). Addiction as a negative state was hardly a concern in Bwiti—in today's terms its adepts might be described as consonant addicts—though as much as possible, and for the most part, iboga(ine) dosage was carefully controlled by leaders of the religion (*3*). Outside Bwiti, of course, colonial and postcolonial administrators, missionaries, and religious elite, Christian priests and pastors, regularly attacked Bwiti as an "addictive religion," both in the metaphoric sense, in that its rituals and philosophy were irresistibly attractive and absorbing, and in the literal sense, that iboga use was the heart of that addiction.

We are, in any event, grateful for this invitation to "fast forward" to this turn of the millennium and bring ourselves up to date on the present relevance of iboga. We were, as our ethnography *Bwiti* makes clear, greatly impressed with the creative achievement of Bwiti as a religion of the deep forest. We sometimes speculated that the psychoaesthetic attractions of this religion might well catch hold elsewhere in Africa—or even in Europe or the United States (as has happened with other African religions). Little did we suspect that the religion's energizing alkaloid would become the object of intense laboratory and clinical study in the interests of helping those trapped in an unreasonable and maladaptive life way—the prison house of addiction—to come back into a more reasonable and socially acceptable world of interactions.

How can our research and knowledge be helpful to the professional and social concerns of this conference? Can we retrieve—in an act of time binding ourselves—from data taken, and experiences undergone, nearly half a century ago anything of value to our concerns here at this millennial moment? Perhaps we can. We have spent many years thinking about Bwiti as a center of religious gravity in the equatorial forest and on the nature of its gravitational attractions, particularly the role of iboga(ine) in its ability to construct an overarching and sheltering ritual culture for its adepts. We would like to suggest that some of the answers to our questions over the years, and perhaps to some of this conference's questions about this bridge from the opiates to a less chemically dependent life way, lies in this notion of "time (and space) binding."

III. Pathologies of the Colonial World and the Role of Iboga in Religious Movements

Iboga use in Equatorial Africa must be placed in the context of the colonial situation that exerted a set of pressures on native peoples, and was accompanied by characteristic pathologies to which the leaders of Fang Bwiti were sensitive and to which, through cult activities and the use of iboga(ine), they attempted to respond. Fang Bwiti is the particular religious movement under discussion here, although other peoples and other cults—the Mitsogo people, for example, and the women's cult of Mbiri most particularly—made extensive use of the plant. There is evidence that iboga(ine) has long been employed in that part of Africa as a stimulant and because of its mind-changing properties. Because of its capacity to enable sustained effort by suppressing symptoms of fatigue, it was allowed, if not actively encouraged, by the German colonial authorities on work gangs in the German Cameroun before the First World War, quite outside any context of religious movement. This use may have been influential in these various religious movements, which made their appearance at about the same time.

The literature on the "colonial situation" is now extensive in anthropology. Moreover, there has recently developed a renewed interest in colonialism and postcolonialism as it has affected subaltern peoples the world around, and as they have creatively responded to its particular set of conditions of life. Fang Bwiti and its iboga usage is most particularly a response to the colonial situation, although iboga use in cult life in this part of Africa surely preceded it, and has also followed it, into the postcolonial world. We cannot, here, review the considerable insightful discussion that has appeared in this literature (see endnote 3). Rather, we will briefly summarize what we know about the colonial world. We are talking here about the French Colonial world, which differed significantly from the British colonial world (the one employing direct rather than indirect rule), as these differed from the Portuguese and the Spanish. This colonial world was characterized until its last and liberalizing preindependence decade, the decade of the 1950s, by an adamant and exacting paternalistic administration of the native population. For the building of the colonies, *corvée* labor and work gangs were a standard practice and, to force this labor and other forms of manual labor in the colonial enterprise, taxation was imposed and heavy penalties exacted for nonpayment. It was supposed that the natives were naturally dilatory and desultory in their work habits, and one paternalistic object was to coerce them away from their endemic idleness. Because French colonization engaged in direct rule and entertained the ideal of eventual assimilation of the natives into French culture, there was, in contrast to indirect rule, strong pressure toward cultural assimilation and correspondingly strong condemnation of native cultural, particularly religious, practices.

While many Africans reacted positively to this paternalistic regime, many did not. The latter experienced a loss of affective allegiance to the norms and projects of their traditional life ways—a state of normlessness known as anomie in the literature—and, because of administrative and missionary attacks on the ancestor cult, a sense of being cut off from this sustaining relationship. Colonial controls were pervasive and effective, and many Africans had the experience of placelessness, of being uprooted, and of being alien in their own land. Being forced by a taxation system to work in colonial projects, either on road projects, in new plantations, or in forest camp labor, they rapidly developed a sense of their own subordination and, on the part of colonials, a set of deprecatory attitudes toward their work habits. The sexual life of men in plantation or forest camp labor also greatly increased venereal disease, and, because of spreading infection, resulted in a high incidence of fetal wastage and a drastic drop in fertility of both men and infected women. A severely declining birthrate, moreover, brought into question the perpetuation of generations over time, the claim of Fang upon the future.

Recruitment to the Bwiti religion and the use of iboga must be seen in relation to these social pathologies. For religious leaders promised, among other things, that the ingestion of iboga(ine) through the visionary excursions it promoted would reestablish contact with the ancestors, would restore the fertility of women and the coming into being of the next generation, and would enable nights of long hard work, and the pride associated with that work, of productive ritual worship. Bwiti chapels and Bwiti villages would give Fang a place to live in the forest free of the capricious, arbitrary, and precipitate commandments and controls of the colonial world. These promises were attractive to a minority of Fang (membership in Bwiti never exceeded 8 to 10% of the Fang population), who found in Bwiti, and through iboga(ine), solace, energy, and community with both the living and the dead. We can summarize the pathologies of the colonial world as pathologies having to do with a diminution or loss of one's sense of place in time and space, and hence we speak of the "time binding" (contact with the past of the ancestors' guarantee of the future in fertility) and "space binding" (creation of a ritual workspace free of alienating colonial commands) accomplished in Fang Bwiti through its use of iboga. This time and space binding was conceived of in metaphorical terms as returning the initiate to "the path of life and death," for he had obviously lost his way and no longer knew how to "work." Bwiti would tie him into the past and into the future, would give him a settled place to live, and give him the most satisfying work to do, the work of the ancestors, the work of their descendants! The promise was that Bwiti, through iboga(ine), would restore directionality to the lost and the wandering, and, to their desultory lives, would restore the purposefulness of a ritual project.

IV. A Comparison of the Introduction and Maintenance of Iboga Use in Bwiti and in Therapeutic Settings (INTASH)

We would now like to compare iboga use in the African context and in a European therapeutic setting, that of the International Addict Self Help (INTASH) organiztion as presented to us in a lucid discussion by Geerte Frenken (*4*). Both in Bwiti and in INTASH the hope is that the taking of iboga, along with associated procedures (in anthropology understood as relatively precise normative commitments or rituals), will enable a change in life way. In Bwiti this is understood, as we know, as the discovery of the "path of life and death," a discovery that saves one from a confused and wandering state within the deep equatorial forest without beginning and without end and without project, that is to say without the capacity or ability for useful work in this world, understood as "the work of the ancestors" (endnote 1). In INTASH, this change of life way is understood as a "breaking of the cycle of self-destructive behavior and the finding of new positive ways to approach life and its problems. The object is commitment to a new way of living," something that could also be said about Bwiti doctrine, although that new way of living is much more clearly defined and enforced in Bwiti as a faithful long-term allegiance to Bwiti doctrine and regular adherence in collectivity to Bwiti ritual practice. Neither approach thinks of itself as a "quick fix" (the searching for which is a characteristic behavior of the previous states of desultory confusion (Bwiti) or addiction (INTASH), but seeks a profound change of attitude toward physical, mental, and emotional well-being. As Frenken pointed out, however, this profound change in the end is left up to the individual both because, in a highly individualistic cultural tradition like the Western one, this is almost inevitable, and because of lack of follow-up funding.

Bwiti, as we have said, operates in a collectivist tradition, although Fang culture is more individualistic than most Equatorial African societies and was appreciated by colonizers for that very reason. Still, it might well be argued that Bwiti can be seen as a collectivist response to the excessive individualism and the existential anxieties and pathologies of subordination and isolation brought on by colonialism and its individualistic bias and through taxation, labor migration, and active suppression of the ancestor cult. It was a response that is, on the part of Bwiti, designed to restore the sociality of familial relations and particularly relations with the dead. Bwiti as a collectivity guarantees a perduring support network and enforces long-term aftercare in the sense of providing repeated opportunities for mild doses of iboga and continuing ritualization of contact with the ancestors. Most especially, Bwiti intensifies sociality through ritually coordinated contact with other cult members. Bwiti weekly, all-night ritual practice, and the morning after follow-up meal of communion, provided an enduring "focus group," as it were, to employ the term used in respect to detoxification groups such as Alcoholics Anonymous.

In INTASH, the fact that iboga is a substance with mind-altering properties has to be explained carefully to those undergoing treatment. The general suspicion of mind alteration and hallucination of any kind in pragmatic, production, and consumption-oriented societies like that in the West requires such explanation; moreover, it helps prepare the candidate for the release of unconscious materials that might be troubling to addicts who often use drugs to suppress unwelcome thoughts and memories and to anesthetize sensitive personal psychic material. In Bwiti it is just these mind alterations and visions that are anticipated, indeed counted on, to facilitate the profound change desired and to convince the adept of the virtuality of the new dispensation into which he or she is entering.

It is of interest to compare the care offered to the initiate or addict in the two practices. A particularly important part of INTASH procedure seeks to provide a supportive, nonjudgmental, and trusting treatment environment, the "pro-choice" attitude as it is called. This trusting environment is carefully cultivated by thoroughly informing the addict about the drug and its possible bodily and psychic consequences, and by medical and psychological screening. Most important is the supportive presence of a team of iboga(ine) experienced ex-addicts, plus clinical personnel who accompany the addict through his or her experience, as a support group. These are never, or almost never, members of the addict's own family, since tensions and problems with family members may, indeed, be part of the etiology of the addiction. The actual administration of the drug in INTASH can employ two different approaches, or a combination of these: a preliminary or initiatory session (not to be confused with Bwiti initiation) in which low doses of ibogaine are used for its spiritual, visionary impact and addiction interruption sessions in which high doses of ibogaine are employed, adjusted both to the weight of the subject and to the duration and intensity of the addiction being treated. The latter, of course, are more difficult to manage than the former. The dose administered gradually over many hours to a Bwiti initiate in full initiation, as observed by us in the 1950s, was estimated to be as high at 35 mg/kg (for a chart of calculated dosage, see Table I) more than a third larger than the 20 mg/kg capsule self-help addiction interruption dose currently reported by Alper *et al.* (5) and others. Many, if not most, Bwiti initiation doses were much lower. The death by intoxication feared by Western health investigators and officials, feared and guarded against in Bwiti, if indeed death did actually occur among them, would very likely much have exceeded 40 mg/kg, which supposes a very large ingestion, indeed, of freshly scraped iboga(ine) root (endnote 2).

The "pick-up" dose taken routinely by Bwiti adepts in order to "lighten the body" and participate without fatigue in the weekly, night-long ceremony is well below one-tenth of the dose administered in the Bwiti initiation, and below the dose administered to facilitate "personal growth and change" by self-help networks. These comparisons suggest that a very low pick-up dose, in combination with regular social support, might be a useful adjunct in the aftercare of those who undergo addiction interruption with iboga(ine) in a clinical setting.

In Bwiti, though here again we want to emphasize the variability among cult groups, great importance is given to providing accompaniment to the initiate undergoing the heavy initiatory course, the "basketful of iboga(ine)" as it is said. Close companions on the visionary trips are the "mother and father of iboga," themselves members of Bwiti who have undergone initiation and thus are iboga users. They are almost never members of the actual family of the initiate, but members of the Bwiti "family" she or he is about to join. Because on occasion the heavy initiatory dose is said to have been fatal—that is, the initiate has "gone to the land of the dead and not returned"—the mother and father of iboga must constantly be attentive to the state of the initiate to make sure that he or she does "return." This father and mother team will henceforth continue to be especially close fellow members of Bwiti.

INTASH reports suggest two different approaches to the administration of iboga(ine) and, we infer, sometimes a combined approach. There is first the initiatory session of low dosage designed to acquaint the addict with ibogaine's "spiritual impact" as it is said—that is to say, with its psychoactive and vision-producing qualities. Here, in the matter of comparison, there is variation from one branch of Bwiti to another. Some branches allow preinitiation members to

TABLE I

COMPARISON OF IBOGAINE DOSAGES REPORTED BY FRENKEN AND ALPER ET AL., WITH THOSE OF BWITI AS REPORTED BY FERNANDEZ, MEDIATED BY CALCULATIONS BY LOTSOF

Frenken, INTASH (*4*)	
Ibogaine low dose for spiritual impact:	nd
Ibogaine full dose for addiction interruption session, in capsule form:	nd
Alper et al. (*5*)	
Ibogaine dose to facilitate personal growth and change:	10 mg/kg
Ibogaine single dose in self-help network for addiction interruption:	20 mg/kg
Animal studies for neurotoxicity; alternate ibogaine doses over 60 days [no toxicity]:	10 mg/kg
Ibogaine dose associated with no evidence of toxicity [but decrease in drug self-administration]:	40 mg/kg
Ibogaine dose associated with cerebellar damage:	100 mg/kg
Lotsof	
(*personally communicated in preparation for ibogaine conference*)	
Ibogaine dosage causing modest psychoactivity with euphoria, altered perception of time:	90-120 mg
Amount of ibogaine ingested by an adept that would allow remaining centered enough to assist in initiation ritual:	200 to 300 mg
Ratio of fresh root scrapings to dry root bark:	15/1
Proportion of *iboga* alkaloids in dry root bark:	2 to 3% (50% ibogaine)
Rounded teaspoon of dry root bark:	3 to 4 g
Amount of *iboga* alkaloid yielded by rounded teaspoon of dry root bark, according to above range of approximations:	60 to 120 mg
Fernandez (*1,2*)	
Pick-up dose, *iboga* alkaloid content of 1 rounded teaspoon of dry root bark:	60 to 120 mg
Large dose for initiation into Bwiti, gradual intake of fresh root scrapings, maximal dose observed:	1000 g [one kilo]
Dose recalculated as dry scrapings [1000/15]:	67 g
Content of *iboga* alkaloids of the above quantity of root scrapings,assuming an average 2.5% *iboga* alkaloid content:	1.675 g
Total maximal Bwiti *iboga* alkaloid dose in mg:	1,675 mg
Maximal Bwiti *iboga* alkaloid dose calculated per kilo of body weight in a small initiate weighing 50 kilos [hence a high estimate]:	33.5 mg/kg

partake of the one or two teaspoons of iboga shared out to Bwiti members before the start of the all-night services. But others reserve the first usage of this "sacred substance" to the several day initiatory sessions—that is, to that moment when a large dose (a basketful) of iboga will be taken. Once initiated, the weekly all-night services for all members begin with the ingestion of a teaspoon or two of iboga. We should also remark, because INTASH treatment carries a several thousand dollar fee to allay costs, that Bwiti initiation carries a mild requirement of monetary, or payment in kind, contribution to the treasury of the chapel and its leaders. As annual per capita income in Gabon at the time was very low in relation to Western standards, though of the middle range in comparison to other African states and territories, these contributions were in the several thousand franc (CFA) range (perhaps 10 to 25 dollars). They were an earnest of the initiate's commitment to his or her initiation.

We see in this comparison of the two courses of iboga(ine) treatment, a similarity of concern about the hazards of the full initiation and the need that it be supervised by practiced and previously initiated older people. We see a similarity of distinction between low-dose usage and high-dose usage, where the latter is intended to "break" the initiates out of a former life way and lead them into a subsequent one that is better adapted to prevailing cultural definitions of useful work. In terms of continuing supervision and the providing of a "focus group," Bwiti has, of course, the greater advantage: It is an established religious collectivity with continuing weekly ritual celebrations that offer the social setting in which such supervision and support is routinely provided.

V. The Contemporary Ethnography of Iboga Use

We have been engaging here in comparison of ethnographies already available and made with different purposes in mind. One (of Bwiti) is made with the interests in understanding processes of cultural decline and cultural revitalization—that is, with processes of cultural construction in mind. The other, Frenken's very helpful, if brief, account of the therapeutic use of iboga(ine), is made with the interests of providing precise and useful information for others interested in iboga(ine) therapy and in the interests of exposing problems needful of correction for better future practice. Frenken's ethnography is thus more clinical and much more patient oriented in the sense that its objective is the use of iboga(ine) in facilitating change and personal growth in addicts desirous of leaving their addiction (Elster's dissonant addicts) (*3*). These two ethnographies—in respect to the use of *a priori* comparable categories of analysis and interpretation—hardly make for a controlled comparison.

The comparisons that can be made, however, show us similarities and differences useful to our understanding: similarities in respect to patient or initiate care during the initiation process, two kinds of iboga(ine) use—mild and heavy, and differences in respect to notions of "personal growth and change" and the meanings of the recent and the long-term past of the given culture. In this regard, Bwiti is much more collectivity oriented than contemporary medicine, whether corporeal or psychiatric; putting a patient back on the right path means *binding* him back into the collectivity in time and space. Reasoning, in contrast to Elster, is collective and the rationale is socially constituted (*3*). In Western culture, the reality of the individual is such, and one's self-realization so important, that therapy necessarily aims at personal change and growth. The rationale is an individual challenge and an individual achievement. The whole notion of "binding," though not irrelevant to Euro-American understandings of the human condition, may seem too restrictive in Western contexts. In any event, what differences these variations in cultural orientation make in the effectiveness of iboga(ine) therapy is an important, but open, question until we have ethnographies more specifically aimed at providing an answer.

In the past quarter century there have been in anthropology—and in related disciplines that practice ethnography—a number of important studies of deviancy and addiction from alcoholism, to AIDS, to drug-related vagrancy and homelessness (endnote 3). There is reason to plan and hope for a more thorough ethnography of iboga(ine) use in the contemporary world, in the context, that is, of Western individualism and the Western sense of the realities of real-world production and project orientation. We should make known to the ethnographic community the pertinence of this subject of study, particularly as we now have some ethnographic basis for understanding iboga(ine) use in an African equatorial forest society, the milieu from which iboga(ine) derives. We should encourage the support of the long-term study of the consequences of iboga(ine) therapy in such societies. We should also encourage the recording of more thorough life histories of Western addicts who have chosen this therapy, a standard and fundamental technique in ethnography. An interesting subcategory of ethnographic attention for anthropology would be the phenomenon of adventure drug seekers who go to Equatorial Africa in search of iboga(ine) mind alteration. The Internet has become an important site of their often quite elaborated reports of such experience. While such seekers are rarely interested in binding themselves into Bwiti collectivities in any significant long-term way, their use of iboga(ine) for personal growth provides highly useful ethnographic materials and is worthy of extended ethnographic study. We do not doubt, therefore, that if awareness of this therapy in its various manifestations and its intentions is made more broadly present in the academic research community in medicine and in social science, that these more revelatory and problem-focused ethnographies will be forthcoming.

VI. Conclusions

As should be clear from the discussion, our study of iboga(ine) took place in a specific cultural context that categorized reality in a significantly different way than is customary in the West. The members of the several religious movements studied regarded iboga(ine) as a sacred substance that powerfully facilitated access to a greater reality than the reality of everyday life in this workaday world—a reality that bound together past, present, and future and life and death. We worked hard in our ethnography to avoid, or at least suspend, the "observer effect," known as ethnocentric judgementalism. We sought to respect the Bwiti categories and what they regarded as significant work in this world and the next. Such is the ethnographic attitude. Despite these differences between an Equatorial African and a Western perspective, there are suggestive similarities between what we in this conference are interested in and what the members of Bwiti were interested in. We are both interested in iboga(ine)'s power to work a transition from one burdensome state of existence to another more workable. We in the West, by and large, live in hard-working and fast-moving cultures where drug addiction is an impediment to effective project planning and project completion and where iboga(ine) shows a suggestive potential for transiting the addict back into such a workaday world. That common purpose and our comparative studies, those begun and those to come, should not be impeded, but rather informed, by significantly different views of the categorizations of reality and its challenges.

One important issue in this comparison concerns the categorization of the relation between life and death. This is a particularly important issue because of the toxicity/fatality factor in gaining governmental support for research and extended clinical practice. Members of Bwiti, in our experience of them, credited iboga(ine) with the power of enabling the initiate and adept to visit the land of the dead and to know, if only briefly, the death side of things. Leaders felt it their solemn responsibility to enable that visitation, through iboga(ine)'s power, without it turning into a permanent sojourn. But this hope and promise also had the effect of introducing a good deal of death and near-death talk into Bwiti, talk about those who had or had not returned but had stayed on the other side. Such talk was an ever-present undercurrent in our fieldwork. But here we have to be careful of differences in categories. For despite all the talk, we ourselves had no firsthand evidence of an actual death by iboga(ine) ingestion in Bwiti. Our discussion of this issue in our ethnography, while accurate and important as ethnographic reporting, should not be read as based on objective measure. The categories of life and death are simply significantly different.

The Bwitists, those of the several branches with which we worked, talked about their work as the work of the ancestors. We took their use of that word, *work*, very seriously, as will be noted, although it is a different kind of work than can be easily appreciated in the practical minded, self-interested West. But one aspect of that work, which all of those working together to solve one part of the problem of addiction—the problem of leaving it—is the dedication of this religion to a long-term follow-up or aftercare, we might call it, of the initiate become adept. Bwiti, in the best of cases, was concerned to maintain a persistent social communion among its adepts, a communion based on binding them together into repeated all-night ritual coordinations.

In this social-religious support lay Bwiti's success in maintaining the fidelity of its membership. This attention to continuing communion is something worth pondering, and indeed it is being pondered in clinical medicine with the recent emphasis on the formation of support groups (*5*). As we know, the Banzie (the initiates of Bwiti) use the trope of the path of life *and* death and its implied meaning of having lost the communal path and of rediscovering it through initiation—an experience usually abetted by the initiatory images induced by iboga(ine) and the sense of path-like excursion that they provide. All support groups need their organizing tropes, and the path trope, perhaps more than the trip trope, seems a particularly apt image for committing the beneficiary of iboga(ine) to enduring change and improvement. In any event, the interest in Bwiti, through a regular pattern of ritual practice and small stimulative use of iboga, in maintaining the member in communion and on the right path, is a challenge to current iboga(ine) therapy taking place in dispersive highly individualistic societies, a cultural condition that at once challenges effective therapy as it is probably part of the cause of addiction itself.

We will end on this note of the challenge to us all of long-term care, a care perilously left to the individual. The Bwiti do not leave it to the initiates to help themselves. They do not live, after all, in a self-help society. Bwiti provides a continuing program of ritual practices with social accompaniments whose purpose is to engage the collectivity in path maintenance for the new initiate and all subsequent initiates. We may perhaps still learn something from the collectivism and communalism of what is rather offhandedly called "the tribal societies." But that learning would be much enhanced by careful comparative ethnography of present day iboga(ine) interventions and therapy.

Endnotes

1. It is of interest that Frenken also employs the journey metaphor, a very widespread one, of course, for encounters with psychoactive substances or "drug trips." The metaphor is probably associated with the sense of movement in space, either experienced directly by the initiate, or taking place before him or her, that is provided by the drug. Frenken speaks of "a journey into one's DNA."

2. Since this issue of the possibly serious, if not fatal, consequences of taking iboga has become a major issue and source of resistance by bureaucrats to government-sponsored trials and support use, we must say that we never—in more than 2 years of study and perhaps a dozen initiations—were ourselves either present at, nor could with certainty (that is, by direct interview of eye witnesses)

confirm and thus attest to, fatal consequences (5). But such consequences were common lore in and around Bwiti circles and made for the anxiety experienced by many undertaking initiation. It was also a reason for many chapel leaders to provide assurance to prospective initiates that they had "never left any initiate on the other side in the land of the dead!" Of course, it was lore used against Bwiti by its administrative and missionary opponents, who frequently spoke of these putative fatal results.

3. We might span a 25-year time period, 1970 to 1995, by mentioning several works in this ethnographic connection: James P. Spradley's "You Owe Yourself a Drunk: An Ethnography of Urban Nomads" (1970), a study of alcoholics and alcoholism on Seattle's notorious Skid Row, and Philippe Bourgois's recent "In Search of Respect: Selling Crack in El Barrio" (1995), an ethnography of the world of Puerto Rican crack merchants and addicts in New York. But this time period would also include the collection of brief ethnographies edited by Peter Furst, "The Flesh of the Gods" (1972 and 1990) and Furst's "Hallucinogens and Culture" (1976); Griffith Edwards, Awni Arif, and Jerome Jaffe's "Drug Use and Misuse: Cultural Perspectives" (1983); Shelagh Weir's "Qat in Yemen: Consumption and Social Change" (1985); John G. Kennedy's "The Flower of Paradise" (1987); Luis Luna and Pablo Amaringo's "Ayahuasca Visions: The Religious Ethnography of a Peruvian Shaman" (1991); Dan Waldorf, Craig Reinarman, and Sheigla Murphy's "Cocaine Changes: The Experience of Using and Quitting" (1991); and Reyes Ramos's "An Ethnographic Study of Heroin Abuse in Mexican Americans in San Antonio Texas" (1995). More recent work would include Stacy Schaefer and Peter Furst's, "People of the Peyote" (1996); and Kevin Rushby's "Eating the Flowers of Paradise" (1999).

Acknowledgments

The Ford Foundation, the American Council of Learned Societies, the Social Science Research Council, and the National Science Foundation supported our work on Revitalization Movements in Africa in the 1950s and 1960s. We are grateful for the invitation extended by Kenneth Alper, M.D., and Howard Lotsof to participate in the Ibogaine Conference and to reexamine our original research in the light of contemporary interests in ibogaine as a transitional alkaloid.

References

1. J.W. Fernandez, "Tabernanthe Iboga: Narcotic Ecstasis and the Work of the Ancestors." *in* "The Flesh of the Gods: The Ritual Use of Hallucinogens." (P.Furst, ed.), p. 237 Praeger, New York, NY. 1972.
2. J.W. Fernandez, "Bwiti: An Ethnography of the Religious Imagination in Africa," Princeton University Press, Princeton, NJ, 1982.
3. J. Elster and O. Skog, "Rationality and Addiction," Cambridge University Press, New York, NY, 1999.
4. G.M.N. Frenken, "Lindesmith Presentation," INTASH, March 2, 1998. (Web document: *http://www.ibogaine.co.uk.geerte.htm).*
5. K.R. Alper, H.S. Lotsof, G.M.N. Frenken, D.J. Luciano, and J. Bastiaans,. *Am. J. Addict.* **8,** 234 (1999).

—CHAPTER 14—

A CONTEMPORARY HISTORY OF IBOGAINE IN THE UNITED STATES AND EUROPE

KENNETH R. ALPER

Departments of Psychiatry and Neurology
New York University School of Medicine
New York, NY 10016 USA

DANA BEAL

Cures-not-Wars
New York, NY 10012 USA

CHARLES D. KAPLAN

Department of Psychiatry and Neuropsychology
Maastricht University
6200MD Maastricht, The Netherlands

THE ALKALOIDS, Vol.56
0099-9598/01 $35.00

I. Introduction

In 1995, Dr. Curtis Wright, then the U.S. Food and Drug Administration (FDA) ibogaine project officer, wrote "What's clear is that a significant portion of the public we serve believes the drug merits investigation" (*1*). Wright's statement intimates a relationship of public opinion to regulatory scientific policy. The statement was made at a time when the FDA, partly in response to highly motivated and organized public advocacy, was modifying its drug development process to accommodate the more rapid evaluation and approval of agents used to treat the human immunodeficiency virus (HIV) (*2,3*). As with treatments for HIV, ibogaine has been associated with a vocal activist subculture, which has viewed its mission as advocating the availability of a controversial treatment to a stigmatized and marginalized minority group of patients who suffer from a life-threatening illness.

Wright's perception of significant public interest in ibogaine was derived from two related subcultural contexts. One such context is the medical subculture of the informal ibogaine treatment scene, and the other is the political subculture of advocacy for the development and availability of ibogaine. This chapter focuses on a contemporary history and description of the medical subculture of the informal treatment scenes of the United States and Europe, and the political subculture of ibogaine advocacy. The period of time spanned by the history presented in this chapter extends from the early 1960s to the present, and it is thus termed a "contemporary" history of ibogaine. Ibogaine has a long history of use as a ritual hallucinogen in Africa. However, the early 1960s marked the advent of the attempt to develop ibogaine as a treatment for substance dependence in the United States and Europe.

II. Ibogaine Treatment Scenes: History and Description

A. Overview

Ibogaine is classified as a Schedule I drug in the United States, with a similar restricted status in Belgium and Switzerland. As a naturally occurring alkaloid that can only be patented with regard to use and not with regard to its structure, and with a mechanism of action that is unknown, ibogaine has been relatively unattractive to the pharmaceutical industry as a potential compound for clinical development. This has led to the existence of a distinctive, unofficial network involving lay individuals conducting ibogaine treatments in nonmedical settings.

Two general types of ibogaine treatment exist. One type of treatment is oriented toward addiction, most commonly heroin dependence, and typically involves dosages in the range of 15 to 25 mg/ kg (*4-7*). The other type of treatment, sometimes referred to as "initiatory," involves a dosage on the order of 8 to 12 mg/kg, or approximately one half of that which is used for addiction, and is oriented toward the goal of facilitating psychotherapeutic or spiritual insight (*8-10*). The existence of these two types of treatment orientations indicates the importance of considering two aspects of ibogaine treatment, namely the pharmacological effects, and the psychological process and social setting.

B. The United States

1. H. Lotsof, NDA International, and ICASH

a. Initial Observations of the Effects of Ibogaine. The setting in which the putative effect of ibogaine on opioid withdrawal was first noted was distinctly marginal with respect to the mainstream culture of medicine. Howard Lotsof serendipitously observed ibogaine's effect on his own heroin dependence at age 19 in 1962. As described in the chapter by Lotsof *et al.* in this volume, he then administered ibogaine to a total of 20 individuals who were part of a lay network of drug users that ingested hallucinogens and somewhat systematically noted and compared their effects. This network appears to have been mainly motivated by curiosity and an interest regarding the subjective effects and possible psychotherapeutic applications of hallucinogens. The intellectual influences of the group included Aldous Huxley's "Doors of Perception" (*11*) and the work of Timothy Leary (*12*). In 1962 and 1963, when the group was active, hallucinogens were not yet regulated (*13,14*). Lotsof described this period in an interview with the Yippie periodical, "Overthrow" (*13*):

"The original work was conducted in 1962 and 1963 in New York, and this was prior to ibogaine being classified as a Schedule I drug. It was the early 60's, there were no restrictions on LSD or mescaline or psilocybin. . . . The regulations pertaining to the restrictions on these drugs occurred in 1966 and 1967. We were not looking for a cure for heroin addiction or cocaine abuse. I was interested in psychoactive compounds and I established a research laboratory, S&L Laboratories, to procure drugs and administer them to interested persons . . .

. . . It was what was sweeping the country in the early 60's, this enormous interest in psychoactive substances...

. . . We received ibogaine from both the United States and England, and what happened was this: I was given a dose of ibogaine from an associate of mine who was actually a chemist involved in a 1950's LSD scene—most people don't even know that there was a '50s LSD scene...

. . . It just happened at that time the FDA was beginning to investigate

hallucinogens, and they realized that our laboratory was ordering large amounts of hallucinogens for experimentation, and they cut off our supplies. So we were not able to continue with this work."

Lotsof gave the initial dose of ibogaine to a friend. The group obtained more ibogaine and gave it at various dosage levels up to 19 mg/kg to a total of 20 individuals. A subset of 7 of these 20 individuals were heroin dependent and noted the alleviation of the symptoms of physical dependence and craving after taking ibogaine (*4,7*). Furthermore, 5 of these 7 individuals reportedly remained free of heroin for 6 months or longer following their treatment with ibogaine. All 7 individuals reported the alleviation of physical withdrawal from heroin, and the 2 individuals who returned rapidly to heroin use ascribed it to their continued identification with the role of the heroin addict, and not to persistent withdrawal symptoms. The activity of the group eventually ceased in 1963 when FDA and law-enforcement agencies eliminated the ability of S&L Laboratories, as Lotsof put it, "to procure drugs and administer them to interested persons" (*13*).

The setting in which the observation of the possible efficacy of ibogaine was made was unconventional, which led to its closure by federal regulatory agencies. Nonetheless, this was the methodological setting of the observation of the putative efficacy of ibogaine in opioid withdrawal and possibly other drug dependence syndromes. An important question is whether clinical observations regarding opioid withdrawal made in this setting can be methodologically valid. The idea of a lay psychotropic user's group as a means of drug discovery is quite unusual. However, there was a systematic aspect of Lotsof and the group's observation and documentation of effects at various dosage levels. An effect of ibogaine on the physical manifestations of opioid withdrawal, if present, might reasonably be expected to have been apparent to the members of this network of experienced users. Subsequently, there has been significant concordance between the observations reported within this original cohort and the ensuing clinical reports on ibogaine for the indication of opioid withdrawal (*4-6*).

b. The Attempt to Develop Ibogaine. In 1982, Lotsof formed a nonprofit corporation, the Dora Weiner Foundation, whose purpose was to promote the development of ibogaine. The nonprofit corporation was unsuccessful in attracting significant financial support. In 1984 it commissioned a review (unpublished) of the ibogaine research literature by Dr. Doris Clouet, who was eventually appointed Chief of the Opioid Pharmacology Branch at the National Institute on Drug Abuse (NIDA). On a psychological level, enlisting Clouet's involvement in project also represented the tactic of "normalization," reaching out to those with scientific credibility in order to offset the controversial and unusual image that ibogaine would surely have as a medication development project. Clouet concluded that the mechanism of action of ibogaine did not appear to significantly involve direct opioid agonist or antagonist activity. The pharmaceutical industry responded unenthusiastically to Lotsof's attempts to develop interest in ibogaine. In addition to ibogaine being a naturally occurring plant hallucinogen originating from a highly irregular developmental setting, the pharmaceutical industry also generally viewed addiction as an economically unattractive area for medication development (*15*).

In 1986, Lotsof formed NDA International, a private company that issued shares of stock. He filed patents for the use of ibogaine in treating multiple dependence syndromes, including opioid (*16*), cocaine and amphetamine (*17*), alcohol (*18*), nicotine (*19*), and polysubstance dependence (*20*). He organized an international conference in Paris in 1987 and traveled to Europe and Gabon to secure sources of ibogaine. In the late 1980s, NDA established contact and provided ibogaine to researchers at the Department of Pharmacology at Erasmus University in Rotterdam, which resulted in the first paper indicating efficacy in an animal model of opiate withdrawal (*21*). The contact of Lotsof with the Erasmus University was provided by the Commercial Attaché of the U.S. embassy in Den Hague, who was aware of the newly formed Center for Addiction Research at the Erasmus University, whose mission was to develop creative approaches to addiction problems. As discussed below, the Erasmus collaboration also led to contact with the Dutch Junkiebond.

NDA International provided a small contract to the laboratory of Dr. Stanley Glick at Albany Medical College in 1989, which stimulated some initial pilot work with ibogaine. Subsequent research by Dr. Glick indicated the efficacy of ibogaine in animal models of opioid dependence (*22,23*). This marked the beginning of what is now a decade of work on *iboga* alkaloids in that laboratory. Lotsof reached out tirelessly to public and private sector scientists who might take an interest in the ibogaine project. A valuable contact was made with Dr. Charles Grudzinskas, who was provided with information about ibogaine while an executive at Lederle in 1990 and subsequently became Director of the Medications Development Division (MDD) at NIDA in 1991.

Ibogaine treatment could not legally be provided to humans in the United States, and patients recruited by NDA International were treated in the Netherlands starting late in 1991 until mid-1993. Prior to the advent of the NDA International treatments, an organization known as the International Coalition for Addict Self-Help (ICASH) began to provide treatments in the Netherlands in 1989. ICASH was an addict advocacy organization founded by Robert Sisko which described itself as having a self-help orientation in the tradition of European user self-help organizations, such as the Junkiebond in the Netherlands (*14,24,25*). ICASH made ibogaine treatment available to a network of heroin users in the Netherlands, among them the late Nico Adriaans, a cofounder of the Dutch Junkiebond (see below). This network of users included Dutch Addict Self-Help (DASH), later known as International Addict Self-Help (INTASH), which provided approximately 10 ibogaine treatments to other Dutch addicts in 1990. Boaz Wachtel, a former medic in the Israeli Army, participated in the ICASH and DASH/INTASH treatments, as well as the treatments conducted by NDA International in the Netherlands. The results of the DASH/INTASH treatments were presented in publications from the Erasmus University group and a squatter's collective in Amsterdam (*26,27*).

A total of approximately 40 to 45 individuals were treated between 1989 and 1993 in the Netherlands involving NDA International and/or ICASH or DASH. The data from these treatments, together with the 20 subjects treated in the United States by Lotsof between 1962 and 1963, provide the principal source of the case study evidence that has been presented to the National Institute on Drug Abuse (NIDA) and the FDA (*4*). As indicated below, this total of approximately 60 cases may be a fraction of the total number of humans treated with ibogaine in diverse informal contexts in the United States and Europe.

The death of a patient in the Netherlands in June 1993 brought an end to treatments there by NDA International. The official Dutch inquiry was not conclusive regarding a causal role of ibogaine in the death (*28*). At the time of the death, the Dutch government was engaged in a protracted inquiry into the existing knowledge on experimental treatments for heroin dependence. The inquiry had been launched to provide background information for evaluating the experiment in heroin prescription (*29*). The death significantly decreased the enthusiasm to investigate ibogaine that did exist in official Dutch circles and stopped the development of a clinical trial protocol that had been quietly in progress at the Center for Addiction Research at Erasmus University. A small number of treatments were subsequently conducted by NDA International in Panama (*30,31*).

NIDA developed its own Phase I/II clinical trial protocol from October 1993 to December 1994. The resulting draft protocol involved single administration of fixed dosages of ibogaine of 150 and 300 mg versus placebo, for the indication of cocaine dependence. In March 1995, NIDA decided not to fund the implementation of the protocol it had developed. Opinions of consultants from the pharmaceutical industry were a significant influence in the NIDA decision not to fund human efficacy trials. The decision was naturally disappointing to the network of advocacy of ibogaine and indicated that the prospect for a U.S. clinical trial of ibogaine in the near future was no longer likely.

2. *Deborah Mash, the University of Miami, and St. Kitts*

Dr. Deborah Mash, Professor of Neurology at the University of Miami School of Medicine, became interested in ibogaine in 1991, after hearing presentations at a conference. In the context of the collaboration between NDA International and the University of Miami that began in 1992, she witnessed treatments arranged by NDA International in the Netherlands. Mash undertook organizing a clinical trial of ibogaine, and in August 1993 she received approval of an Investigational New Drug Application from the FDA Advisory Panel chaired by medical review officer Curtis Wright. The Phase I dose escalation study protocol initially included only individuals with histories of having previously received ibogaine and called for these individuals to receive dosage levels of 1, 2, and 5 mg/kg. The study began in December 1993, but activity was eventually suspended, in the context of the unavailability of grant support. The relationship with NDA International had become strained and eventually involved litigation.

In 1996, Dr. Mash began an ibogaine treatment program in the Caribbean Island of St. Kitts. Treatments are oriented to the indication of addiction, mainly to heroin and cocaine. The program features medical supervision of the intake treatment process and includes cardiac monitoring, systematic collection of pharmacokinetic data, and the use of structured psychiatric interviews and rating instruments. The program attempts to remain in contact with patients for up to a year after the treatment by telephone or questionnaires, but it has not yet presented data relating to longer-term follow-up. As of the time of the writing of this volume, more than 150 patients have reportedly been treated in St. Kitts. In patients treated for the indication of acute opioid withdrawal with single dosages of ibogaine ranging from 600 to 1200 mg, physician ratings of withdrawal signs and symptoms appeared to indicate efficacy (*5,32*, see the chapter by Mash *et al.* in this volume), consistent with the case study literature (*4,7,16,25-27,30,31*).

3. Eric Taub

Eric Taub states he has arranged treatments conducted in the Caribbean since 1992 (*33,34*) and claims an estimated total of 310 such treatments, approximately 130 of which were sought for the treatment of addiction, predominantly heroin dependence. The remaining majority of treatments have been "initiatory," involving nonaddicts seeking psychological or, as Taub puts it, "psychospiritual" insight as the therapeutic goal. He uses an overall range of doses of 9 to 12 mg/kg for an initiatory treatment, 12 to 18 mg/kg for cocaine dependence, and 20 to 24 mg/kg for dependence on heroin or methadone.

The majority of Taub's treatments have been of the initiatory type, and he conceptualizes the personal work that his clients seek as an attempt to change a pattern of "reactive" or subconsciously determined behavior (*33,34*). In this regard he appears to be interested in conventional psychotherapeutic goals. A commonly pursued goal of conventional psychotherapy is to achieve insight as to how the processing of prior experience affects behavior, so as to allow greater conscious flexibility and spontaneity in adapting to the demands of the present, rather than subconsciously determined reaction.

Taub regards obsessive behavioral tendencies as particularly responsive to ibogaine treatment. He appears to identify a common attribute of pathologically acquired or "learned" associations in addiction, obsession, and forms of "reactive" behavior that might conventionally be described as neurotic. A common attribute of these behaviors is an intense subjective motivational state elicited by an environmental cue or an internal representation, and a subsequent repetitive, inflexible behavioral response. Similar to the conceptualization suggested by Goutarel (35), ibogaine's putative effect in facilitating psychotherapeutic change, as well as treating addiction, is viewed as involving attenuation of pathologically acquired or "learned" linkages that associate cues or internal representations with corresponding motivational states and behavior.

Taub reported that it is common for individuals who have taken ibogaine to reflect on and process the content of their experience for months afterward. This does not imply some sort of "flashback" or reexperience of the subjective state of being on ibogaine, but a continued interest in the meaning of the material that emerged during the ibogaine experience. He also noted that women seem to experience visual phenomena from ibogaine to a greater extent than men, possibly reflecting the greater bioavailability of ibogaine in females versus males noted in animal models.

Taub estimated that about 70% of addicts treated in their 20s or 30s, and 30% of those treated in their 40s or 50s, eventually return to their prior drug of dependence. He views ibogaine as creating a "window of opportunity" of diminished craving for several months after treatment during which time involvement in aftercare is essential to prevent eventual relapse. The idea of the post-treatment period as critical for processing the content of the ibogaine experience, gaining an appreciation of the psychological aspects of one's addiction, and establishing participation in aftercare is shared by others who have been involved in the treatment of drug dependence with ibogaine (*7,36*).

4. *The West Coast and Other North American Scenes*

There is apparently a network involving the use of hallucinogens, including ibogaine, as psychotherapeutic agents that has existed on the West Coast of the United States since the 1950s (*8-10,14*). This network has operated with a low profile, and there is no Web site. The focus of ibogaine use in this setting has mainly been psychotherapeutic and not the treatment of addiction, and it apparently has included the significant involvement of credentialed health professionals.

According to Stolaroff (*10*) in his book *The Secret Chief: Conversations with a Pioneer of the Underground Psychedelic Psychotherapy Movement*, the dosage of ibogaine administered by his informant was reportedly in the range of a total of 150 to 300 mg, as a single fixed dose, which did not frequently cause hallucinations. Some of the expectations and beliefs regarding the use of ibogaine in this setting appear similar those regarding LSD and other serotonergic hallucinogens (*37-39*), although ibogaine appears to be attributed with some distinctive features. In agreement with the account of Taub above (*33,34*), Stolaroff's informant also described the phenomenon of patients reflecting on and processing the material that emerges during an ibogaine experience continuing for months after the session. This phenomenon is described as a relatively distinctive attribute of ibogaine relative to other psychedelics. Stolaroff's informant also attributed ibogaine with particular efficacy in "blocks":

"This is a place where people will confront blocks, as a matter of fact, if they have been unwilling to accept them or believe them or often try to deny them. A truth that you're trying to deny. Ibogaine won't let you do that."

The effect attributed to ibogaine as a psychotherapeutic agent in the above statement appears to involve particular efficacy in individuals with a behavioral set that is overly constrained, and has a general resonance with attributions of efficacy in changing "reactive" or obsessive behavioral patterns (*33,34*). Another putative ibogaine effect is the existence of a "window of opportunity" following ibogaine treatment, during which there may be a heightened interest and capacity for engagement in the psychotherapeutic process (*7,10,33,34,36*). Operationalizing and studying such effects, although challenging from a research perspective, could be relevant to optimizing patient management, if ibogaine were to be approved for clinical use.

C. The Netherlands

1. DASH/INTASH and the Center for Addiction Research at the Erasmus University

Following his treatment in May 1990, ICASH provided Nico Adriaans with enough ibogaine to treat approximately 10 other addicts. These addicts were treated in the summer of 1990 by a group that referred to itself as DASH, and subsequently INTASH, which at that time consisted mainly of three individuals: Adriaans and his girlfriend, Josien Harms, and Geerte Frenken. The treatments were conducted at Nico Adriaans's apartment in Rotterdam. A description by Frenken, whose chapter is included in this volume, of these treatments reveals an attempt to create ritual (*14*):

"We'd have the person come over the night before his treatment and introduce ourselves as Mamma and Papa Iboga, because the addict is the child going through rebirth. We would tell them about the African ritual the night before the treatment so that they'd totally know what they were going to face. We'd tell them that the Africans use it basically to guide people into adulthood so that they become more responsible people, and that's what we'd be doing. Making a person more responsible in life, so that he'd have control over drugs instead of drugs over the person.

If the person was a shooter, we would let him or her take their last shot, then let her or him destroy their syringe, symbolizing the destruction of their addiction . . .

. . . We had a room with two doors, which was perfect. We'd move someone in through a door with the moon painted on it, into the room where they would have the experience, and then after treatment guide them out the doorway decorated with a sun."

The treatments, mainly for heroin dependence, reportedly went smoothly and were successful in achieving the goal of alleviating the symptoms of acute withdrawal, and those treated denied craving and generally avoided heroin for weeks or months afterward. A focus group was formed of those who had been treated in an investigation conducted in collaboration with the Center for Addiction research at Erasmus University. Eva Ketzer, a physician intern attached to the center, facilitated the groups and compiled data for scientific publication (*26*). However, as time passed, the pull of the social network of the users, and possibly the gradual attenuation of the apparent effect on craving became increasingly evident.

"[T]hey basically all slowly but surely fell back into their addiction. The people who were dealing the heroin were very powerful, socially. They had friends in the focus group going back ten, fifteen years—they went all the way back to hippie times together. They had been through all kinds of shit and they had this intense bond.

The leader of this group was a heroin dealer . . . In the end he won. After the ibogaine people realized they were alone, and discovered that the one thing they had in common was that every day, they used to sit at this guy's table and get together—and that this contact would happen because of smack (which was his income). In the end they all fell back into heroin to maintain the social contact, and the ibogaine was wearing off."

The focus group eventually dissolved. The focus group experience appeared to indicate that it would be necessary to integrate ibogaine into conventional treatment settings, such as the therapeutic community, in order to produce lasting success. It was envisioned that in such a formal treatment setting, the addict patient could be involved in a therapeutic mileu that could protect against the social isolation that appeared to be a major determinant of relapse in the focus group cohort.

The dissolution of the focus group marked a stormy time at the Center for Addiction Research at Erasmus University. Adriaans and Harms were both employed as community field workers at the center. Their function was to maintain contacts with the heroin addict populations and recruit their participation in center projects. The director of the center at that time, Dr. Charles Kaplan (a co-author of this chapter), had been introduced to Lotsof, and the ibogaine issue provided a unique opportunity for investigating innovative treatments. The idea was to initiate an ibogaine medication development project at Erasmus University that would start with animal Phase I studies and progress to clinical trials. However, the activist spirit of Nico Adriaans and Lotsof found common ground, and they undertook the initiative of providing informal treatment to Dutch addicts, an action that they felt was legitimized by the AIDS epidemic. (Nico Adriaans eventually died of AIDS in 1995 (*40*), and a foundation was established in Rotterdam that bears his name and continues his work.) Although Kaplan was sympathetic to this development, it went beyond what was officially permissible with regard to the ibogaine medication development project. Nonetheless, his assistant, Eva Ketzer, had been authorized to work with Adriaans to study the ongoing underground treatments. This ex post facto approach had been considered ethically acceptable since it simply organized a systematic way of observing the self-help initiative of the addicts themselves that already existed outside of the boundaries of formal addiction treatment in the Netherlands.

The tension between official institutional policy and the collective sentiments of the Addiction Research Center intensified with the appearance of sensationalist "success" stories in the international press, which focused an uncomfortably large amount of publicity on the center. This wave of publicity precipitated a controversy and displeased some of the members of the board of the Erasmus University and the private foundation that supported the Addiction Research Center. The underground informal ibogaine treatments were perceived as irresponsible and potentially dangerous, and they raised concerns with the publicity, which appeared to associate the center with the unauthorized ibogaine treatments. A backlash developed in official circles concerned with addiction and medical policy in the Netherlands, which ultimately resulted in the removal of Kaplan as director of the center in 1991. The staff at the Addiction Research Center persisted in attempting to develop a protocol for an ibogaine clinical trial for some time after the departure of its director, but it eventually abandoned the project.

DASH members, mindful of the apparent limitations of their treatment setting, became more focused on attempting to involve the professional drug addiction treatment community. In 1992, in conjunction with NDA International, four treatments were conducted that were observed by Dutch physicians affiliated with the Nederlands Instituut voor Alcohol en Drugs (NIAD) and by American physicians, including Dr. Lester Grinspoon of Harvard Medical School and Dr. Carlo Contoreggi of NIDA. The treatments went smoothly and impressed those present, but did not in themselves succeed in introducing ibogaine into the conventional addiction treatment community in the Netherlands.

2. Jan Bastiaans

Dr. Jan Bastiaans was a major figure in the history of the psychotherapeutic use of hallucinogens (*41-43*). He was a professor at the State University of Leiden from 1963 to 1985. Active with the Dutch resistance during the Second World War, Bastiaans had a long-standing interest in the incorporation of pharmacologic methods into the psychotherapy of war-related trauma. In 1961, he began utilizing hallucinogens for this purpose and eventually treated approximately 300 patients with LSD and psilocybin. As with a number of other researchers who were interested in the possible medical use of hallucinogens, Bastiaans's work became increasingly stigmatized with the rise of the drug culture, and he eventually became the last remaining Dutch psychiatrist with official authorization to use hallucinogens clinically.

The individuals he treated included the Israeli writer Yehiel De-Nur, a survivor of Auschwitz who wrote under the name Ka-Tzetnik 135633 and published an account of his treatment (*44*). It appears that Bastiaan's clinical skills were well respected by his peers, but the controversy over hallucinogens left him isolated politically in his profession. Doubts existed as to the safety of hallucinogens and whether his methods were sufficiently well systematized to be as effective when utilized by others. A report published in 1987 on behalf of the State Department of Public Health concluded that the data was insufficient to systematically evaluate the effects of Bastiaan's treatment approach.

Bastiaans began working with NDA International in 1992 and was present at a total of 18 treatments from 1992 to 1993 (*4*). In June 1993, a patient died during a treatment, and Bastiaans, as supervising medical doctor, was asked to end his therapeutic practice by the Medische Tuchtraad, the Dutch supervisory board of the medical profession. The official Dutch inquiry into the case did not substantiate any charges of wrongdoing by Bastiaans, and noted uncertainty in determining the cause of death due to the lack of information regarding the toxicological significance of post-mortem ibogaine or noribogaine levels, as well as the possible involvement of surreptitious heroin use during the treatment (*28*). However, the adverse action of the Medische Tuchtraad apparently related to an administrative failure in obtaining official permission regarding the use of an experimental treatment. Bastiaans died in October 1997, and in his last years he had become bitter over the lack of recognition for his methods in the medical profession.

D. Other European Scenes

1. Slovenia

According to Marko Resinovic (*45*), founder of the Slovenian Iboga Foundation (*46*), a treatment scene has existed in Slovenia since 1995. He has estimated that approximately 150 people have been treated within multiple networks in Slovenia. The majority of those taking ibogaine have done so for the indication of opioid dependence, with a minority for psychotherapeutic reasons. The foundation's role is to provide information about the use of ibogaine and to assist individuals and their social or family networks in arranging the treatment situation. The foundation is also a source from which ibogaine can be purchased, and the price of the ibogaine for the treatment of a heroin addict is currently the equivalent of approximately US $200 per gram. The scene in Slovenia seems loosely organized, with addicts often setting their treatments up individually. Outcomes in this scene have not been systematically tracked, and it is difficult to assess what the collective clinical experience with ibogaine has been in Slovenia. Dr. Andrej Kastelic, the head of the Center for Treatment of Drug Addiction in Ljubljana, Slovenia's main addiction treatment center, described variable results in the patients with whom he came in contact. He reported the apparent efficacy of ibogaine in acute withdrawal, but commonly with relapse to heroin use within days to weeks (*47*).

2. Denmark

A relatively large ibogaine scene is alleged to exist, or to have existed in Christiana, a former military barracks located in a suburb of Copenhagen turned squat and currently occupied by approximately 1200 people. This information was furnished to one of the authors of this chapter (K.A.) by Carl Waltenburg, a 45-year-old Danish man who describes a lifelong interest in hallucinogens as a psychological or spiritual paradigm (*48*). He has also apparently been involved with hallucinogens as an entrepreneur. He claims to have been associated with Christiana since its inception in 1972, and he first encountered ibogaine in the context of general interest in the psychotherapeutic or sacramental use of hallucinogens. He believes that ibogaine and serotonergic hallucinogens, such as LSD, share a common attribute of facilitating insight, but that ibogaine is distinctive with respect to its efficacy against the physical symptoms of opioid withdrawal. He also has described the use of low doses by nonaddicts for an apparent stimulant-like effect, similar to the use of tablets containing 8 mg of ibogaine which were marketed as Lambarène in France before being banned in 1970.

Waltenburg stated that in 1981, a European industrial manufacturer took an interest in ibogaine and created a 44 kg supply of alkaloid extract from 500 kg of *Tabernanthe iboga* root bark. He eventually obtained possession of this supply of the alkaloid extract, which he refers to as "Indra," and estimated that the total number of heroin dependent individuals who have received it through the Christiana network is approximately 1000. The Indra ibogaine extract is said to have been available through the Christiana network and more recently has been sold via the Internet (*49*). The current price of a gram of extract is approximately US $25. The alkaloid content of the Indra extract is said to be approximately five times that of dried, unextracted root bark, and total *iboga* alkaloid content is claimed to be on the order of 20%. The assertion that this alkaloid extract supply dates back to 1981, and the uncertainty regarding the conditions under which it has been stored raises questions regarding its content, stability, and whether samples taken from various physical locations within this supply are uniform. In January 2000, e-mails posted to the Ibogaine Mailing List (*50*) alleged that in London, a 40-year-old heroin addict who died had taken 5 g of Indra extract 40 hours prior to his death. At the time of the writing of this book, the official British inquest into the matter is in progress.

Waltenburg stated that he began to guide sessions in 1982 that involved a total of about 100 heroin users that year, with some treatments conducted simultaneously on groups of up to eight patients. The duration of sessions was typically 2 days. Subsequent to 1982, he guided about 70 more treatments until the present time, with the remainder of the total of 1000 treatments involving other guides who had acquired their experience in the Christiana network. Approximately half of these heroin dependent patients predominantly used the intravenous route, with the remainder smoking or snorting.

This account, while certainly interesting, conflicts significantly with accounts provided by other informants (*51,52*). The authors of this chapter have not been able to find any independent informant that confirms the existence of the putatively extensive Christiana ibogaine scene. The account of an informant referred to the authors of this chapter by the Danish Drug User's Union (*51*) is inconsistent with that of Waltenburg. Richard T. Lionheart, a Christiana resident of 28 years from the time of its inception, does not recall ever meeting Waltenburg. He was not aware of the use of ibogaine in Christiana, and certainly not on anywhere near the scale Waltenburg claimed. Lionheart provided a different account of how Christiana came to terms with its heroin problem. By 1979, Christiana was afflicted with a severe heroin crisis, and over one-third of its adult population was addicted (*52*). A policy termed "blockade" was adopted, under which all addicts were forcibly removed from Christiana and forbidden to return until they had been off heroin for 6 months. Detoxified heroin addicts were often sent to communes in the North of Denmark and were then permitted to return to Christiana after 6 months. After the imposition of the blockade policy, a "zero tolerance" policy toward hard drugs was imposed in Christiana. The cannabis dealers of Christiana were permitted, but not their counterparts who attempted to sell harder drugs. Lionheart stated that a number of "alternative" treatments had been considered for detoxification, including ibogaine, and were eschewed in favor of the "cold turkey" approach, apparently mainly on philosophical grounds.

"We decided instead to use 'cold turkey' because it seemed to last the longest. There might be an aspect of difficult come difficult go.

A reason not to use ibogaine or acupuncture was to not send a signal of 'Drugs are not so risky. If you get addicted you can always get to the reset button through ibogaine or acupuncture.' It was a period of time when a lot of people in Denmark and especially in Christiana got addicted.

So we did things the hard way."

The above conflicting accounts exemplify some of the methodological difficulties and uncertainties of conducting a contemporary historical study. There simply is not enough distance from the historical phenomenon to assemble and sort through all the relevant evidence. Nevertheless, we can tentatively conclude that while the lack of confirmation raises doubts regarding Waltenburg's account, it appears possible, given the apparent philosophical and power differences, that the use of ibogaine in Christiana could conceivably have occurred in a secretive context not revealed to those who imposed the "blockade" policy.

3. The Netherlands

While the addict self-help scene of the late 1980s and early 1990s no longer exists in the Netherlands, a newer scene emerged in late 1999. "Sara," a mother of five children without a formal medical background, described treatments with ibogaine in her home located in the countryside outside of Amsterdam. She stated that as of November 2000, she had treated 15 individuals for the primary indication of opioid dependence, 5 for cocaine dependence, 2 for alcohol dependence, and 4 individuals whose motivation was "psychospiritual". Most patients are from the United States. Her interest in ibogaine began with Eric Taub and Dan Lieberman's visits to the Netherlands in 1999, and she herself took ibogaine in an "initiatory" session before beginning to treat others in December 1999. She usually used 3 to 6 g of "Indra" (*49*), an extract of *T. iboga* root bark. She has also occasionally used ibogaine hydrochloride at dosages of 300 to 500 mg to bring about a more rapid onset of action in patients who are already evidencing signs of opioid withdrawal. Her reason for combining the root bark extract and the purer hydrochloride appear to be pharmacokinetic. She described the hydrochloride as having a faster onset of action, which is desirable in patients who are verging on acute withdrawal, whereas the extract is described as having a more sustained peak effect and a slower onset of action.

The cost of the treatment at Sara's home is approximately US $600 to $1000, and she described her main compensation for her role as "not a great income, but a great pleasure." She has placed great importance on empathic contact between the treatment guide and the patient, and stated she sometimes takes a small dose of approximately 1/2 g of the extract to strengthen the bond with the patient and to maintain her own wakefulness through the all-night session. She believes that her home setting is reassuring to many of her patients, who receive the treatment in a darkened room in her house. She noted that some patients prefer to speak about issues such as the negative self-image of the addict or their hopes for the future, while others prefer simply to be left alone to attend to their inner experience. She noted the importance of aftercare and for treated addicts to form new nondrug-centered social networks after treatment, and she believes that social isolation is a major factor involved in relapse. She has attempted to encourage treated addicts to maintain contact with one another in the form of support groups, but noted that the interest in this among the addicts she has treated appears to have been limited.

4. The United Kingdom

An ibogaine scene began in the United Kingdom in 1998, apparently stimulated by among other things, the appearance of a lengthy article on ibogaine in the *London Times* Saturday Edition (*53*), distribution of the book *The Ibogaine Story* (*14*), and an ibogaine video by the ibogaine-advocating organization Cures-not-Wars (*54*) (see below) that circulated among members of the Green Political Party and others interested in drug policy reform. In 1998, activists Chris Sanders and Nick Sandberg formed their respective organizations, The Ibogaine Project (*55*) and Ibogaine.co.uk (*56*). They journeyed to Slovenia to purchase a supply of ibogaine, and treated an addict in the United Kingdom in December of that year.

The ibogaine treatment scene in the United Kingdom is loosely organized and not apparently centered around any single organization or individual. Ibogaine has been available in certain small shops in the United Kingdom and can be purchased via the Internet as well. It is estimated that as of late 2000, underground treatments were proceeding in London with a frequency of approximately one per week (*57,58*). The Ibogaine Project has been involved with activities related to political organizing such as influencing the Greens to officially include advocacy of ibogaine research and treatment in their agenda and lobbying the Lord Mayor of London to support an ibogaine treatment and research project involving local physicians. Ibogaine.co.uk has mainly been involved in advocacy in the form of activities such as distributing pamphlets and organizing lectures.

5. Italy

A small scene has existed since 1994 in Italy involving a chemist who made local contacts in Africa and developed his own method of extracting ibogaine from root bark shavings (*33*). Approximately 35 to 40 people have been treated, predominantly for heroin dependence.

6. Czech Republic

A treatment scene began in the Czech Republic in January 2000 (*59*). A total of approximately eighteen opiate-dependent addicts and 60 nonaddicts were reportedly treated. The cost of the treatments was the equivalent of about US

$500. Nonaddicts received dosages ranging from 13 to 18 mg/kg, and addicts 20 to 22 mg/kg. The treatment provider terminated this activity approximately a year after beginning it, due to his impression that offering ibogaine treatment alone without systematic aftercare appeared to yield limited benefit.

7. France

G., who is discussed here anonymously due to uncertainty regarding the position of French authorities on the use of ibogaine in that country, was initiated into the Bwiti cult in 1980, and has conducted ibogaine sessions in France and Cameroon since 1998. In France, he has treated 11 people, one for heroin dependence, 1 for alcoholism, 6 for psychotherapeutic purposes, and 3 "out of curiosity." The sessions have generally been at no cost. G. has usually conducted the sessions on his own, although a Gabonese Nganga (guide) did participate in one individual's treatment, at a cost of approximately US $1000. G. noted that drug and alcohol use seemed to be decreased for a variable length of time following ibogaine, regardless of the indication for which the session was sought. G.'s sessions in Africa are described further below.

E. Africa

This chapter focuses on ibogaine use in the United States and Europe. However, the work of two individuals is discussed here who have been involved in arranging African Bwiti ceremonies predominantly on behalf of Europeans or Americans.

Dan Lieberman (*60*), an ethnobotanist based in South Africa and a Bwiti initiate, had been involved in arranging initiations for Europeans and Americans in Gabon, and also apparently treated heroin addicts in South Africa. He unfortunately died in an automobile accident in August 2000. It is estimated that Lieberman arranged approximately 20 ibogaine experiences in Africa (*33*). The following description emphasizes his respect for, and his experience of, the sense of community and empathy in the Bwiti ceremony (*61*):

"The plant tends to throw one directly into a state of 'sacred knowing,' a state unclouded by thought. The more profound the stillness experienced, the greater will the initiates insights be, and the more fundamental and abiding will the changes thus be... The Bwiti, who understand this state perfectly, have developed a culture of healing around this plant, which is embodied in their initiatory ceremonies. These are musical outpourings of ritualized compassion and caring, which show the indivisibility of healing, art, music, dance and religion in a most perfect and sublime manner.

These initiations are held for those people who request it, and only Bwiti initiates (those who have eaten of the plant) are able to initiate others. The Bwiti practitioners' entire lives are suffused with a consideration and an awareness of the interconnectedness of all things. They live in and around the temples, which are the center of village life, and have a respect for the sanctity of healing, and for the inherent potential for goodness in every human being. . . . The entire village takes part in the ritual, each individual having a specific role to perform, instrument to play, part to sing, and they know the ceremony intimately. It is a combined effort by the whole village, rich in cultural nuance and significance, and that you, as initiate, as neophyte, are the centre of all their focus, that you are the 'Banzie,' makes this a very powerful experience indeed."

Lieberman clearly believed that the set and setting of ibogaine administration are important, and that the Bwiti ceremony, as he has experienced it in Africa, has significant implications for optimizing the use of ibogaine for the indication of addiction:

"In personal communications and e-mail correspondence with various addicts who had undergone the iboga/ibogaine treatments, I found that inadequate or incompetent caregiving was often cited as a barrier to greater insight. (To the Bwiti this is the heart of the matter—the sanctity of the temple, it's members, rites and accordances are the foundation of a successful and pure initiation.) For a more enduring effect, particularly for those staving off addictions, it seems that the setting is critical to the longer-term efficacy of the experience."

G., whose activities in France have been previously discussed, has also arranged Bwiti initiations in Cameroon for a total of seven individuals, three of whom sought help for heroin or alcohol dependence and four who sought a "mystical experience." The cost of these experiences was approximately US $1200 to $2000. G. stated that he no longer arranges these ceremonies, because "for Westerners, staying with local people in their home and adhering strictly to the traditions is too hard" and that "Although it is bearable for someone who comes back cured of addiction, such is not the case for someone searching for a mystical experience and who discovers the infernal (and real) life of Africans. I discontinued initiations . . . I expect to have them in a context more suited to our poor little asepticized bodies."

A problem with arranging "iboga tourism" is the lack of control over the conditions and circumstances of the ceremony, and there can be problems such as malaria, or unexpected negotiations regarding the fee at inopportune times (*56*). It should be kept in mind that there is considerable variability with respect to any number of factors such as experience, interest in initiates, and the manner in which the Bwiti ceremony is practiced by its adherents.

III. The Political Subculture of Advocacy

There has been energetic public sector activism on behalf of ibogaine, much of which is associated with the lay NDA/addict self-help treatment network that originally accumulated the clinical case study evidence that was presented to the FDA and NIDA. In addition to interactions with NIDA and the FDA, ibogaine activism has encompassed activities such as participating in harm reduction or drug policy oriented conferences and engaging the media. Ibogaine advocates did network with activist organizations that had a history of utilizing aggressive and provocative tactics of civil disobedience that resulted in arrests of demonstrators. However, no arrests ever resulted from the generally calm and orderly actions and demonstrations that were staged on behalf of ibogaine.

Activists were regularly included in major NIDA and FDA meetings on the development of human research with ibogaine in the years of NIDA's ibogaine project from 1991 to 1995. Ibogaine advocacy sought to influence the public sector, which controlled policy and budgetary decisions of great importance for ibogaine's possible development. NIDA underwrote most of the research expenditures for preclinical and toxicological work on ibogaine's development as a pharmaceutical agent and weighed the possibility of funding its own human clinical trial. The FDA provided the regulatory authorization to permit going forward with clinical trials.

In the United States and Europe, there is a history of groups afflicted with an illness organizing in order to influence public policy through political activism. Examples of this include the health activists who have organized around AIDS or breast cancer (*2,3,62,63*). Such groups are bound by a sense of group identity, share a feeling of being marginalized, and have formed their disease specific agenda in the context of a preexisting political agenda. AIDS activists organized in the context of a preexisting gay political movement, and breast cancer activism in the context of feminism. The preexisting political base for the advocacy of ibogaine was that associated with drug addiction, and it included the European drug-user self-help unions, and harm reduction and AIDS activism communities in the United States.

The use of ibogaine originated in a medical subculture, involving individuals without formal medical training or credentials and the treatment of a stigmatized condition. Identification with marginalization and stigmatization is a resonant theme in the contemporary history of ibogaine and an important element of the group identity of those involved in ibogaine's advocacy. Advocates of ibogaine experienced themselves as having witnessed strong evidence of efficacy in themselves or others, but felt that their observations were discounted and their views excluded due to their lack of credentials. Many of its advocates believed ibogaine had not been developed because society has marginalized and stigmatized addicts and did not want to expend resources to develop treatment, and furthermore discounted evidence of the efficacy of ibogaine because it originated from a nonmedical network involving addicts. Activism was a response to the perception of marginalization, and providing ibogaine treatment was viewed as a legitimate act of civil disobedience.

The ibogaine project drew a constituency from other activist groups whose collective identity also featured a sense of marginalization and whose agenda might accommodate ibogaine advocacy. The AIDS Coalition to Unleash Power (ACT UP) was the most tactically important alliance with a group possessing a related agenda. The shared agenda of ACT UP and ibogaine was to accelerate the development of treatment for drug dependence as a disorder causing an increased rate of transmission of HIV. The prospect of user self-help in the development of an effective, alternative and noncoercive treatment approach interested the Dutch Junkiebond. Some common ground existed with drug policy reformers who generally favored the development of treatment instead of incarceration and the drug war and because the scheduling of ibogaine and its stigmatization as a "psychedelic" arguably impeded ibogaine's development. Issues of potential interest from an African-American perspective involved the community impact of the drug problem, the ethical and outreach aspects of clinical research in the community, and the Afrocentric theme of the retrieval of a "lost" traditional medicine.

A. Origins and Actions

1. *Activist Tactics*

The most significant activist tactic employed on behalf of ibogaine was simply to spread information. The advocates of ibogaine felt they had a strong case, with the apparent congruence of the human case reports and animal work indicating evidence for efficacy, and their convictions drew on their own personal experience and the collective experience of the network. The advocates of ibogaine were energetic with respect to the distribution of scientific literature, seeking contact with the media and networking with the drug policy and harm reduction activists. Arrangements were made for physicians or professional researchers to witness treatments, sometimes discreetly and confidentially. Individuals who had been involved in the treatments shared their first person perspective at conferences and meetings, and in interviews with the media. NDA International frequently mailed collections of reprints from the scientific literature on ibogaine to various individuals in industry, academia, or government, and also distributed them at conferences and meetings related to drug policy or harm reduction.

2. *Civil Disobedience*

The classic model of nonviolent civil disobedience advocated and exemplified by Mohandas K. Gandhi or Martin Luther King, Jr., involved the violation of the law in such a way as to challenge the onlooker to question the morality of the law or policy at which the action was directed. On the New York ACT UP Web page a piece titled "Why We Get Arrested" (*64*) bluntly summarized that activists who make use of civil disobedience to attempt to achieve change must "1) make absolutely clear what change is desired, usually by listing specific demands; 2) target a group or individual with the power to bring about the desired change; and 3) design actions so that the cost of resisting change is perceived by the person/group in power to be greater than the cost of giving in." This approach appears to have been effective, and ACT UP employed these techniques repeatedly with significant success in gaining specific agenda items such as lowering the price of some medications from pharmaceutical manufacturers or accelerating the regulatory process of medication development at the FDA (*3*).

Advocates of ibogaine did not engage in this sort of civil disobedience, although some of the organizations with which they networked did. The classic paradigm of civil disobedience involves the deliberate, open violation of an unjust law. When advocating the development of ibogaine as a treatment, the advocates of ibogaine did not break the law and never were arrested in any action or demonstration. Treatments such as those arranged in the Netherlands or St. Kitts could not have legally been conducted in the United States, but could be conducted legally in accordance with local law. Also, the treatments for the most part were not "open" and subject to the bounds of confidentiality. Nonetheless, the perceived apprehension of the FDA or NIDA regarding possible civil disobedience on behalf of ibogaine may have been useful in gaining activists access and possibly influencing the development process of the ibogaine project, as for example the "Storm NIDA" action of July 1991, which is described further below.

3. *The Tactics of Political Theater and Confrontation*

Health activists, who have become accomplished practitioners of political theater, have drawn from a heritage of tactics once previously associated with the radical left. The Dutch Provos (as in provocateur), which originated in Amsterdam in 1965 (*65*), were a seminal influence in the tide of countercultural protest that eventually peaked in Berkeley, Mexico City, Prague, and at the Democratic Convention in Chicago in the summer of 1968. The Provos embraced an attitude of antiauthoritarianism, a sense of "no going back" to the past, and a general tactical approach of attempting to provoke furious, embarrassed reaction from the authorities with highly theatrical, and often absurd actions (*66*).

The Provos were a major constituent from which the American Yippies formed in 1967. In the Netherlands, the Provos also gave rise to the Kabouters (translated

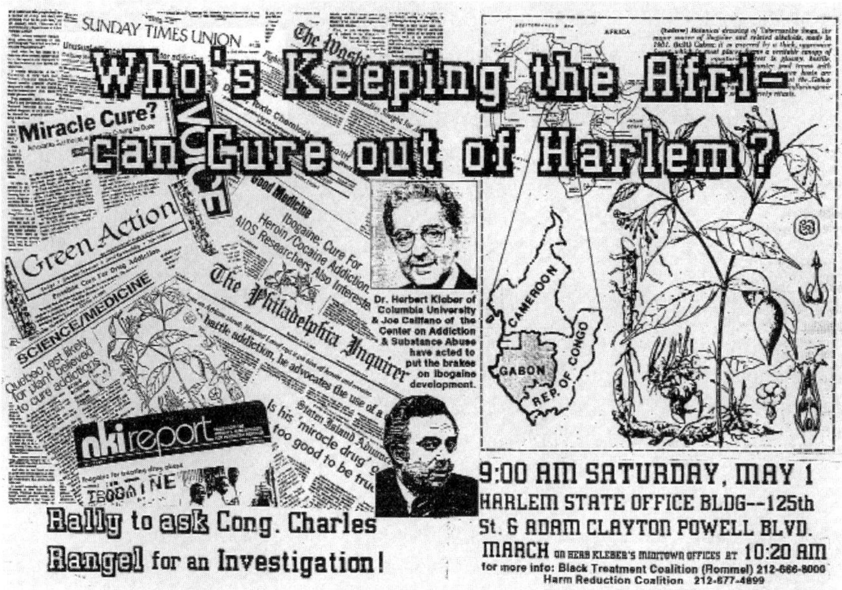

FIGURE 1. An example of the tactics of provocation, this 1993 poster asks, "Who's keeping the African cure out of Harlem?"

as "gnome," "elf," or "dwarf"), which formed from old Provo factions and the squatters movement in 1970, and they became an important political force in the Dutch drug policy of separation of cannabis from hard drugs in 1976. This emergent strength of the drug policy movement contributed to the future political base of drug-user self-help unions and harm reduction.

Some prominent actions on the part of the Yippies, the Junkiebond, and ACT UP exemplify a common related tactical heritage of the use of political theater to shock, inform, and engage. In 1967, the Diggers, a group which was eventually subsumed within the Yippies, engaged in an action in which they threw dollar bills from the balcony of the New York Stock Exchange. As political theater, individuals lunging for dollar bills momentarily disrupted the normal business activities of the Exchange. The action succeeded in its goal of exposing and drawing attention to the elemental display of the psychology of greed, as the trading of vast sums of capital was momentarily disrupted by individuals lunging for dollar bills. A decade later, activists of the Dutch Junkiebond and squatter's movement scattered black and white photocopies of the colorful Dutch currency ("black money"), to expose the role, eventually verified by state prosecution, of a bank using real estate transactions to launder money made by drug dealers (*14*).

From their first action in 1987, ACT UP repeatedly targeted the New York Stock Exchange as a symbol of greed and corporate exploitation of the AIDS crisis (*3*).

In addition to attempting to establish a shared agenda with organizations that used such techniques, advocates of ibogaine attempted to make some use of confrontation and political theater, sometimes with questionable success. As illustrated in Figure 1, a 1993 poster inquired, "Who's keeping the African cure out of Harlem?" and implied that Columbia Professor Herbert Kleber had "acted to put the brakes on ibogaine's development." Such an action was intended to raise the profile of ibogaine and contribute to an impression of a broader base of interest than likely existed at the time. However, Dr. Kleber is not known to have played a direct role in any decision regarding ibogaine at NIDA or the FDA. The action may well have been more related to ACT UP's differences with Kleber's Center on Addiction and Substance Abuse (CASA), regarding the issue of needle exchange, than to ibogaine. The incident illustrates a potential hazard of the diffusion of the focus of ibogaine activists in their attempt to combine their agenda with ACT UP.

4. The Junkiebond and European Addict Self-Help

The Dutch Junkiebond (Junkie "league" or "union") formed in 1980 around the central focus of promoting the "normalization" of the image and status of drug users in society (*40,67,68*). The Junkiebond was effective in advocating for the involvement of drug users in policy making and improving the accuracy of their image in the media. The Junkiebond started the first needle exchange in responsc to hepatitis B in Rotterdam in 1981, and advocated additional harm-reduction policy innovations such as low-threshold methadone maintenance and the "safe stroll" or "tipplezone" for street sex workers (*69*). Drug-user self-help unions now exist in a number of other European Union countries such as Germany, Denmark, and Britain, and in Australia. Drug-user unions have not been as significant a development in the United States as they have in Europe.

Nico Adriaans, a cofounder of the Dutch Junkiebond, was treated with ibogaine in May 1990. His research work at Erasmus University had a strong policy-making objective, that of helping to provide access to the user community in research relevant to harm reduction policy development (*70,71*). He described his role in facilitating access of the addict community to researchers as "the tribesman who helps the scientist access and understand tribal culture" (*40*). This understanding is, in turn, essential for assessing the suitability of a given policy innovation for the target population. Adriaans believed that he had found a powerful policy transformation tool in ibogaine. He advocated ibogaine to other European user groups, but it had a limited impact, apparently due their greater interest in agenda items such as methadone maintenance or the legalization of heroin, as opposed to the development of an experimental treatment (*14*).

5. Cures-not-Wars

Cures-not-Wars (*54*) is an advocacy organization whose agenda includes ibogaine, harm reduction, and drug policy reform. As the name of the organization indicates, it advocates the development and availability of treatment over pursuit of the "Drug War". Specific Cures-not-Wars agenda items include the decriminalization and medical use of cannabis. Cures-not-Wars and like-minded counterculture ibogaine activists contend that ibogaine is a victim of the "Drug War" mentality, which has impeded the development of an effective treatment for chemical dependence because of its status as a scheduled drug and a hallucinogen. Dana Beal, a coauthor of this chapter, is the founder of Cures-not-Wars. Beal's interest in ibogaine began with meeting Howard Lotsof in 1973 and intensified due to the upsurge in heroin use the early 1980s. A former Yippie with experience in grass roots activism, Beal pursued alliances with other activists whose agendas would accommodate ibogaine. These alliances involved a core activist group, which evolved through multiple affiliations with other activist organizations with whom a shared agenda that included ibogaine could be established. Ibogaine activists in this group were successful in gaining access to major meetings and decision makers of the ibogaine project. They were an active presence at the NIDA and FDA ibogaine project meetings from 1991 to 1995 and were regularly included in meetings with officials in the FDA, NIDA, and the Office of National Drug Control Policy.

The alliances pursued by this group of ibogaine advocates included the Yippies; Rock against Racism, a movement of rock musicians and fans that began in Britain in 1978; the Green political party; and various factions of the drug policy reform movement. The alliances also included Jon Parker, an AIDS activist who played a pioneering role, often involving civil disobedience, in helping to establish needle exchange in the United States and who organized the first needle exchange in the United States in New Haven, Connecticut. Another was with African-American community leader Dhoruba Bin Wahad, who had been a leader of the East Coast Black Panthers. The most tactically important of these alliances with respect to NIDA's ibogaine project involved ACT UP, as described in the following paragraphs.

6. ACT UP

ACT UP, formed in 1987, utilized tactics of civil disobedience and political theater that were effective in influencing the process of decision making regarding the development and availability of treatments for HIV (*2,3*). From 1991 to 1994, ACT UP supported ibogaine as an item on their agenda, taking the view that ibogaine's putative effectiveness in heroin and cocaine dependence could possibly reduce the spread of HIV. This support may have been significant for the ibogaine project at NIDA, as evidenced by the apparent effect of the "Storm NIDA" action. In May 1990, ACT UP had organized a national action to

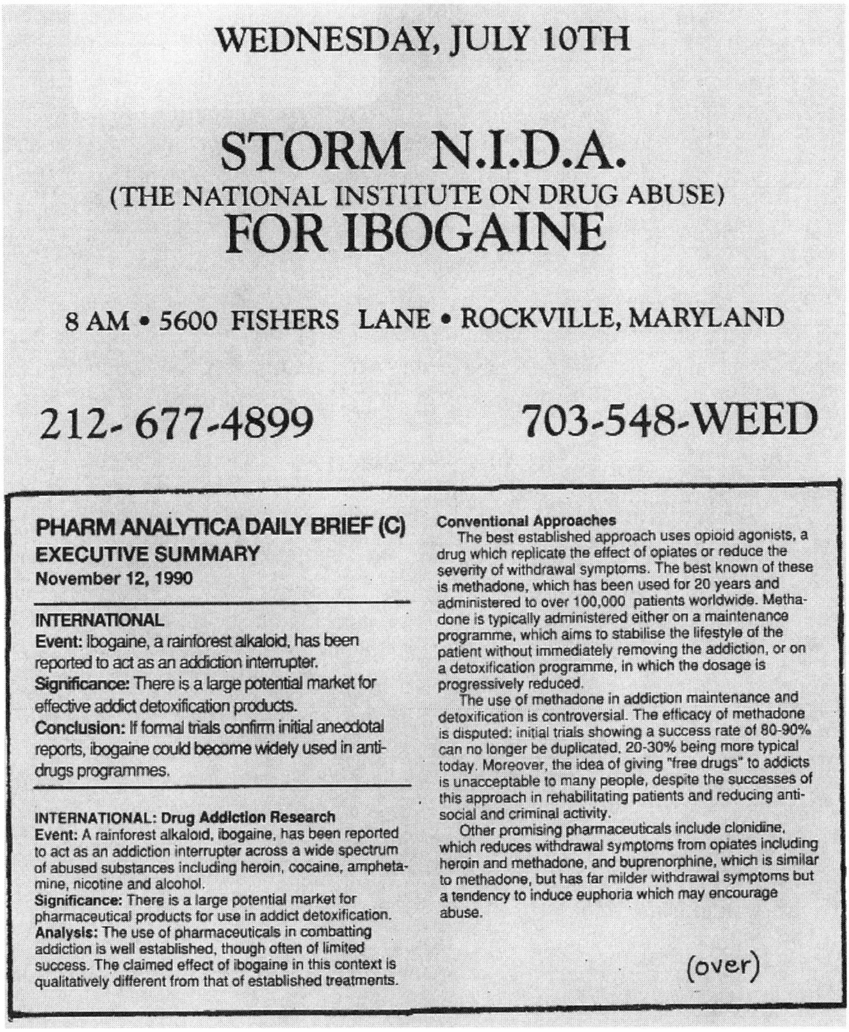

WEDNESDAY, JULY 10TH

STORM N.I.D.A.
(THE NATIONAL INSTITUTE ON DRUG ABUSE)
FOR IBOGAINE

8 AM • 5600 FISHERS LANE • ROCKVILLE, MARYLAND

212-677-4899 703-548-WEED

PHARM ANALYTICA DAILY BRIEF (C)
EXECUTIVE SUMMARY
November 12, 1990

INTERNATIONAL
Event: Ibogaine, a rainforest alkaloid, has been reported to act as an addiction interrupter.
Significance: There is a large potential market for effective addict detoxification products.
Conclusion: If formal trials confirm initial anecdotal reports, ibogaine could become widely used in anti-drugs programmes.

INTERNATIONAL: Drug Addiction Research
Event: A rainforest alkaloid, ibogaine, has been reported to act as an addiction interrupter across a wide spectrum of abused substances including heroin, cocaine, amphetamine, nicotine and alcohol.
Significance: There is a large potential market for pharmaceutical products for use in addict detoxification.
Analysis: The use of pharmaceuticals in combatting addiction is well established, though often of limited success. The claimed effect of ibogaine in this context is qualitatively different from that of established treatments.

Conventional Approaches
The best established approach uses opioid agonists, a drug which replicate the effect of opiates or reduce the severity of withdrawal symptoms. The best known of these is methadone, which has been used for 20 years and administered to over 100,000 patients worldwide. Methadone is typically administered either on a maintenance programme, which aims to stabilise the lifestyle of the patient without immediately removing the addiction, or on a detoxification programme, in which the dosage is progressively reduced.

The use of methadone in addiction maintenance and detoxification is controversial. The efficacy of methadone is disputed: initial trials showing a success rate of 80-90% can no longer be duplicated, 20-30% being more typical today. Moreover, the idea of giving "free drugs" to addicts is unacceptable to many people, despite the successes of this approach in rehabilitating patients and reducing anti-social and criminal activity.

Other promising pharmaceuticals include clonidine, which reduces withdrawal symptoms from opiates including heroin and methadone, and buprenorphine, which is similar to methadone, but has far milder withdrawal symptoms but a tendency to induce euphoria which may encourage abuse.

(over)

FIGURE 2. This 1991 "Storm NIDA" leaflet capitalized on the notoriety of a much larger ACT UP action that had taken place previously at the NIH.

"Storm the NIH" (National Institutes of Health) (*3*). One thousand protesters attended the action and demanded more AIDS treatments and improved representation of women and minorities in the planning and implementation of clinical trials. It was a substantial action, and was duly noted at the NIH.

The following year, ibogaine activists working within ACT UP created a "Storm NIDA" action on behalf of ibogaine and scheduled it for July 10, 1991. The event was advertised one to two weeks in advance with thousands of leaflets (see Figure 2), which were extensively distributed throughout the offices of the FDA and NIDA and intentionally leaked to the police, creating the impression of a large pending action, such as that of the previous year. Ibogaine did not, in fact, have a constituency that would draw a number of protestors anywhere near the size of the original 1990 Storm NIH event. As it turned out, the action was calm and involved a total of fewer than 20 demonstrators. However, NIDA on hearing of the pending action, immediately scheduled a meeting with the participants, and on July 9, the day before the action, notified the protestors that ibogaine had now been placed on the list of drugs to be evaluated as treatment for addiction (*14,72*). This effectively marked the beginning of the NIDA ibogaine project, which was active from 1991 to 1995.

Another action involving ACT UP participation was the summer of 1993, when about 40 people, including Philadelphia ACT UP, joined the New York contingent to picket FDA/NIDA at the Parklawn Building in Rockville on July 5. Activists believe that the picket and spirited activist participation in the August 1993 FDA review panel may have influenced the decision to resume Phase I clinical research that had been paused due to concerns about neurotoxicity.

The placement of ibogaine on the ACT UP agenda was a result of the participation and influence of ibogaine activists in the ACT UP Treatment and Data Committee. The Treatment and Data Committee served to track the medications development process and to interface with government and industry in trying to accelerate the development of drugs to treat HIV. A political/ideological division within ACT UP involving, among other issues, the relative priority of development of medications versus immediate social and healthcare needs led to a major realignment of ACT UP in 1994, with many who had served with the Treatment and Data Committee leaving ACT UP to form a separate organization, the Treatment Action Group. It was in the context of this political realignment that ibogaine was concluded to be nonessential to the ACT UP agenda, from which it was dropped in 1994.

B. Media Coverage of Ibogaine

The earliest media coverage for ibogaine as a treatment for drug dependence began in the alternative press with an article in the Yippie periodical *Overthrow* in 1985 (*13*). Subsequently, ibogaine has been covered in the mainstream media since the late 1980s. In general, the media coverage of ibogaine has emphasized the scientific story of the development of an agent originating from the African rain forest as treatment for addiction (*73-75*), with relatively less emphasis on the subcultural aspect of the informal treatment and advocacy networks. The inclusion of ibogaine-treated addicts in the coverage mainly focused on clinical validation and human interest, and not the nonconforming nature of their activities as a medical or advocacy subculture. The reference to support for ibogaine's development by a "significant portion" of the public (*1*) at the beginning of this chapter is due in part to the coverage that ibogaine has received in the media. This was particularly true during the period of the NIDA ibogaine project from 1991 to 1995.

In addition to describing the aims and scope of the NIDA ibogaine project, the media have apparently been engaged by aspects of scientific theoretical thinking about ibogaine. In 1993, the French chemist Robert Goutarel published a monograph that suggested the hypothesis of a possible functional analogy between the action of ibogaine and rapid eye movement (REM) sleep. Goutarel suggested that ibogaine might work by diminishing the pathological linkages of drug-related stimuli to feelings of valuation and salience, and he suggested a process analogous to the reconsolidation of memory during REM sleep (*35*). Goutarel's hypothesis has a certain intuitive appeal, and apparently influenced media stories that emphasized the "dreamlike" nature of the ibogaine experience (*14,73,76*).

The Ibogaine Story (*14*) published in 1997, provided a social and cultural history of the development of ibogaine and the political subculture of ibogaine advocacy. Although intended for a general lay audience, the book also surveyed the topics of major significance from the scientific literature on ibogaine. The book attempted to engage interest by including creative scientific speculation such as Goutarel's work (*35*), or theoretical discussion regarding a possible relationship of quantum physics to a model of consciousness. A video also titled *The Ibogaine Story* included excerpts of television news productions, as well as the testimony of patients and treatment guides. Cures-not-Wars estimates it has distributed some 4,000 copies of the book in its final form, 2,000 copies of various preliminary versions of the book in the form of photocopies, and approximately 5,000 videos in the effort to promote and popularize ibogaine.

In a television interview in 1993 Frank Vocci, the NIDA official in charge of the ibogaine project stated, "I think that the bad press that LSD got in the 1960s is not going to help ibogaine, because I think the American people will expect that this is something gone awry, that the people in Washington are off their rockers, they really lost it." However, the media tended to place relatively greater emphasis on the scientific controversy regarding the question of ibogaine's effectiveness than on the potential social or political controversy of ibogaine as a hallucinogen. The media's coverage suggested the possibility that if the ibogaine project were to have produced favorable clinical results, its developers would not have been perceived as "off their rockers" solely because it involved the use of a hallucinogen.

C. The Internet

The Internet has played an important role in the increasing the globalization of the medical subculture of ibogaine. Various providers of treatment and sources of ibogaine offer Web sites devoted to ibogaine for the purpose of disseminating information. A Web site (*58*) summarizes the current availability of ibogaine and offers links to providers of treatment/sessions or sources of ibogaine, most of whom maintain their own Web sites.

With the exception of some countries such as the United States, Belgium, or Switzerland, it is presently generally possible to purchase ibogaine through contact with sources via the Internet in much of the world. The Internet functions as a sort of "underground railroad," linking suppliers and treatment guides with those seeking ibogaine treatment, and it has likely been a factor in the increased availability and the decline of the price of ibogaine over the past several years.

Other Internet activity includes the Ibogaine Mailing List (*50*), which began in March 1997 and has had a significant reach in accessing and "binding" the networks associated with ibogaine. The correspondence of its contributors is archived, and is an informative resource regarding the collective ideas, beliefs, and expectations regarding ibogaine in the United States and European subcultures. Some of the more prevalent topics include the availability of treatment and the experiences of patients. The list also offers lively discourse that includes more esoteric and theoretical themes that illustrate ibogaine's role as a subcultural intellectual phenomenon. Threads of discussions on the list include topics such as the sacramental aspect of ibogaine as an "entheogen," linkages of the ibogaine concept to world historical religious cultures such as Gnosticism or Buddism, and references to Bwiti, the sacred culture of ibogaine in Africa. The tone of some of the list correspondence suggests that ibogaine has become associated with a sacred or transcendent status in the subculture in the United States and Europe, much as it is associated with a sacred culture in Africa. Some excerpts taken from the list illustrate the transcendent or sacramental quality that some in the subculture have attributed to ibogaine:

"I believe both NDE's (near death experiences) and ibogaine can induce certain forms of archetypal experience relating to the creation of physical existence."

"I did have some preconceived notions about ibogaine experience . . . like the Bwiti notion of 'meeting your ancestors,' what that became for me was to 'see' my past lives: the actual faces of them . . . Having done many psychedelics, I am convinced that ibogaine is a sacred substance . . . no hit of LSD or mushroom or cactus came close to where ibogaine took me. Period. Everything I experienced during the ibogaine I had already experienced; the ibogaine 'trip' was just what

I call the 'remembering'."

"The Bwiti cover all bases. Their religion is so new and so utterly synchretic, I'm sure there isn't a religious belief structure in the world that isn't represented in it somewhere . . . and, as such, it does I guess stand a good chance of being adopted extensively, certainly in the States."

IV. Conclusions

The unusual treatment and advocacy subculture in which ibogaine exists raises several issues. Participation in, or identification with, this subculture involves a belief in the efficacy of ibogaine, which appears to be sustained by word of mouth and Internet communication, in addition to published clinical reports. The ibogaine belief system is also an expression of the demand for alternatives for existing treatments for addiction. The sense of marginalization and stigmatization, with which some of ibogaine's advocates appear to have identified, is a shared general feature of the drug subculture. An awareness of why individuals choose to use ibogaine, and what they appear to have derived from it, could be useful in accessing the treatment-seeking behavior and motivations of hidden populations. Such knowledge could also be useful in designing the milieu of present treatment settings, and if ibogaine were to be approved, in devising manual-driven behavioral approaches.

The successes and failures of group activism for the development of a new pharmacotherapy of addiction provide an example of the problems, issues, and factors associated with the tactics of addiction health policy advocacy. To a significant extent, activism on behalf of ibogaine can be viewed as disease-specific advocacy, such as that associated with AIDS or breast cancer. Ibogaine is a theoretically interesting drug intended for a disorder with high collective societal cost and morbidity, but it is relatively unattractive to the private sector because it presents a limited apparent profit potential and significant developmental expense. The private sector's weighting of costs and long-term incentives can be irrational from the larger standpoint of society, a situation that public sector spending is intended to remedy. Because of limited private sector enthusiasm, the development of new treatments for addiction is particularly dependent on public sector spending relative to the development of treatments for other conditions. Public sector spending is subject to political factors, which may be influenced by disease-specific activism.

Activists viewed advocacy on behalf of ibogaine's development as necessary because the disorder it is intended to treat afflicts a marginalized group.

Preexisting movements, such as the drug user's unions in Europe and harm reduction and AIDS activists in the United States, contributed to the political base. A stronger user's movement in the United States might have benefited the political base for the development of innovative and experimental treatment for addiction, including ibogaine. Such an American user's movement might have provided a significant countervailing force for the development of new pharmacotherapy for addiction in academic research, government, and the pharmaceutical industry. Greater insurance reimbursement for the treatment of substance dependence could also provide an incentive for development of pharmacotherapy for addiction in the private sector, just as insurance reimbursement has motivated the pharmaceutical industry to develop drugs for HIV.

Ibogaine's advocates gained significant access to public sector decision makers involved in the development of addiction treatment. It is possible that they exerted significant influence, and that they still might do so in the future. It has been suggested that the reactions elicited by ibogaine's advocates may have been counterproductive in some respects (see the foreword by Kleber in this volume). As an example, AIDS advocacy apparently accelerated the development and release of drugs to treat HIV, but also been suggested to have discouraged companies and individuals from involving themselves with HIV (*2*). The current status of the lack of official approval can be cited as evidence that advocacy on behalf of ibogaine worked at cross-purposes to its development. On the other hand, it can also be argued that the political activist subculture may have positively influenced support for the allocation of public resources toward ibogaine's development. Perhaps more importantly however, the medical subculture of ibogaine did yield human experience that influenced the decision to pursue the significant body of preclinical work that was NIDA's ibogaine project. This medical subculture presently remains as a persistent, and an apparently growing phenomenon.

Acknowledgments

The authors would like to thank Howard S. Lotsof, Dr. Stephen J. Sifaneck, and Aivia Monitto-Angel for their assistance with this chapter.

References

1. C.R. Wright, *Memorandum to Franck Vocci: Ibogaine*. March 10, 1995.
2. R.M. Wachter, *N. Eng. J. Med.* **326**, 128 (1992).
3. AIDS Coalition to Unleash Power (ACT UP), "Capsule History," *http://www.actupny.org/documents/capsule-home.html*, 2000.
4. K.R. Alper, H.S. Lotsof, G.M.N. Frenken, D.J. Luciano, and J. Bastiaans, *Am. J. Addict.* **8**, 234 (1999).
5. D.C. Mash, C.A. Kovera, B.E. Buck, M.D. Norenberg, P. Shapshak, W.L. Hearn, and J. Sanchez-Ramos, *Ann. N. Y. Acad. Sci.* **844**, 274 (1998).
6. D.C. Mash and C.A. Kovera, "Ibogaine Is Effective in Blocking Opiate Withdrawal Symptoms: Role of the Metabolite Noribogaine," *Sixty-Second Annual Scientific Meeting of the College of Problems on Drug Dependence*, San Juan, Puerto Rico, June 2000.
7. H.S. Lotsof, *Multidiscip. Assoc. Psyched. Stud. Bull.* **5**, 16 (1995).
8. C. Naranjo, *Clin. Toxicol.* **2**, 209 (1969).
9. C. Naranjo, "The Healing Journey," Pantheon Books, New York, 1973.
10. M.J. Stolaroff, "The Secret Chief," Multidisciplinary Association for Psychedelic Studies (MAPS), Charlotte, NC, 1997.
11. A. Huxley, "The Doors of Perception," Harper, New York, 1954.
12. T. Leary, G. Litwin, and R. Metzner, *J Nerv. Ment. Dis.* **137**, 561 (1963).
13. S. McCrosky, *Overthrow* **7**, 11 (1985).
14. P. DeRienzo and D. Beal, "The Ibogaine Story," Autonomedia, New York, 1997.
15. W.K. Schmidt, C.W. Gorodetzky, N.L. Teti, F. Vocci, S.A. Grossman, and R.S. Mansbach, "Drug Abuse Pharmacotherapy Development in the Private Sector," *in* Proceedings of the 60th Annual Meeting, The College of Problems on Drug Dependence, p. 20. U.S. Department of Health & Human Services, Bethesda, MD, 1999.
16. H.S. Lotsof, *U.S. Patent* **4,499,096**; *Chem. Abstr.* **102**,160426w (1985).
17. H.S. Lotsof, *U.S. Patent* **4,587,243**; *Chem. Abstr.* **106**,12967r (1986).
18. H.S. Lotsof, *U.S. Patent* **4,857,523**; *Chem. Abstr.* **112**,32041m (1989).
19. H.S. Lotsof, *U.S. Patent* **5,026,697**; *Chem. Abstr.* **116**.17031x (1991).
20. H.S. Lotsof, *U.S. Patent* **5,152,994**; *Chem. Abstr.* **116**,100980b (1992).
21. E.D. Dzoljic, C.D. Kaplan, and M.R. Dzoljic, *Arch. Int. Pharmacodyn. Ther.* **294**, 64 (1988).
22. S.D. Glick, K. Rossman, S. Steindorf, I.M. Maisonneuve, and J.N. Carlson, *Eur. J. Pharmacol.* **195**, 341 (1991).
23. I.M. Maisonneuve, R.W.J. Keller, and S.D. Glick, *Eur. J. Pharmacol.* **199**, 35 (1991).
24. R. Rand, *The Truth Seeker* **117**, 27 (1990).
25. B. Sisko, *Multidiscip. Assoc. Psychedel. Stud. Newsl.* **4**, 15 (1993).
26. C.D. Kaplan, E. Ketzer, J. DeJong, and M. DeVires, *Social Neurosci. Bull.* **6**, 6 (1993).
27. S.G. Sheppard, *J. Subst. Abuse Treat.* **11**, 379 (1994).
28. Court of Appeal at the Hague, *Order of the Court.* Cause List Number: 97179K09, 1999.
29. W. van den Brink, V.M. Hendriks, P. Blanken, and J.M. van Ree, *Ned Tjdschr Geneeskd* **144**, 108 (2000).
30. D. Luciano, *Am. J. Addict.* **7**, 89 (1998).
31. D. Luciano, E.A. Della Sera, and E.G. Jethmal, *Multidiscip. Assoc. Psychedel. Stud. Bull.* **9**, 27 (2000).
32. Healing Visions, "Healing Visions—Narcotic Addiction Recovery Using Ibogaine," *http://ibogaine.net,* 2000.
33. E. Taub, *Personal communication*, August 2000
34. J. Kent, "An Interview with Eric Taub," *http://www.xs4all.nl/~knehav/taubinterview.html*, 1999.
35. R. Goutarel, O. Gollnhofer, and R. Sillans, *Psychedel. Mono. Essays* **6**, 70 (1993).

36. B. Judd, "Ibogaine, Psychotherapy, and the Treatment of Substance-Related Disorders," *The Eighth International Conference on Drug Related Harm,* Washington, DC, November 19, 1994.
37. S. Grof, *LSD Psychotherapy*, Hunter House, Pomona, CA, 1980.
38. R. Yensen, *J. Psychoact. Drugs* **17**, 267 (1985).
39. R.J. Strassman, *J. Nerv. Ment. Dis.* **183**, 127 (1995).
40. J.P. Grund, *Intl. J. Drug Policy* **6**, 65 (1995).
41. J. Bastiaans, *Rev. Med. Chir. Soc. Med. Nat. Iasi* **59**, 573 (1974).
42. J. Bastiaans, "Mental Liberation Facilitated by the Use of Hallucinogenic Drugs: Psychedelic Reflections," Human Sciences Press, New York, 1983.
43. S. Snelders, *Multidiscip. Assoc. Psychedel. Stud. Bull.* **8**, 18 (1998).
44. Ka-tzetnik 135633, "Shivitti," Harper and Row, New York, 1989.
45. M. Resinovic, *Personal communication*, March 27, 2000.
46. Ustanova Iboga, "Ustanova Iboga," *http://www.ustanova-iboga.si*, 2000.
47. A. Kastelic, "Self-Help Use of Ibogaine in Slovenia," Presented at the First International Conference on Ibogaine, New York, November 5, 1999.
48. C. Waltenburg, *Personal communication*, March 2, 2000.
49. Indra, "Indra Shamanic Herbarium," *http://www.indra.dk*, 2000.
50. egroups, "Ibogaine Mailing List," *http://www.egroups.com/group/ibogaine*, 2000.
51. Danish Drug User's Union, "Drug User's Union," *http://www.brugerforeningen.dk*, 1999.
52. R.T. Lionheart, *Personal communication*, September 2000.
53. S. Witter, *The Times Magazine* (London), July 11, (1998).
54. Cures-not-Wars, "Cures-not-Wars," *http://www.cures-not-wars.org*, 2000.
55. C. Sanders, "Schmoo Connection: Ibogaine," *http://www.schmoo.co.uk/iboa.htm*, 1998.
56. N. Sandberg, "Ibogaine.co.uk" *http://www.ibogaine.co.uk*, 1998.
57. C. Sanders, Personal communication, August/November 2000.
58. N. Sandberg, "Ibogaine Supply and Treatment Options," *http://www.ibogaine.co.uk/options.htm, 2000.*
59. B. Mariano, *Personal communication*, October 2000.
60. D. Lieberman, "African Ethnobotanical Homepage," *http://ibogaine.lycaeum.org*, 1998.
61. D. Lieberman, "Technology Transfer and *Tabernanthe Iboga* Treatments: The Application of Relevant Traditional Bwiti Techniques to a Western Medical Setting," Presented at the First International Conference on Ibogaine, November 6, 1999.
62. M.J. Brendtro, *Adv. Prac. Nurs. Quart.* **4**, 54 (1998).
63. M.K. Anglin, *Soc. Sci. Med.* **44**, 1403 (1997).
64. A. McKean, "Why We Get Arrested," *http://www.actupny.org/documents/whywe%20get.html*, 2000.
65. P.S. Stansill and D.Z. Mairowitz, *in* "BAMN-Outlaw Manifestos and Ephemera 1965-1970," Autonomedia, New York, 1999.
66. R. de Jong, *in* "Anarchism Today" (D.E. Apter and J. Joll, eds.), p. 164. Macmillian, London, 1971.
67. G.F. van de Wijngaart, *Am. J. Drug Alcohol Abuse* **14**, 125 (1988).
68. S.R. Friedman, W.M. de Jong, and D.C. Des Jarlais, *in* "The Effectiveness of Drug Abuse Treatment: Dutch and American Perspectives" (J.J. Platt, ed.), p. 315, 1990.
69. C.D. Kaplan, M. Morival, and C. Sterk, Needle-Exchange IV Drug Users and Street IV Drug Users: A Comparison of Background Characteristics, Needle, Sex Practices and AIDS Attitudes, in Division of Epidemiology and Statistical Analysis, ed. Community Epidemiology Work Group Proceedings: Selected Issues: Acquired Immunodeficiency (AIDS), Drug Abuse and Crime, Ethnographic Research, Rockville, MD. National Institute on Drug Abuse, p. IV-16-25, 1986.
70. J.P. Grund, C.D. Kaplan, and N.F.P. Adriaans, *Am. J. Public Health* **81**, 1602 (1991).
71. J.P. Grund, P. Blanken, N.F.P. Adriaans, C.D. Kaplan, C. Barendregt, and M. Meeusen, *J. Psychoact. Drugs* **24**, 41 (1992).

72. H.S. Lotsof, *Personal communication*, July 2000.
73. S. Blakeslee, *New York Times*, October 27, C-11 (1993).
74. P. Park, Miami *New Times* **12**, 16 (1997).
75. B.I. Koerner, *US News & World Report* **127**, 82 (1999).
76. J. Bors, *The Baltimore Sun*, August 9, 1 (1993).

—— CHAPTER 15 ——

FROM THE ROOTS UP: IBOGAINE AND ADDICT SELF-HELP

GEERTE FRENKEN

INTASH
San Francisco, CA 94142

I. Introduction of Ibogaine and Addict Self-Help in the Netherlands

The subject of this chapter is the formation of International Addict Self-Help (INTASH), the introduction of ibogaine in the addict self-help scene in Holland, and the contribution of the addict self-help movement to the development of ibogaine treatment. It also describes the INTASH intake, treatment, and aftercare procedures and the importance of addict self-help involvement in future developments with the clinical use of ibogaine.

Ibogaine was introduced to the addict community in 1990 by Howard Lotsof and Bob Sisko from the International Coalition for Addict Self Help (ICASH) (*1,2*). The late Nico Adriaans, Josien Harms, and myself formed an informal organization, Dutch Addict Self Help (DASH), which today is called INTASH, in order to treat addicts with ibogaine. Nico Adriaans was one of the founders of the first addict self-help movement in Europe, the Junkiebond, which promoted innovations such as needle exchanges, decriminalized prostitution strolls, and the general attitude of humanization of the addict within the institutions of Dutch society, and inspired many other initiatives throughout the world (*3-6*).

INTASH performed four initial treatments with ibogaine on opiate-dependent

THE ALKALOIDS, Vol.56
0099-9598/01 $35.00

poly-drug users, some of whom had been in methadone maintenance programs for many years. The successful results of these treatments were impressive. "Successful" is, of course, a relative and somewhat subjective term. An outcome was regarded as successful if our subjects acutely experienced a complete elimination of withdrawal symptoms, or were left with only minor residual symptoms such as mild chills or some yawning, and elimination of cravings from approximately 5 months to 2 years. As Nico Adriaans often pointed out, "there is no substance known in the world today, besides ibogaine, that can eliminate withdrawal of high maintenance doses of methadone without causing extreme discomfort."

The goal of our self-help organization was to provide treatment with ibogaine in a nonjudgmental and trusting environment. Treatments with ibogaine were provided over a period of several months for a group of eight Dutch addicts, most of whom originated from the same town and social network. After being thoroughly informed during a month-long intake, all participants were treated in the same private setting. All subjects received a physical examination by a medical doctor and nobody suffered any adverse physical or mental consequences due to ibogaine treatment.

The philosophical approach to these treatments was "pro-choice" and not based on an absolute opposition to drug use; instead, the goal was to provide alternative treatment options to people that wanted to quit using drugs in an obsessive way. For example, in one particular case, a subject came through the treatment without any withdrawal symptoms, but nevertheless expressed the need to use heroin. In response to questions regarding his motivation to resume heroin use and whether he was still feeling withdrawal symptoms, he responded that he felt fine, but that the lifestyle of heroin use was still appealing to him. Since he requested to use heroin and he was not in his hometown, we actually helped him cop. The major concern was with the welfare of the subject, to make sure that he did not wander around town or get bad product. In addition, it was important to assure that he would only use a very conservative amount, since ibogaine sets the subject back to a pre-addictive state, which creates risk for accidental overdose.

All of the other subjects in the group stayed clean for an average period of six months. During this period, Dr. Charles Kaplan, an experienced international sociologist and drug researcher, suggested a focus group for those who had undergone treatment that would meet on a weekly basis. His German colleague, Eva Ketzer coordinated these meetings in order to collect data and to provide the subjects with an opportunity to share their experiences (*7*). Unfortunately, due to practical difficulties and very limited financial resources, the focus group dissolved after a several sessions.

The following 2 years focused on processing the data of these treatments, which were published in the *Journal of Substance Abuse Treatment* (*8*). An effort was also made to inform therapeutic communities and drug-abuse treatment institutions in Holland about the existence of ibogaine and to request further research. Representatives of INTASH traveled the world to participate in drug-related conferences to spread awareness of the existence of ibogaine. The reception of both professionals and the international addict community was skeptical, a kind of wait-and-see attitude and often a frankly uninterested reaction. It seemed that the professionals within the drug-treatment community in Holland viewed ibogaine as a politically difficult issue. Holland was already under a lot of pressure from the newly formed European Community to change its progressive drug policy. Taking on ibogaine, a hallucinogen, was considered too much of a leap, and the attitude seemed to be one of "let some other country take the lead this time." With the exception of a Russian and a German group, the international addict community at that time was more interested in establishing legalized methadone programs and needle exchanges in their individual countries.

It was possible, however, to interest a few key people in Holland to observe treatments or review data. The next series of ibogaine treatments in the Netherlands took place in 1992, and they were observed by the late Jan Bastiaans M.D., former chairman of the Department of Psychiatry of the State University of Leiden (9). Dr. Erik Fromberg and Dr. Delano Gerlings from the Netherlands Institute for Alcohol and Drugs (NIAD) also monitored the treatments. INTASH then moved to the United States and integrated more issues around drug abuse. In 1996 we began a Web site called "The Junkie Domain" that covered topics from art and safer injection manuals to personal reports on ibogaine experiences.

II. Intake and Treatment

There are two different types of sessions with ibogaine. People who are not necessarily substance abusers have sought to use ibogaine to benefit from its spiritual impact in what is called an "initiatory" session, in which a low dose of ibogaine is used, typically on the order of 10 mg/kg. The other method involves a larger dose, on the order of 20 mg/kg, for an addiction interruption session. The addiction interruption session is more difficult in a medical sense, because many addicts have physical and psychiatric conditions related to their substance abuse. INTASH members designed a protocol, a modification of the Lotsof procedure (*1*) to provide sessions for addiction-interruption in a semi-clinical setting. They reached the conclusion that a thorough professional physical and psychological screening was needed. The experience of observing most people ultimately relapse at different time intervals after their treatments indicated the need to provide some type of aftercare.

The intake procedure consisted of establishing a preliminary process during which the addict requesting treatment was gradually prepared, while a relationship of trust developed. The addict was thoroughly informed of the physical and psychological consequences of a treatment with ibogaine. Each person who seriously considered treatment was well informed about ibogaine, and went through an initial screening, which included a general physical exam history, a visit to a psychiatrist, and an optional visit to a psychotherapist. Laboratory evaluation included liver function tests, blood count, and an electrocardiogram.

Once the subject had passed this screening, INTASH did an unstructured life and drug history interview with the subject, which included information about the treatment procedure in order to prepare the subject as thoroughly as possible. It is hard for many hard-core substance abusers to imagine or comprehend that ibogaine will cause them to be clean from one day to the next without the pain and agony of withdrawal, especially for those who have been using high doses of methadone daily. Therefore the information given during intake encompassed many aspects and started with a clear and firm warning of the danger of using drugs, in particular heroin, during and immediately after the treatment. This warning was repeated on the day of the treatment, and is important because subjects undergoing the treatment need to be aware that ibogaine may potentiate opiates that are still in the system. Opiate use during treatment can lead to overdose.

The subjects were then told what happens during a treatment. Early in the morning, 10 hours after the last use of food and drugs, the subject took the ibogaine orally in capsule form. Sometimes the ibogaine was mixed with a digestive aid. This took place in the morning, when the subjects normally would have used their "wake-up" dose of drugs. An hour after administration the subjects usually noticed that their familiar morning withdrawal symptoms had disappeared and expressed a desire to lay down and get comfortable. A quiet, darkened room, specially prepared in a personalized, though non-distracting manner was made available for this purpose. The room was darkened because light bothers most subjects on ibogaine. The room was quiet because sound is usually experienced in an amplified and oscillating way. During movement, subjects generally experienced ataxia, which is loss of muscular coordination similar to that which occurs with drunkenness. Since the ataxia was sometimes accompanied by vomiting, he or she was asked to lay still with the least amount of motion possible. When closing their eyes, approximately 75% of subjects experienced vivid dreamlike visions. However, when subjects opened their eyes they usually reported no real visual or auditory hallucinations, and some level of communication was possible, but usually not preferred by the subject. Many subjects perspire heavily and were advised to wear comfortable shirts/pants that could be easily replaced. The first stage takes place for about 4 to 8 hours, during which the subject was regularly checked by the treatment team, and members of this team are constantly available on request. During the first stage, subjects generally do not complain about any withdrawal symptoms.

In the second stage, some individuals still experience a dreamlike period, although it is often reported as less intense. There is time to evaluate the visionary experiences, which can bring about profound insight into life and death and the reasons behind addictive behavior. Some subjects request something to drink and/or light food, such as fruit. The subject usually stays awake most of the time. During this phase some individuals complain about exhaustion, which some of them interpret to be withdrawal symptoms. It is at this stage that the presence of ibogaine-experienced ex-addicts is crucial. The trusting relationship previously established with the subject gives the guide the opportunity to assure the subject that this is a common stage and that all that is needed is some sleep. Patients and ex-addicts can relate on the basis of shared experiences, which has proven to be very effective and very important in order to prevent the subject from using any drugs that they might have concealed. In many of these cases, the subject is calmed down and sleep medication can be requested and is often advised by the team.

After approximately 20, and up to 30 or 40, hours after taking ibogaine, most subjects fall asleep for a couple of hours, with or without the help of some sleep medication. They generally awaken from this sleep feeling rested, very hungry, and in need to wash up. In the course of this day, most people are able to resume normal activities. Many subjects need to spend more time in or around the treatment environment to process what has happened to them and to adjust. Some people request to talk about their experience, others prefer privacy. Most subjects experienced complete resolution of withdrawal symptoms after treatment, while a few experienced some minor residual symptoms such as minor chills or a little yawning. An increased amount of energy and appetite and a decreased sleep requirement then continues over a 3- to 4-month period, diminishing slowly. Subjects usually stay free of cravings for several months.

III. Psychological Aspects of the Ibogaine Treatment Experience

The visual and auditory experiences that may occur during ibogaine treatment potentially include the release of repressed memories. These visions are relevant to the addiction interruption process because they seem to help the individual to develop an understanding of the underlying reasons for their addictive behavior. Subjects were asked what their expectations are around these possible ibogaine visions. Since many addicts use drugs for their consciousness-suppressing qualities, some of them express fear of ibogaine's mind-altering effects. It is then explained to them that the visions do not seem to be experienced as intensely emotional. That is to say, there is no element of release of emotions like laughing or crying as is reported with many hallucinogens. Besides, many of the repressed memories that are released are positive, since most addicts have been dwelling on the ones that are negative.

It is important to explain the similarities between an addiction interruption session and the use of ibogaine in the African tribal tradition. Some West African tribes have used ibogaine for centuries as a form of initiation that occurs once in a lifetime when a young person is to make their transformation into adulthood by reviewing their past and to "restore communication with the ancestors." (see the chapter by Fernandez and Fernandez in this volume). People taking ibogaine for addiction-interruption purposes describe the visionary and auditory elements of the experience as a state of "dreaming wide awake." Visions can occur in a repetitive mode. They often report visualizing a rapid run-through of their lives and/or the lives of family members, even of those who have already passed away. They have noted the experience of going both backward and forward in time and being able to come to an understanding of their spiritual roots. The term "spiritual" does not necessarily mean religious, but a heightened level of awareness. I like to call the experience a "journey into one's DNA."

The possible amount and intensity of released material can be so overwhelming, that people have said that they simply could not remember everything they had seen, or that it took months to remember certain visions. Therefore, the processing of released material, and the ability to verbalize these matters and learn to interpret their often symbolic content, can take extended amounts of time and continue over years. Subjects have reported experiencing a mental or spiritual transformation due to the ibogaine, which they compare to 10 years of therapy in 2 days, or taking a "truth serum." Whatever people report on their experiences, they have been observed returning from their ibogaine experiences with a greater understanding of choices they have previously made. However, this does not mean that the ibogaine experience offers them the skills to interpret and approach this material in a constructive manner that can lead to positive and productive solutions and changes in the life after treatment.

For many people, ibogaine treatment itself is not enough to maintain a substance abuse-free life. Most subjects require some type of aftercare in which these and other matters are addressed. Psychotherapist Barbara Judd, who has been working with substance abusers for more than 15 years and who has treated people before, during, and after ibogaine treatments for more than 6 years, is of the opinion that a person treated with ibogaine is more ready and willing to undergo therapy sessions compared to the average recovering drug abuser (*10*). Many addicts who have used ibogaine seem to be able to access sensitive material that lies at the core of their addictive behavior without the usual feelings of trauma and fear, and the need to anesthetize these feelings with drugs as a way of defense. Their newly acquired knowledge and attitude can save the therapist time in terms of confronting the individual with possibly painful issues. In case there are traumatic issues, they need to be worked through in order to break through the cycles of self-destructive behavior and find new, positive ways to approach life and its problems. Subjects are stimulated to seek out or create support networks, which could range from attending Narcotics Anonymous meetings to organizing ibogaine support groups of their own.

IV. Aftercare

The aftercare strategy is defined through collaboration with each subject during the intake phase and after the treatment. Individualized aftercare plans are based on the life and drug history taken earlier in the interview and the subject's present situation. Any form of aftercare is, of course, optional and it's up to the subject to follow through in whatever way the subject feels is necessary. Motivation to design an aftercare strategy and intentions to follow through on such plans are taken into account when reviewing the eligibility of each individual requesting treatment. Some people might need a therapeutic community, others a halfway house, and yet others just manage on their own. What we try to do is make people aware before the treatment that taking ibogaine involves a commitment to a new way of living, that ibogaine is not just a "quick fix," and that staying clean is based on a profound change of attitude toward physical, mental, and emotional wellbeing.

Crucial aspects of aftercare that need to be considered include housing, education, jobs, and the psychological consequences of assimilation back into relationships, the family, and the community. If unanswered, these matters could otherwise ultimately cause relapse to old behavioral patterns. Based on the psychiatric evaluation, some subjects need to be made aware of options like antidepressants, nonaddictive antianxiety medications, and so on. Subjects are made aware of the availability of some fairly new antidepressants that are currently on the market, which are particularly suitable for recovering addicts. They may be helpful in mitigating possible chemical imbalances in the brain produced by the extensive use of hard drugs. Stabilizing dysregulated neurotransmitters is not only important in terms of treating depression, anxiety, and other symptoms caused by extensive drug addictions, it is also crucial in terms of dealing with psychotherapy in an effective way.

All subjects receive a list with important issues relevant to any recovering substance user/abuser, including suggestions regarding physical well-being, such as how to eat healthy, the need for exercise, how to deal with hypoglycemia, information on vitamins, the benefit of sauna, and so on.

V. Conclusions

Having covered the intake and aftercare protocol used by INTASH, the need is emphasized to include ibogaine-experienced ex-addicts in the process of treating substance abusers with ibogaine. Even if ibogaine should be made available through the medical establishment, it is crucial to do so in cooperation with ibogaine-experienced addict representatives. The presence of these peer counselors is very important because there is a possibility of a trust relationship, which reduces possible risks and optimizes the chances of a successful outcome. The use of peer counselors is a convention that is widely used in the field of treatment and harm reduction as pointed out and applied by people like Dr. Vincent Dole and Nico Adriaans (*3,11*). Most substance abusers prefer to receive treatment in the presence of former addicts who are experienced with the treatment procedure, because they can relate to each other through similar experiences. In a world where addicts have constantly been submitted to rejection and secrecy, many have developed low self-esteem. We therefore find it crucial to provide a treatment environment that is nonjudgmental, in which addicts feel respected and free to express themselves, and where the right to choose is always present. For example, most addicts will not change their behavioral patterns if they are being pushed into treatment by family, friends, or the judicial system. Being prepared for treatment with ibogaine means being ready and willing to take a physical and spiritual leap forward. It is therefore important that the treatment team includes ibogaine-experienced ex-addicts in order to provide loving and understanding support and guidance, in which mutual trust is the central issue. When the treatment is completed, a process of self-discovery and self-realization can start to develop, in which it is vital that the former addict can relate to others with a similar experience in order to prevent feelings of alienation from his or her environment. This has been done by creating focus groups where people can share this common ground, or by treating several members of one particular scene of drug users.

In my opinion, there is no substance in the world today as effective as ibogaine in combating addiction to opiates and other drugs of abuse including methadone, cocaine, amphetamine, alcohol, and nicotine. However, ibogaine is not a "cure" in itself, but potentially a very effective part of a larger treatment scheme. It can therefore also play a role in the prevention of the spread of drug-related infectious diseases, like the HIV virus amongst i.v. users. Even if people do decide to return to drug use after treatment, they usually find that they need less of the drugs to get high, because they have less tolerance, and because ibogaine seems to diminish the need to use drugs. Ibogaine is a potentially powerful harm reduction and relapse prevention tool. A clinical argument can be made that ibogaine is safer and more effective than ultra-rapid detoxification with naltrexone or naloxone.

Ibogaine is potentially cost-effective and should be available to the estimated 200 million addicts in the world today. Are different countries around the world playing the waiting game as to who is going to test and market ibogaine first, as seems to be the case with Holland? Is the United States waiting for another country to take the lead? Are we going to let it? While we wait, let's consider the outcome in, for example, Russia or Eastern Europe where the prevalence of substance abuse and HIV infection has taken on epidemic proportions (*12*). Let's not forget that the heroin epidemic in Eastern Europe began only a few years ago and that both clean needles and treatment centers are not widely available. You only have to guess the statistics in a few years from now to know that the results are going to be very tragic. All too often ibogaine has been approached from a political perspective instead of one of medical necessity. The question of how to make ibogaine widely available as soon as possible, and in the most effective way, should not be decided by politics.

Ibogaine already is available in the currently existing international black market where it is being bought and sold in the streets and used without the proper medical screening and attention that is needed, which can lead to possibly hazardous situations. For example, there are reports of a French organization that actually takes addicts to West Africa to chew on the root in the bush, or people in Europe who sell ibogaine on the street. Ibogaine should not become just another illegal street drug with an antisocial stigma attached to it. Issues like the quality of product that is used, or the dose range, or undiagnosed physical or mental health conditions of the people who choose to take it in an irresponsible, unprepared manner can lead to negative consequences. The main concern should be the safety of the substance abusers. Possibly hazardous outcomes might lead to further delay in proper testing of ibogaine by the appropriate authorities. Only through adequate testing through Food and Drug Administration (FDA) medication testing procedures can the safety and dose range of ibogaine be established and the compound produced by pharmaceutical companies. It should then be implemented in clinical settings and the currently existing detox and treatment centers. Since there has been no funding for an FDA-approved trial, there is now the need for other sources of funding to finish these testing procedures. Spare change anybody?

References

1. H.S. Lotsof, *Multidiscip. Assoc. Psyched. Stud.* **5,** 16 (1995).
2. B. Sisko, *Multidiscip. Assoc. Psyched. Stud.* **4,** 15 (1993).
3. J.P. Grund, P. Blanken, N.F.P. Adriaans, C.D. Kaplan, C. Barendregt, and M. Meeusen, J. *Psychoactive Drugs* **24,** 41 (1992).
4. J.P. Grund, C.D. Kaplan, and N.F.P. Adriaans, *Am. J. Public Health* **81,** 1602 (1991).
5. J.P. Grund, *Intl. J. Drug Policy* **6,** 65 (1995).
6. G.F. van de Wijngaart, *Am. J. Drug Alcohol Abuse* **14,** 125 (1988).
7. C.D. Kaplan, E. Ketzer, J. DeJong, and M. DeVires, *Social Neurosci. Bull.* **6,** 6 (1993).
8. S.G. Sheppard, *J. Subst. Abuse Treat.* **11,** 379 (1994).
9. K.R. Alper, H.S. Lotsof, G.M.N. Frenken, D.J. Luciano, and J. Bastiaans, *Am. J. Addict.* **8,** 234 (1999).
10. B. Judd, "Ibogaine, Psychotherapy, and the Treatment of Substance-Related Disorders", *The Eighth International Conference on Drug Related Harm*, Washington, DC, 1994.
11. V.P. Dole and M.E. Nyswander, *NIDA Research Monograph No. 43* 5, 1983.
12. K.L. Dehne, J.P. Grund, L. Khodakevich, and Y. Kobyshcha, *J. Drug Issues* **29,** 729 (1999).

—CHAPTER 16—

CASE STUDIES OF IBOGAINE TREATMENT: IMPLICATIONS FOR PATIENT MANAGEMENT STRATEGIES

HOWARD S. LOTSOF
NORMA E. ALEXANDER

NDA International, Inc.
POB 100506
Staten Island, NY 10310-0506

I. Introduction

Ibogaine's development as a putative treatment of substance-use disorders may certainly be described as unusual. In Western Europe and the United States, the use of ibogaine originated among individuals using drugs for the purpose of altering consciousness in the early 1960s, a period identified historically with the widespread introduction, and ethnographic and ethnobotanical studies of hallucinogens (*1,2*). The work of Timothy Leary and R. Gordon Wasson (*3-5*), as

THE ALKALOIDS, Vol.56
0099-9598/01 $35.00

well as the media attention focused on psychedelics, led to the establishment of the group, in which ibogaine's apparent utility to treat opiate, cocaine, and amphetamine dependence was first described. Historically, the lines of this research can be traced to Aldous Huxley and Lewis Lewin and, with some liberty, into prehistory (*6,7*).

This chapter reviews the use of ibogaine in three different contexts; a drug user group, a self-help organization, and a clinical research setting. Each section is followed by the self-report of a subject who has taken ibogaine within the setting being reviewed. These settings contrast with the use of iboga in Gabon, Africa as a practice of the Bwiti, an African religion sometimes referred to as an initiation society, as documented by Fernandez and Gollnhofer, during which ibogaine-containing plants are provided in rites to assist in the transition from adolescence into adulthood, for psychiatric healing, or for other purposes (*8-10*).

II. Drug User Group

Research on ibogaine as a putative treatment for addiction began in the United States with a group of lay drug experimenters organized in New York in 1962 by the first author (HSL), who was 19 at that time. The group had formed to study a topic of common interest, namely, the evaluation of the subjective effects of psychoactive drugs. It consisted of a core of 20 individuals, who were mainly Caucasian, in their late teens and early 20s, and were attending or had attended college. Seven of these individuals were heroin dependent. During this period, which preceded the scheduling of "psychedelic" and other psychoactive drugs, the group attempted to evaluate a variety of agents obtained from botanical and chemical supply houses. The method of "evaluation" consisted primarily of discussion comparing the subjective effects of various categories of drugs, such as stimulants and hallucinogens. The group shared an interest in the psychotherapeutic potential of hallucinogens.

The effects of ibogaine were entirely unknown to this group's participants, who ingested it in escalating dosages ranging from 0.14 mg/kg to 19 mg/kg. An apparent effect on opioid withdrawal was noted by the heroin-dependent subgroup, and these observations became the basis for subsequent research on ibogaine in the United States. The oversight provided by the group consisted of a single individual taking notes on the experiences reported by the person taking ibogaine during the period of the strongest visualization effects, which was generally the first 4 hours. Post-treatment discussions would take place among the group members after their ibogaine experiences, the interpretations of which tended to reflect the group's common interest in psychotherapy. The absence of any philosophical conflict over drug use provided a permissive, nonjudgmental atmosphere in which individuals could disclose information relating to effects on craving. Those who did return to heroin use offered a candid explanation that they were doing so because they identified, and accepted themselves, as heroin addicts.

The following self-report of a heroin addict treated with ibogaine is comprised of posts to a virtual Internet drug user's group. This particular individual took an initial dose of ibogaine, followed by a second dose 17 days later, and then took a third dose 56 hours following the second dose. He then reports on his state 41 days following the third dose.

A. Self-Report (Anonymous)

I was lucky enough to be able to obtain some ibogaine to try and help treat my heroin addiction.

Briefly here is how it went. I took the ibogaine and laid down. It took effect within a half hour or so. I had a sensation as though I were rocking, gently forward and backward. The rocking sensation would slowly increase in intensity, but not in speed. Then I would have this weird image of a twisted stick or root being shook rapidly, accompanied by a deep sound, something like a didgeridoo. When I had this image the rocking would stop. The image lasted about 3 to 5 seconds. After the image left my mind, the rocking would start again, slowly forward and back. The rocking would go on, increasing in intensity for about 20 minutes or so, then the image of the shaking root and the sound while the rocking stopped, and then started all over again. I kept feeling like something else was going to happen although nothing else did. This went on for what seemed about 5 or 6 hours.

Then I fell asleep. I had already closed my eyes after about 3 hours of this, and I don't know if it was the rocking sensation or what, but I simply fell asleep. Now I am very disappointed that I did because I feel like I missed out on most of the beneficial effects of the drug. I had no astounding revelations, nor did I really even get to analyze any part of my drug history or motivations or anything.

The next 18 hours or so consisted of intermittent waking from bizarre dreams. Every time I woke, I would be very confused and had trouble discerning reality from the dreams. None of the content of any of the dreams is memorable to me now, though. Each time I woke, I was extremely exhausted and fell back asleep within minutes. At one point I got up to go to the bathroom and was very ataxic. By the time I woke and didn't feel too exhausted to stay awake, the experience was pretty much over—just about 24 hours, which is somewhat shorter than what I understand to be the normal duration (36-48 hours). There were other slight lingering effects for another 6 hours maybe, but they were very minor.

I was extremely disappointed in myself for sleeping through the drug's effect, as I did not get any of the mental resolve and self-insight or other effects that many ibogaine users report. After waking, I had a runny nose and watery eyes and I was truly afraid that it had not attenuated my withdrawals. However, those symptoms resolved within 3 to 4 hours, and no other withdrawal symptoms at all occurred. I was very happy about that.

So that was it. I don't feel as though I have gained anything in the way of mental, emotional, or spiritual help in staying off of drugs, which makes me fear that it didn't really work other than to get me off of the dope for right now. So I am hoping that AA will help me stay clean. I am going to try and get a hold of a second dose, but I doubt I will be able to.

I kind of feel like I blew a once-in-a-lifetime opportunity. But I didn't think that sleeping on that stuff was even possible as I have never been able to fall asleep on other psychedelics, so when I closed my eyes, I made no attempt to try and remain awake. And after I had already slept on it, the exhaustion I felt each time I woke was overwhelming.

. . . A while back I wrote about the disappointment of my first ibogaine experience. Well I have much better news this time. Overwhelming success is what I would call it.

My second experience was very different from my first and came 2 weeks and 3 days after my first experience. I had already decided to try and not use heroin for a day before taking the second dose to try and avoid sleeping through the experience like I did last time. I made it about 20 hours and was already pretty sick from withdrawal by the time I took the ibogaine. My second and third doses were both smaller than my first dose being about 14mg/kg.

The initial few hours were similar to my first experience, with a strong feeling of being, well, I guess, drugged. After the first few hours I spent the next 20 hours or longer, I really can't remember how long but it was, thinking about everything under the sun; actually I spent some time thinking about the sun itself too. I had lingering effects this time also, mostly visual effects in dim lights, but also an almost complete inability to sleep, and occasional waves of warmth or chill across my skin, especially my scalp.

I had initially planned on taking my third dose somewhere between a week and a month after the second, but I had time off from work and decided to take it during this period since it is unusual for me to have even 2 days off in a row. This may have been a mistake, but it doesn't seem to matter to me much now. So I took the third dose a little over 56 hours or so after the second. The reason I say it may have been a mistake is because the lack of sleep has lasted long enough to grow uncomfortable.

My third experience was pretty much the same as the second, both in effect and duration. It has now been about 60 hours since I took the third dose and I still have the same lingering effects I listed above.

I have had absolutely zero desire to touch heroin since I took the second dose.

It has no temptation for me at all. So despite the inability to sleep I am very pleased with the results.

Well, there you have it. It does work. And it is also true that multiple doses can be effective when a single dose is not. Even though it is 4 a.m. and I haven't slept more than about 6 hours in 4 days, I am extremely happy with how things turned out.

. . . I fell off the wagon, got back on that horse, went back out and whatever other euphemisms you can think of. I had 41 days off the shit today. I went out to buy some freaking CDs with no intention of using, and somehow I came home with smack. I had to go to four places to find the two CDs I was looking for, and somehow while I was driving around the thought to use occurred, which then turned into a craving, and finally a deep need. Or so it seemed anyway. I am seriously bummed by this turn of events. At least I was smart enough to know my tolerance would be low, but damn I am flying off the smallest piece. So now I am seriously considering flushing the other 95% of what I bought. Sorry to say that in here, I know just about everyone who reads that will wince at it, but there it is. I only know where to get black tar around here, so I had to buy a $25 blob (the smallest amount you can get) and I don't want to have any of this shit left over tomorrow or I will use it. And I doubt I could finish it tonight without getting really sick

. . . Well I got sick as a dog and flushed the rest of the shit. Man, I didn't even enjoy the buzz. My lack of tolerance pretty much made it so that I have just been puking with a throbbing headache. I am back on the wagon, off that horse, and whatever other euphemisms you can think of for quitting dope.

I have heard from some people that their past is played out something like a movie, but I never experienced it quite like that. I did have certain things that I had more or less repressed, things dealing with my parents and my childhood, that came to the surface and I was forced to deal with them. Some of that stuff is quite unpleasant but I always feel better later. Someone put it like this and I found it fairly accurate. He said it was like taking out your emotional carpet and beating it with a broom so that all the dirt comes out. And you do feel somewhat "cleansed" afterward . . .

As far as it wearing off, you're just talking about the relief from cravings. Damn, getting off heroin with no withdrawals is reason enough to take it, the relief from cravings is really just a bonus. And a heroin addict really can't expect that just taking some other drug is going to cure them of their addiction. I mean, there are consequences of becoming an addict. Besides, the longer you go without using, the less you crave. So if you got 6 months free of cravings, you should be able to resist the urge to use if it did reappear. I had a month free of cravings and then when they did come back, I was so unaccustomed to them I think I was actually surprised to have them. Now I crave now and then, but I know that if I can resist for a few hours at the most, they will go away. And there were some mental changes (as far as attitudes and stuff) that I think were probably permanent. No acid trip I ever had made any permanent changes in my attitudes. That to me makes ibogaine seem like a very powerful drug.

III. Self-Help Organizations

In the late 1980s, The International Coalition for Addict Self-Help (ICASH) initiated a small number of ibogaine treatments among heroin dependent drug users, which appeared to confirm that ibogaine would eliminate narcotic withdrawal signs and interrupt drug craving (11). More than 25 years had passed between the original observation of ibogaine's ability to interrupt opioid and cocaine dependence and the revisiting of those findings by a second group of drug users. The work of ICASH can be viewed historically in a context of AIDS patient activism and the activities of advocacy groups such as ACT UP.

ICASH sought out countries with drug policies that were not hostile to drug users and made contact with drug user activists in The Netherlands. The first and most influential of these contacts was with Nico Adriaans, a fieldworker for the European Addiction Research Institute (IVO) at Erasmus University in Rotterdam, and one of the founders of the Junkie Bond or Junkie's Union as it was popularly called (*12*). Shortly after contact between ICASH and Adriaans, Adriaans was treated with ibogaine and thereafter established Dutch Addict Self-Help (DASH) with G. Frenken and others. DASH's principal goals were to supply ibogaine to Dutch heroin users and to influence the Dutch government to authorize ibogaine as a medicine.

After a year of discussion concerning the background and merits of ibogaine with the late Professor Dr. Jan Bastiaans of Leiden, The Netherlands, Bastiaans agreed to attend and observe treatments for patients recommended by NDA International, a U.S. corporation involved in the attempt to develop ibogaine. NDA informed DASH of this agreement, and DASH, in turn, referred people who wished to be treated to Bastiaans for ibogaine therapy. Other Dutch physicians and researchers became interested, and in some cases they observed ibogaine therapies. The self-help organizations, during their work in the late 1980s and early 1990s, provided further case study evidence of the antiaddictive effects of ibogaine.

A. Self-Report (Anonymous)

The following article was written in 1990, 6 months after I underwent treatment with ibogaine in an attempt to curtail my heroin use. I was born and raised in Amsterdam and was 26 years old at the time.

I heard about ibogaine from a friend in New York, and then contacted the International Coalition for Addict Self-Help (ICASH) to request treatment for me and my boyfriend. We were the first people to be treated in Holland. My ibogaine treatment took place on October 25, 1989, in a hotel room in Amsterdam. My boyfriend had been successfully treated the day before.

The night before my treatment, I was given a small oral dose of 100 mg of Ibogaine to see if I would have an allergic reaction, which I didn't have. After an hour, I had strong memories of my childhood. I was walking through the house I was raised in. This kind of memory was a new experience in the sense that I actually viewed the interior of the house at the visual height of a child at age 4. While walking, I recalled all kinds of details in the house, which I never expected to be relevant. I experienced how my parents must have seen me when I was a child.

Before the treatment I was told that, like in a movie, I would relive certain events in my life and I would experience repressed memories. In my experience, it didn't happen in a chronological way. At 10 o'clock in the morning I take 1200 mg of ibogaine in capsule form with some tea on an empty stomach.

I wait for a flow of memories. Twelve hours have passed since I took my last dose of heroin, therefore I am experiencing withdrawal symptoms.

After about an hour, I start visualizing pink diamond shapes. My body feels quite heavy, but I am still able to coordinate my functioning. For about an hour I am being checked on by the person who is guiding me through the treatment. To me, his appearance now resembles a pygmy. He wants to see if I start walking wobbly, one of the symptoms that ibogaine is taking effect. I am told to walk through the room several times. This request bothers me; I don't want to be disturbed. Even though the ibogaine is affecting my coordination, I keep walking straight lines. I want to show that through willpower, no drug has to influence you if you don't want it to. Through this experience I realize that the same goes for all the other drugs you can take. There is one eternal aspect in yourself that is unchangeably present. My conclusion was, "why take drugs to suppress this state of consciousness?" I also realized the enormous possibilities of a mind that is crystal clear.

In the following four hours, stroboscopic flashes of remembrance happen to me in visions and sounds. Sounds are particularly irritating to me because they echo back loudly in an oscillating way. There is a constant zooming in the room, as if there is a gigantic fly in the top corner of the room behind me. It makes me think of the writer Carlos Castaneda when he described the fly as a guard "between two worlds." I resent the idea of experiencing "this older world." In the meantime, I have already reached its vast planes.

I see several rolls of film unfold from my head through the room, displaying cartoons. I notice that the humor in these cartoons is mostly based on violent interactions, and realize that these are the first imprints in the mind of a twentieth-century child like myself.

There are sounds in the back of my head. The more I concentrate, the louder they become. There are sounds of African drums, and immediately I have visions of walking through the jungle of dark Africa. I hear a neighboring village transmit a message on hollow tree-trunks and I play them a message back. It's like the rhythms have always been in my head, they just needed to be relived. Totally realistically, I'm walking barefoot through the jungle and dew from the leaves drops on my skin. I'm scared of the possibility of getting attacked by wild animals or stepping on a snake. This is why the visions of Africa make me uncomfortable. For the first time in my life, I am grateful to be born white in the 20th century in Northern Europe. I realize the enormous comfort and technological possibilities I am able to enjoy in my current life.

I sit on my bed and bend my hands in front of my mouth, blowing through them as if it is a large tube or pipe directed to the ground. I am producing vibrating tunes and hear wailing sounds in the back of my head, but I don't understand why. Months after the treatment, while on a trip in India, I met a German man in Delhi, who played an Australian instrument called a "Didgeridoo." This is the first time I learned of the existence of such an instrument. It's a tube-like branch of a tree, 6 feet long, that is hollowed out by termites and rubbed with beeswax. The instrument is played by putting the top part against your mouth and pointing the bottom down, in order to create a vibrating, wailing sound while blowing through it.

Two hours have passed since I have taken ibogaine. My stomach is upset. I throw up a little. I have a vision of walking through my brain, as if walking in a giant computer-like file cabinet. There are long narrow drawers with selected, collective information, to be opened on request. Somebody in the hotel turns on a radio and commercials are playing. Immediately a drawer in my brain is opened and all the jingles I have heard in my life come out as one long song. I realize the incredible amount of bullshit that is taking the place of more important information.

Four hours after taking ibogaine, I throw up twice. The person that is supposed to guide me through the treatment has fallen asleep, and his snoring disturbs me. On top of that, my boyfriend keeps interrupting me in an effort to share his enthusiasm of being clean. Where the hotel had provided a quiet setting for him the previous day, activities have broken loose on this Monday. Maintenance people are washing windows, vacuum cleaning hallways, and cutting trees in front of the hotel. I decide to go home to Utrecht. I leave my sleeping guide a thank-you note, in which I wish him fun exploring new, freaky cultures in Africa with his friends. On the train home I see a lot of people who I experience as being "dead in the head." I feel intensely connected with black people.

At home, I don't feel good. This gives me the feeling I have been cheated. I am not supposed to feel withdrawal now. Everything seems very awkward and I try to throw up. Throwing up helps me feel relief, but everything tastes and smells bitter from the ibogaine. In an attempt to get rid of ibogaine's effects, I decide to cop some smack (heroin), after which I feel less anxious and more relaxed, though still trippy. Later that night the effect of the heroin finishes and I lay down on the couch and fall into a dreamlike half-awake state.

I see myself laying on the couch, which is followed by a vision of myself as a fetus in my mother's womb. I am actually, very rapidly, going through a rebirthing process. I feel an incredible devotional love coming from my parents. This memory enables me to accept the mistakes my parents have made in raising me. For the first time I can feel respect for my parents, which shapes our whole relationship into a more harmonious reality.

Many other dream flashes appear. The next morning I awake fully refreshed, newborn, and hungry as a wolf. I give my heroin away to my roommate. My boyfriend and I start to evaluate our experiences. It is as if new things keep falling into place. It's as if all information in your brain file cabinet is shaken out of it's drawers on to one big pile, looked at "objectively" and put back in, untwisted from emotional trauma.

It takes time to realize that we're not getting sick. There is no need to arrange money to run to the dealer anymore. That time can now be used to prepare our planned trip to India and Nepal. The following days go by in an up and down rhythm. One day is incredibly energetic and active; the next one is needed to relax. We both feel very positive, joyful and enthusiastic. A withdrawal never took place just some occasional yawning and minor chills. In a normal withdrawal, you need all your motivated energy to go through being sick, which burns you out completely. This time, the motivated energy is reinforced, and together with all of the visual experiences, it puts you on the path directed toward your goal.

Initially my junkie friends were very skeptical, until they realized that my boyfriend was selling his daily portion of 65 mg methadone every day, for weeks in a row, and he was not spending his money on smack, cocaine, or alcohol, but on traveling gear. Some of them thought our enthusiasm was irritating. Others wanted to experience ibogaine too, and it felt very frustrating that I couldn't give it to them.

The presence of hard drugs in my environment after the treatment was not threatening in any way. It didn't seem particularly positive or negative. It just didn't matter anymore. I did use some smack to see if I would still like it, but I didn't care for the effect anymore. It actually seemed like it reactivated the ibogaine. Up until 4 months after the treatment, I experienced colors and light very intensely. I never experienced any negative side effects, mentally or physically, after ibogaine. I've noticed that I'm not sensitive to the influence of drugs as I used to be. I lost a great deal of interest in drugs in general, because the effect of ibogaine goes far beyond their effect, though not necessarily in a pleasant way. Ibogaine is quite an ordeal; therefore I hope I don't ever need to use it again. It is not possible for me to resume the same addictive personality, unless it would be my conscious choice. Ibogaine has given me this choice. Heroin never did. Momentarily, I can use any drug without being used by them.

It is through the treatment that I don't experience events in the past as problematic anymore. I experience the present with the past as reflection. The past therefore is no longer perceived as an obstacle, but as a source of collective information. The realization of the collective consciousness is a mystical, religious experience. It confirms a unity with all living beings and old feelings of separation between "you and the outside world" disappear.

Ibogaine was a mental process for me, a form of spiritual purification and a truth serum. I had to experience its results through time. It's only now after 6 months that I can say I'm not addicted anymore. It takes time to admit there is no way back. Ibogaine is not the solution in itself, although it takes withdrawal away completely and gave me clues that made it possible to figure out why I got strung out in the first place. Ibogaine made it possible for me to accept life on its own terms and access the willpower inside myself that I needed to pick up where I had left off.

After the treatment I was clean for about a year. I got retreated, but relapsed in a matter of weeks as a result of lack in aftercare. In that time treatments were still very experimental. As I treated many other addicts, I realized that in order to stay clean, most people need some kind of therapy. Besides a quick and effective detox, ibogaine can offer a lot of information to the underlying reasons for becoming a junkie, which can be helpful in working with a therapist. I eventually quit my addiction the "old-fashioned way," with the use of some methadone and pills. After my first treatment with ibogaine, I was so impressed that I started treating other addicts. Together with Nico Adriaans and Josien H., I set up an addict self-help group and we treated many people. We learned how important it is to provide treatments in the presence of ibogaine experienced ex-addicts and to provide aftercare.

Today we are called INTASH (International Addict Self-Help) and work to establish worldwide ibogaine treatments.

IV. Regulatory Development and Clinical Research Settings

The Dora Weiner Foundation (DWF), a philanthropic organization, was established in 1983 for the purpose of promoting the formal regulatory development of ibogaine. The fundraising efforts of the foundation were not successful. The general public appeared to have little more than an adversarial interest in drug users, whose condition they felt was self-induced and were deserving of little or no comfort. The pharmaceutical industry as a whole had little interest in the treatment of chemical dependence, due to the liability associated with the patient population of drug users and the desire to avoid the stigma attached to chemical dependence. Additionally, there appeared to be little profit in a medication like ibogaine that would be provided only a few times to the patient.

DWF found that though sizable grants were being provided for drug use prevention, there was no interest from the major philanthropies to make funding available for the development of medications to treat chemical dependence. The DWF became inactive, and in 1986, a for-profit corporation, NDA International, Inc., was established to raise the necessary financing to meet FDA regulatory requirements for the approval and marketing of ibogaine.

In late 1990, Howard Lotsof, president of NDA International, Inc., contacted Dr. Charles Grudzinskas, then a vice president with Lederle Laboratories. Dr. Grudzinskas asked that he be sent material on ibogaine. In a discussion some weeks later, he informed Lotsof that Lederle was not interested, but that as of January 1991, he would be named as the director of the National Institute on Drug Abuse's (NIDA's) new Medications Development Division (MDD), and to contact him once he was in office. That contact resulted in MDD/NIDA requesting NDA International to submit a Product Profile Review to MDD, and thus began the clinical development stage of ibogaine. This eventually led to FDA approval of an IND, in 1993, for University of Miami personnel under contract to NDA International, Inc., to initiate a Phase I study of ibogaine in human subjects.

The use of ibogaine in a conventional research setting has occurred in the Republic of Panama, on the island of St. Kitts, and with FDA approval at the University of Miami (*13,14*). The hospital setting of these treatments contrasts with the nonhospital environment characteristic of the informal treatments in the Netherlands and elsewhere (*15-17*).

A. Self-Report

The following report is from a hospital setting in the Republic of Panama. The patient is a physician who had become dependent on 600 mg of Demerol a day and had attempted to stop his drug use a number of times without success. One particular interest we had in this subject was the hope that, as a medical doctor, he might provide us with some professional insight into the results of his treatment. He kept notes and provided a report on the four different doses he received, which is presented in its entirety. This subject proved to be more sensitive to ibogaine than any other individual in our studies conducted outside the United States, and he had a full-blown subjective experience from a 10 mg/kg dose. The patient had participated in a research protocol, which called for an intermediate dose of 10 mg/kg of ibogaine, which was administered as part of a pharmacokinetic study and was not expected to have a therapeutic effect, but it did. As part of the protocol, patients would then be administered a known therapeutic dose of 20 mg/kg.

Needless to say, this patient enjoyed certain advantages that affected his treatment outcome. He had a career, was highly motivated, and had available significant psychosocial supports needed by so many others who do not have his background.

(Anonymous)

First Day: 100 mg (test dose #1). I've taken my ibogaine dose, and went to bed, and stayed laying down. I felt nothing, until the medical staff arrived to do the 1-hour tests. I was surprised, because in my mental measurements, I thought I had taken ibogaine about 20 minutes earlier. When I stood up, I felt a little drowsiness, and it was difficult to walk in a straight line. I was feeling photophobia and every little noise seemed to be much louder than in reality. The sounds were very disturbing to me.

During the 2-hour testing, symptoms were worse. It was very difficult to walk in a straight line, and the room seemed to beat, like a heart. I felt very tired, and the only thing I wanted was to rest in bed. Each head movement seemed to make things worse.

When I stood up for the 3-hour test, I felt that the symptoms were disappearing. I was very hungry and ate. After eating, I was a little nauseated. For the following hours I felt nothing, except for a sensation that my mind images were richer in details than before, like a 3-D movie.

I ate with no nausea, slept very well, and awakened in very good condition.

Second Day: 25 mg (test dose #2). After this dose of ibogaine I felt nothing different from my normal state.

Third Day: 10 mg/kg (experimental dose). For the first 2 hours I felt a little different, like I had smoked marijuana. I was very calm and relaxed and all the tension of the beginning of the procedure was gone. The room seemed to be a little different and the colors around me sharper than normal. The lights and sounds were disturbing to me, like the first time. Suddenly, with my eyes closed, I began to see images that appeared in screens, exactly like TV or cinema screens. These screens were appearing in small sizes and then they would get bigger as I focused my attention on them. Sometimes they appeared small and would then begin to grow, like I was walking in their direction, and sometimes they were going from left to right, in a continuous way.

The images on the screens were moving in slow motion and were very sharp and well defined. I saw trees moving with the wind, a man with bells in his hands, various landscapes with mountains and the sunset. At this time I was a little nauseated, and when the doctors asked me to stand up for some tests, I vomited. From all of the hundreds of images I saw this day, I recognized only two: the first, an image of myself as a child, static like a photo. This image began to approach me and get bigger, but something in the room happened and I opened my eyes, losing the image. The second image I recognized was one of some horses dancing in a circus. It was a TV show that I had seen two days before. The time seemed to go very quickly, because after about 4 hours (in my mind), they told me I had taken ibogaine 9 hours earlier! It was very difficult for me to speak in English or in Spanish. I was only able to speak in my native language. At this time the images started to appear at a slower rate and for another 2 hours I saw only screens with no images on them. About 10 to 11 hours after the beginning of the experiment, they disappeared.

I ate very well and stayed awake all night long, falling asleep only about 7 a.m., almost 24 hours after the medication had been administered. During the night I had some insights about my life and about the things I realized I was doing wrong. I stayed all the following day very tired, sleepy, but very happy and relaxed, in a way I never was before.

Fifth day: 20 mg/kg (therapeutic dose). The first 3 hours were similar to the last time: photophobia and a bad sensation with little noises. After that the images began to appear, in a slower rate than the other time. There were less images, but I was recognizing all of them as part of my childhood. I saw myself playing in my father's farm, riding a motorcycle, playing with a cousin, feeding a fish and other things. I saw some recent images, like one of my father, laughing in the living room of my house. This happened about a year ago. I understood that I had a happy childhood, and there was no one to blame for my addiction, only myself. I felt their love coming from my parents and relatives. I was feeling the same time distortion that I felt the other day, and after many hours I suddenly had an insight. It was that my mind and the universe were the same thing, and that all the people in the universe and all things in the universe are only one. I saw many mistakes I was doing in my life, so many attitudes I could not have, and this helped me to decide very strongly that I will never use Demerol again. Now I can see very clearly that I don't need Demerol to live my life. And I feel better if I don't use it. During the first 8 hours after taking the ibogaine I vomited four or five times, always when I tried to move. I was able to sleep about 4 a.m., and to eat only about 9 a.m. the following day. I awakened feeling weak, tired, and drowsy. As the hours were going, I slept a lot and began to feel better and in the morning of the following day I was normal.

Differences in Day-by-Day Life after the Experience. I returned to my normal life with absolutely no cravings, with better appetite than before, and highly self-confident. Now I can see differences in some aspects of my personality, things are changed. For example, I used to avoid driving at night, because it reminded me of a car accident I had years ago. Now I can drive anytime, day or night, without anxiety. I'm sure that this is caused by ibogaine, because now I'm not the same very anxious person I was. I'm not as shy as I used to be, too. It's easier now to contradict people when I think they are wrong, and to make them know what I want and what I think. I used to accept all that other people said only to avoid a discussion, even when I was sure that my point of view was the correct one.

These are the main happenings in my ibogaine experience and the main differences I can perceive in these few days.

Some Months Later. The most important thing I learned with all that happened is that I can never underestimate the power of the addictive personality I have inside. I can never say I'm cured because if I do this, I will forget to protect myself from drug-using thoughts. I must know I have a chronic disease that will be quiet in its place until I decide to give it a chance to grow. This decision, and that's the point, is a conscious decision. If I give in, the disease will be out of control in a few days. But if I could be strong to take real and honest control of my Demerol addiction using thoughts, I will be free forever.

A few days ago, because of professional needs, I had to keep two Demerol doses with me, in my house, all night long. To protect myself, I gave them to my wife. But it was amazing to see how I was not anxious to use them but to give them to the patients that needed them. I clearly felt that Demerol was a strange thing in my environment. I wasn't curious about the place my wife had put them. I wasn't feeling any craving. I was only looking forward to the moment I could give them to the patient and say, I've done it. And I did it, because of all of you from NDA. I don't want to be boring, but I have no words to say how grateful we, my family and I, are. I will remember you for a lifetime.

V. Psychological Aftereffects

The effects of ibogaine in human subjects have been described (*1,8,9,11,13-21*). In brief, ibogaine provides three different phases of effects in most, but not all, patients. In the first phase, the greatest intensity of which lasts approximately 3 hours, the patient appears to experience dreaming with eyes closed while awake. The form of the material experienced during this ibogaine visualization period is as varied as the scope and breadth of material seen in ordinary dreaming, in that it may be realistic or symbolic, in black and white or color, and diverse in subject matter. The visualization will be interrupted if patients open their eyes. It should also be noted that this dreamlike phase tends to end abruptly.

A second phase consisting of cognitive evaluation lasts between 8 and 20 hours. The material reviewed and reported by patients during the cognitive evaluation phase may consist of material from the dreamlike experience, or of other memories, and often concerns traumatic or emotional experiences, personal relationships, and important decisions that the patient has made. The second phase transitions slowly into a third phase of residual stimulation. The third stage may last as long as 36 hours or longer in some patients. The first three phases will run their course in most patients within 48 hours. It is not uncommon for a subset of patients to recover within 24 hours.

Psychological aftereffects of ibogaine may include what appears to be a process of abreaction. Patients report an understanding of their condition associated with traumatic events and resolution of issues in a process somewhat similar to that of psychoanalysis. Whether improvement in anxiety and depression in patients treated with ibogaine is a result of this phenomena, or more readily explainable on the basis of effects on neurotransmitter systems, has yet to be determined (*22-24*).

In the authors' experience, ibogaine-related psychological aftereffects have been apparent either immediately or after an interval of weeks or months. Clinical examples of aftereffects seen either acutely, or after a period of months, are provided below.

A. Immediate Psychological Effects Following Ibogaine Treatment

An example of the acute effects of a single treatment with ibogaine is that of a 39-year-old female patient dependent on 80 mgs per day of oral methadone that she supplemented with between $120 and $250 worth of heroin by IV administration. She had a 20-year history of heroin use.

The patient was HIV-positive and had an adolescent daughter. The patient's reason for seeking treatment was so that she might provide her daughter with quality time before her own anticipated death. Her treatment proceeded unremarkably, and she experienced little discomfort. On the third day after the treatment, being significantly recovered from the physical effects of ibogaine, she met with the psychiatrist who had supervised her treatment. The patient disclosed that she was suffering emotionally from the loss of her husband, who had died of a heroin overdose while she was in prison on a charge of heroin smuggling. The patient stated she had repressed emotions relating to the loss of her husband while in prison, she complained that the feelings of loss were now painfully intense, and she cried. She was angry that she had to feel these emotions. Her psychiatrist suggested that she wait another day and visit him again to see how she felt.

On the following day she appeared to have achieved a significant degree of resolution of the grieving relating to the loss of her husband, and she attributed this to her treatment. She had no desire to use opiates and showed no signs of opiate withdrawal.

B. Delayed Psychological Effects Following Ibogaine Treatment

The patient was a 26-year-old female dependent on heroin for 3 months. Her husband had been using heroin for approximately 10 years. The treatment of the husband and wife overlapped in time. The man was treated on day 1. The woman was very cooperative and assisted in every way with her husband's treatment. The woman was treated the following day. Her husband refused to leave her alone in the treatment room. He informed the treatment team, including four medical doctors observing the treatment, as well as the attending psychiatrist, that if they attempted to keep him out of his wife's room and bed he would disrupt her treatment. The husband further told the treatment staff not to come into the room to care for his wife and only authorized entry to address a lengthy list of somatic complaints concerning his own condition. Eventually, he left the treatment site to go for an extended bike ride. During his absence, his wife cried in the arms of one of the female members of the treatment team. She told of her husband's abuse toward her, of his manipulating her into a ménage-à-trois, which he had planned with an earlier girlfriend, and how he then leaked information about the relationship to his wife's family in order to isolate her from them.

The following day, both husband and wife were recovered enough to return to their apartment. The treatment team's attempts to meet with either the husband or wife over the next few days were not successful.

Six months later, the treatment team was able to meet with the wife. She informed us that within days she and her husband had returned to heroin use, though they were not in withdrawal. She maintained the relationship for approximately 3 months, after which she realized that she no longer wanted the life of heroin addiction. She ceased heroin use, left her husband, filed for divorce, and entered nursing school, while hiding from her husband in the home of one friend or another. She attributed her ability to alter her lifestyle to a catharsis precipitated by her experience with ibogaine.

VI. Discussion

A. The Clinical Significance Of Ibogaine Visions

There is evidence to suggest that ibogaine has been particularly efficacious for addressing physical opioid withdrawal signs (*20,25,26*). Reports over a period of more than three decades also suggest that ibogaine may be effective in reducing the craving that drives opioid (*21*) and cocaine dependence (*27*). The use of ibogaine to attenuate cocaine self-administration has been published in independent preclinical research studies by Cappendijk, and Sershen and Glick, and its ability to attenuate alcohol self-administration has been studied by Rezvani (*28-31*). Preclinical studies showing ibogaine's influence on drug place preference and learning, which are models of craving, are reported by Parker *et al.* (*32*). Double-blind clinical studies are needed at this time to verify the efficacy indicated in informal reports that have come from at least a dozen countries.

The self-reports that have been included in this chapter allow a review of the similarities of ibogaine effects that are consistent to the experience across multiple subjects. The most consistently reported ibogaine effect appears to be that of oneirophrenia, or the experience of apparently dreaming while awake. A question exists as to whether this phenomenon is significant to the antiaddictive action of ibogaine or separable from it. This issue may eventually be resolved should an *iboga* alkaloid devoid of oneirophrenic effects be developed, tested, and proven to have antiaddictive properties similar to ibogaine.

Whether the visualization or dreamlike experience of ibogaine is important or not, it is widely reported in the clinical literature (*1,8,9,11,13-21*). While the reports of dreamlike effects within the ibogaine experience are common, they do not invariably occur. Some ibogaine-treated subjects indicate that they did not experience oneirophrenia and proceeded directly to the stage of cognitive evaluation, which is generally considered to be the second phase. In some cases, visual images are reported during the actual experience, but are not apparently recalled afterward. Patients generally recall only a few of many images that they may have seen, similar to normal dreaming. It is difficult to assess whether the patients denying visualization after the fact simply do not recall the images, or that the images are so personal that the individuals do not wish to share them, or that visualization was simply not experienced.

The self-reports, the first example of which is found in Section II concerning drug user groups, provides an example of limited visualization occurring during the period when the majority of subjects have reported a rich and varied tableau. The self-report in Section II states, "Then I would have this weird image of a twisted stick or root being shook rapidly, accompanied by a deep sound something like a didgeridoo. When I had this image the rocking would stop . . .

I kept feeling like something else was going to happen although nothing else did. This went on for what seemed about 5 or 6 hours." In keeping with normal expectations during the cognitive evaluation or second stage of ibogaine activity, the subject reports "After the first few hours, I spent the next 20 hours or longer, I really can't remember how long but it was long, thinking about everything under the sun."

The above example should be compared with the reports found in Section III, the self-help section, and Section IV, the clinical environment section in which the authors of those reports indicated more varied and significant visual material. Whether the more extensive review of visual material may be responsible for the longer duration of time free of craving and drug use will require additional research. The two individuals reporting greater oneirophrenic effects remained either drug or craving free for approximately a year after a single ibogaine treatment, while the individual reporting minimal visualization, by comparison, had a drug-free period of only months.

B. Optimizing The Ibogaine Treatment Setting

The question of the environment in which ibogaine should be administered is disputed among ibogaine treatment providers. Ibogaine treatment providers come from disciplines as diverse as varieties of shamanism, self-help, clinical research, and African religious practices. To maximize the possibility of success of the ibogaine experience in a hospital setting, certain matters should be addressed in the design of the protocol. It is advisable to include persons who have previously taken ibogaine for a substance-related disorder to work with the treatment team. Patients will relate to these team members and are generally reassured by their presence, knowing that these individuals may uniquely understand what the patient is experiencing during the procedure.

Keeping the treatment site free of distracting sensory stimuli, such as loud noises, discussions or arguments, strong or irritating odors, and bright lights, is strongly recommended during the therapy. Patients should not be compelled to open their eyes or respond to staff anymore than is absolutely necessary during the first 3 to 4 hours of the ibogaine experience, as this may interfere with mental processing and adversely affect the outcome of treatment.

The personal ibogaine experiences of the authors, and the reports of patients in various treatment venues, appear to support the value of the visualization and subsequent abreaction. Ibogaine aftereffects are an area of particular interest to clinicians, therapists, and counselors working with ibogaine-treated patients. The aftereffects appear to involve learning and understanding by ibogaine-treated subjects regarding issues of psychological importance, and the resolution of those issues may lead to attenuation of anxiety, depression, and drug self-administration.

For the treatment provider, therapist, or counselor, it is important to understand that ibogaine-treated subjects may be more open and able to make use of various occupational, educational, or psychiatric forms of therapy. Professionals involved with ibogaine-treated patients should regard the process of visualization and subsequent abreaction as a window of opportunity. This could require more intensive work, but it could also be more rewarding work.

C. Ibogaine Versus Other Treatment Modalities

Four issues relating to ibogaine versus other treatment modalities are suggested for further discussion. The first is the differing philosophies held by the drug users involved in ibogaine and those not so involved. The second issue is the differences between medical professionals and the ibogaine self-help groups. The third matter of discussion concerns the various reasons for resistance to the development of ibogaine, and the fourth issue involves what might be learned from other medication programs for chemical dependence, such as methadone.

Among drug users, one unexpected result of advocacy for ibogaine treatment was a dispute that arose during a conference in Europe attended by Dutch Addict Self-Help (DASH). Nico Adriaans and other DASH members were promoting ibogaine availability and were quite surprised when other user groups indicated strong opposition to ibogaine in that they felt the availability of ibogaine would interfere with the possibility of legal heroin. DASH indicated that the two were not mutually exclusive, but to no avail. This conflict among heroin users over ibogaine was completely unexpected and continues to the present time.

In addressing the opposing philosophies of drug users and medical professionals providing treatment, a struggle relating to control and empowerment is seen in ibogaine therapy between user activists and the medical establishment. This is a result of ibogaine being first established within the self-help group context, and not the conventional medical setting as is usual for most medications. Some self-help groups feel the empowerment allowed by ibogaine should be maintained by drug users and self-help groups and not be turned over to the medical community for administration and control.

Resistance to ibogaine development is understandable. New technologies are often viewed with skepticism, and ibogaine appears to represent a particularly radical paradigm shift. As previously reviewed, the pharmaceutical industry for reasons of liability, perceived lack of profit, a lack of emphasis in the development of medications to treat chemical dependence, and a desire not to be associated with the stigma of the drug user population, chose not to involve itself in ibogaine development.

As to the research community, until the advent of ibogaine, pharmacotherapies to treat chemical dependence tended to be distinct for each form of dependence. Some medications were used to treat opiate dependence, others to treat alcohol dependence, and others to treat stimulant and sedative dependence. Thus, there was an understandable resistance to a single pharmacotherapy reported to have efficacy in multiple forms of substance-related disorders. There was also perceived resistance to the discovery being made by a person with no medical credentials, and promoted principally by drug users equally lacking in academic credentials. Years were spent in attempting to find interested pharmacologists who would perform ibogaine research. Once the first studies were accomplished, promising results accelerated research on the drug.

With regard to the future of ibogaine therapy, a look at methadone therapy may provide an understanding of what might go right and what might go wrong in the development of effective medications to treat chemical dependence. Methadone maintenance was the creation of Dr. Vincent Dole and the late Dr. Marie Nyswander (*33-35*). It consists of providing an opioid agonist in doses high enough to block the effects of heroin. Methadone is long acting, may be provided orally, and has been shown to promote a heroin-free lifestyle, social stability, and to reduce drug-related crime. So, what is wrong with methadone? The answer is nothing is wrong with methadone, but a good deal is wrong with many of the current providers who all too often fail to follow the Dole/Nyswander protocol. Adequate doses are often not given, and the humanitarianism shown patients has been replaced with indifference, animosity, a failure to acknowledge the patient as an individual, and the insistence of continuing to attach stigma to the patient in a punitive clinic environment (*36*). How this bodes for ibogaine therapy where even more skills may be needed, only the future can tell.

If a reduction in chemical dependence is to be accomplished, whether with ibogaine, its analogs, or other modalities, it will require that patients be better treated and better respected. It should not be anticipated that chemically dependent patients will readily remove themselves from that condition while they are marginalized, criminalized, and stigmatized. If the ultimate goal is to reintegrate these individuals into productive society, chemically dependent individuals must be provided with the same level of care and rights as patients who are being treated for other chronic, life-threatening conditions. In addiction medicine, as in other medical disciplines, it is of paramount importance that the physician listen to, respect, and not underestimate the patient. Ibogaine therapy offers a unique opportunity both for the physician and the patient.

References

1. H. S. Lotsof, U.S. Patent **4,499,096;** *Chem. Abstr.* **102,**160426w (1985).
2. P. De Rienzo and D. Beal, "The Ibogaine Story: Report on the Staten Island Project," Autonomedia, Brooklyn, NY, 1997.

3. T. Leary, R. Metzner, and R. Alpert, "The Psychedelic Experience, A Manual Based on the Tibetan Book of the Dead," University Books, New Hyde Park, NY, 1964.
4. R.G. Wasson, "Seeking the Magic Mushroom," *Life Magazine*, May 13, 1957.
5. R.G. Wasson and V.P. Wasson, "Mushrooms, Russia and History," Pantheon Books, New York, NY, 1957.
6. A. Huxley, "The Doors of Perception and Heaven and Hell," Harper Colphon Books, Harper & Row Publishers, New York, NY and Evanston, IL, 1963.
7. L. Lewin, "Phantastica," Park Street Press, Rochester, VT, 1998.
8. J.W. Fernandez, "Bwiti: An Ethnography of the Religious Imagination In Africa," Princeton University Press, Princeton, NJ, 1982.
9. R. Goutarel, O. Gollnhofer, and R. Sillans, *Psychedel. Monogr. Essays*, **6,** 71 (1993).
10. O. Gollnhofer and R. Sillans, *Psychotropes*, **2,** 95 (1985).
11. B. Sisko, *Bull. Multidiscip. Assoc.Psychedel.* Stud., **4,** 15 (1993).
12. J.-P. Grund, Letter for Nico, *Mainline*, Amsterdam, NL (1995).
13. D.J. Luciano, *Am. J. Addictions*, **7,** 89 (1998).
14. D.C. Mash, C.A. Kovera, B.E. Buck, M.D. Norenberg, P. Shapshak, W.L. Hearn, and J. Sanchez-Ramos, *Ann. NY Acad. Sci.*, **844,** 274 (1998).
15. C.D. Kaplan, E. Ketzer, J. DeJong, and M. deVries, *Soc. Neurosci. Bull.*, **6,** 6 (1993).
16. S.G. Sheppard, *J. Subst. Abuse Treat.*, **11,** 379 (1995).
17. A. Kastelic, "Self-Help Ibogaine Use in Slovenia," Presented at the First International Conference on Ibogaine, New York School of Medicine, November 5-6, 1999.
18. D. Pinchbeck, "Tripping on iboga in Gabon, a disenchanted journalist embarks on a hallucinogenic tribal rite," *http://www.Salon.com*, November 3, 1999.
19. C. Naranjo, "The Healing Journey," Pantheon Books, New York, NY, 1973.
20. K.R. Alper, H.S. Lotsof, G.M.N. Frenken, D.L. Luciano, and J. Bastiaans, *Am. J. Addictions*, **8,** 234 (1999).
21. H.S. Lotsof, *Bull. Multidisc. Assoc. Psychedel. Stud.*, **5,** 16 (1995).
22. H. Sershen, A Hashim, and A Lajtha, *Brain Research*, **42,** 161 (1997).
23. I.M. Maisonneuve, K.L. Rossman, R.W. Keller, Jr., and S.D. Glick, *Brain Res.*, **574,** 69 (1992).
24. P. Popik and P. Skolnick, in "The Alkaloids," (G. A. Cordell, ed.), Volume 52, p. 197, Academic Press, San Diego, CA, 1999.
25. E.D. Dzoljic, C.D. Kaplan, and M.R. Dzoljic, *Arch. Int. Pharmacodyn.*, **294,** 64 (1988).
26. S.D. Glick, K. Rossman, N.C. Rao, I.M. Maisonneuve, and J.N. Carlson, *Neuropharmacology*, **31,** 497 (1992).
27. H.S. Lotsof, U.S. Patent, **4,587,243;** *Chem. Abstr.* **106,** 1986.
28. S.L.T. Cappendijk, R. De Vries, and M.R. Dzoljic, *Eur. J. Pharmacol.*, **241,** 261 (1994).
29. H. Sershen, A. Hashim, and A. Lajtha, *Pharmacol. Biochem. Behav.*, **47,** 13 (1994).
30. S.D. Glick, M.E. Kuehne, J. Raucci, T.E. Wilson, D. Larson, R.W. Keller, Jr., and J.N. Carlson, *Brain Research*, **657,** 14 (1994).
31. A.H. Rezvani, D.H.Overstreet, and L. Yue-Wei, *Pharmacol. Biochem. Behav.*, **52,** 615 (1995).
32. L.A. Parker, S. Siegel, and T. Luxton, *Exptl. Clin. Psychopharmacol.*, **3,** 344 (1995).
33. V.P. Dole, *N.Y. State J. Med.*, **66,** 2011 (1966)
34. V.P. Dole, *Sci. Amer.*, **243,** 138 (1980).
35. M. Nyswander, "The Drug Addict as a Patient," Grune & Stratton, New York, NY, 1956.
36. H. Joseph, "Medical Methadone Maintenance: The Further Concealment of a Stigmatized Condition," Ph.D. thesis, Graduate Center, City University of New York, 1995.

THE ALKALOIDS: CUMULATIVE INDEX OF TITLES, 1950–2001

SUBJECT INDEX